ALGEBRAIC COMPUTABILITY AND ENUMERATION MODELS

Recursion Theory and Descriptive Complexity

ALGEBRAIC COMPUTABILITY AND ENUMERATION MODELS

Recursion Theory and Descriptive Complexity

Cyrus F. Nourani, PhD

Apple Academic Press Inc.	Apple Academic Press Inc.
3333 Mistwell Crescent	9 Spinnaker Way
Oakville, ON L6L 0A2	Waretown, NJ 08758
Canada	USA

©2016 by Apple Academic Press, Inc.

First issued in paperback 2021

Exclusive worldwide distribution by CRC Press, a member of Taylor & Francis Group
No claim to original U.S. Government works

ISBN 13: 978-1-77463-575-9 (pbk)
ISBN 13: 978-1-77188-247-7 (hbk)

Library and Archives Canada Cataloguing in Publication

Nourani, Cyrus F, author
Algebraic computability and enumeration models : recursion theory and descriptive complexity / Cyrus F. Nourani, PhD.

Includes bibliographical references and index.
Issued in print and electronic formats.
ISBN 978-1-77188-247-7 (hardcover).--ISBN 978-1-77188-248-4 (pdf)
1. Computer science--Mathematics. 2. Computational complexity. 3. Model theory.
4. Functor theory. 5. Recursion theory. 6. Descriptive set theory. I. Title.

QA76.9.M35N69 2016   004.01'51   C2015-908306-0   C2015-908307-9

Library of Congress Cataloging-in-Publication Data

Names: Nourani, Cyrus F.
Title: Algebraic computability and enumeration models : recursion theory and descriptive complexity / Cyrus F. Nourani, PhD.
Description: Toronto : Apple Academic Press, 2016. | Includes index.
Identifiers: LCCN 2015047105 (print) | LCCN 2016000519 (ebook) | ISBN 9781771882477 (hardcover : alk. paper) | ISBN 9781771882484 ()
Subjects: LCSH: Functor theory. | Model theory. | Computable functions. | Algebra, Homological. | Kleene algebra.
Classification: LCC QA169 .N6935 2016 (print) | LCC QA169 (ebook) | DDC 511.3/5--dc23
LC record available at http://lccn.loc.gov/2015047105

Apple Academic Press also publishes its books in a variety of electronic formats. Some content that appears in print may not be available in electronic format. For information about Apple Academic Press products, visit our website at **www.appleacademicpress.com** and the CRC Press website at **www.crc-press.com**

ABOUT THE AUTHOR

Cyrus F. Nourani, PhD

Dr. Cyrus F. Nourani has a national and international reputation in computer science, artificial intelligence, mathematics, virtual haptic computation, information technology, and management. He has many years of experience in the design and implementation of computing systems. Dr. Nourani's academic experience includes faculty positions at the University of Michigan-Ann Arbor, the University of Pennsylvania, the University of Southern California, UCLA, MIT, and the University of California, Santa Barbara. He was also a research professor at Simon Frasier University in Burnaby, British Columbia, Canada, and TU Berlin, Germany. He was a visiting professor at Edith Cowan University, Perth, Australia, and a lecturer of management science and IT at the University of Auckland, New Zealand.

Dr. Nourani commenced his university degrees at MIT, where he became interested in algebraic semantics. That was pursued with a category theorist at the University of California. Dr. Nourani's dissertation on computing models and categories proved to have intuitionist-forcing developments that were published from his postdoctoral times on at ASL. He has taught AI to the Los Angeles aerospace industry and has authored many R&D and commercial ventures. He has written and co-authored several books. He has over 350 publications in mathematics and computer science and has written on additional topics, such as pure mathematics, AI, EC, and IT management science, decision trees, predictive economics game modeling. In 1987, he founded Ventures for computing R&D. He began independent consulting with clients such as System Development Corporation (SDC), the US Air Force Space Division, and GE Aerospace. Dr. Nourani has designed and developed AI robot planning and reasoning systems at Northrop Research and Technology Center, Palos Verdes, California. He also has comparable AI, software, and computing foundations and R&D experience at GTE Research Labs.

CONTENTS

LIST OF ABBREVIATIONS

AMAST	Algebraic Methodology and Software Technology
ICC	implicit computational complexity
LATA	language and automata theory and applications
LTS	labeled transition system
OI	open induction
PA	Peano arithmetic
PCA	partial combinatory algebra
POPL	principles of programming languages
SDC	System Development Corporation
SO∃	second-order existential formula
TAP	tree amplification principle
UIC	universal inductive closure
UTP	unique termination property

PREFACE

This volume is written with a newer perspective to computability that is based on algebras, model-theory, sets, categories, and recursion hierarchies, deploying concepts from topological structures, set theory, and definability that are the most recent developments on those areas during 2014. The book is intended for advanced undergraduate students to first/second-year graduate students and, naturally, for computability and mathematics faculty and professionals.

This book is written following the author's functorial model theory book in 2014. Natural to the development is newer computability models, from functors to topos, concluding with a chapter on structural realizability for computability questions. The models and sets basis treats all areas from the arithmetic hierarchy, and Gödel's accomplishments, to Turing degrees, enumerability, and reducibility. Turing's accomplishments in computability theory and the Church-Turing's work on the unsolvability of the decision problem for first-order logic are the basic initial foundational areas. While we study formal systems for language computations, "computability" has higher dimensions with infinitary descriptions that are treated here with infinitary language definability, for example. Turing reducibility, sets based reducibility, and the notions of degree of unsolvability since Post are carried forward to the newer post degrees and theory. Newer computing models, based on fragment definability on positive categories and Horn categories with applications to a Positive Process algebraic computing are presented. These are pertaining to Kleene structures and newer considerations for recursion on Kleene algebras. Infinitary positive language categories are defined, and infinitary complements to Robinson consistency from the author's preceding papers are further developed to present new positive omitting types techniques and infinitary fragment higher stratified computing categories. Admissible models for computability on admissible generic diagrams functors to admissible structures are defined. These infinitary computability areas are further developed merging or lifting techniques accomplished by other authors for important enumerability and reducibility applications with ordinals. Basic descriptive computing and admissible computable models are presented, whereby a specific descriptive computing is defined. Generic model computing diagrams and situations compatibility are characterized, and Boolean model mappings are characterized also. A brief on "description logic" brings us closer to what the computer scientists have been developing for the past decade. More practical algebraic computing deploying the recursion theorem on signatures algebraic trees are presented to with morphic tree computing program synthesizers. The areas are augmented with a foundation for computation with algebraic subtree replacement systems for initial or standard models. Positive forcing conditions are defined on algebraic trees such to create canonical term initial models. Descriptive Sets and Infinitary Languages ushers us to the analytic hierarchy towards admissible sets and structures. Admissible languages, Hanf number characterizations, and Boolean valued models are presented. The hyperarithmetical and the analytical hierarchy on subsets of Cantor and Baire spaces are briefed with the basic accomplishments. The alternative on analytical hierarchy on Baire space allows us to define the analytical hierarchy on subsets of Cantor space from the hierarchy on Baire space, towards Set reducibility, admissible tree recursion, towards admissible set reducibility. Positive forcing techniques to computational complexity from the past decade are presented to become intimate with computational complexity on structures. Newer connections to Admissible sets and Kripke-Platek Urelements to computational set genericity are presented next. Applications to concrete and implicit complexity theory are examined with forcing conditions on r.e. sets. Complexity set reducibility based on Diophantine definability is examined. The proof techniques combine major advances in recursion theory, number theory, model theory, sets and forcing defined on structures. A Hilbert like program for viewing computational complexity from the models viewpoint, where the complexity questions are addressed as forcing and number theoretic problems are the highlights that avoids a direct computability comparison with a recursive effective procedure. There the Kripke Platek recursion on admissible sets complexity questions are a revisit to p-generic sets on admissible compact formulas. Further topological structures and more on complexity notions on the arithmetic hierarchy and Turing degrees are treated where applicable to concrete descriptive complexity. Degree theory, arithmetic hierarchy, and enuemarability towards a structure for reducibility on the arithmetic hierarchy is presented based on decidable sets, language recognition, definability degrees, and Turing Jumps enumeration definability. Newer set reducibility based on model theory is presented, for example towards Muchnic reducibility and alternatives. The isomorphism types for sets are examined for the Turing degree spectrums. More structures on Turing degrees are examined based on newer generic pairs, for example, Kallimulin pairs since this author's computational complexity generic pairs, over a decade ago. KPU recursion on urelements with isomorphism types towards lifts on K-pairs are briefed with direct product filters to examine degrees. Enumeration Definability and Turing Jumps are presented with new automorphisms and lifts defined based on compact admissible sets, towards admissible enumeration model theory. Peano Arithmetic Models and Computability are presented where basic foundations from Gödel to Hilbert and von Neumann are briefed. Recursion on Arithmetic Fragments, Peano Arithmetic, Hyting, and Bounded Computations, are treated with respect to the computational reducibility and expressive nature. Reducibility based on product fields since this author's publications during the past several years and newer authors on fragments of Peano arithmetic is briefed. Infinitary theories and countable N models are studied as example areas to address reducibility and Turing degrees. Intuitionistic arithmetic fragments and the Intuitionist Topos are briefed with a glimpse at the intuitionistic arithmetic hierarchy for countable N Models and a categorical Enumeration Model Theory. Further glimpse on reducibility areas on term algebra models on topological structures, admissible Hull, or Hausdorf spaces are newer developments. Ordinal admissible computability models are examined towards the automata theoretic concrete computability areas. Generic computability and filters on genericity and reducibility are stated that compare a newer syntactic notions on genericity, summarizing the newer developments by colleagues since the past several years, concluding with generic Horn filter applications the author developed to the newer developments on degree pair definability for enumerability. Categoricity on degrees are reviewed with glimpses on degrees, determinacy, projective sets, and saturation. The book concludes with Realizability on topos for further insights on computability structures. There are at least five realizability areas that are not all mutually disjoint on that are considered in this volume. Realizing a type with closures omitting extras, realizing a model or type with morphism and filters; realizability on topos, triposes, or sheaves; realizing universal arrows in categories, Kleene and intuitionistic realizability. Newer areas are treated in the author 2014 book on realizability on generic filters, ultrafiters and ultraproduct projective computability on saturated types, Gödel sets, and presheaves. From Longley (2012) we have the glimpse that amongst the preceding schools there is no general definition of what realizability is. Kleene first introduced the idea of realizability in a paper of 1945. This author's volume expounds on new realizability areas based on omitting types, forcing, and adjoint functors, with applications to Kleene structures. The treatments are on further on filters and ultraproduct models, beginning with Hyting and Kleene on intuitionistic logic. All chapters have exercises that are times good starts for dissertations.

— *Cyrus F. Nourani*

Berlin, Germany
January 2016

CHAPTER 1

INTRODUCTION

This volume is written with a perspective to computability that is based on algebras, model-theory, sets, categories, and recursion hierarchies, deploying concepts from topological structures, set theory, and definability. The emphasis is not proof-theoretic, whereas all such areas from the Arithmetic hierarchy Gödel's incompletes to Turing degrees, enumerability, and reducibility are encompassed. The very beginning accomplishments are from Turing in computability theory and the Church-Turing's work on the unsolvability of the decision problem for first-order logic. Turing formalized relativized computation with oracle Turing machines, commencing Turing reducibility. Comparing sets based on reducibility the notions of degree of unsolvability was brought forth by Post in 1944. Studying such poset degrees is called degree theory. That area is on Chapters 7 and 8 in this book, however, the beginning is on Chapter 6.

While we study formal systems for language computations on a chapter or two in this volume, Chapter 2, for example, with finitary natural descriptions, "computability" has higher dimensions with infinitary descriptions. However, since formulas may be identified with natural numbers, for example, "Gödel numbering" there is the possibility for "languages" some of whose formulas would be naturally identified with *infinite sets*: hence the term infinitary languages. Syntax and semantics of infinitary languages were presented since 1967. The expressive power is examined on chapters here with examples. There are infinitary languages with only finite quantifier sequences, therefore more tangible for computations. There are applications to game theory and economic game models not on the author's recent decade publications and other colleagues.

Here is how the chapters are organized. Chapter 2 is on Computing Categories, Language Fragments, and Models. There this author presents the newer algebraic computing model areas starting with Infinite language categories (Nourani, 1995). Newer computing models, based on fragment definability on positive categories and Horn categories (Nourani, 2005–2015) on fragment categories, and applications to a Positive Process algebraic computing, are presented. The areas pertaining to Kleene structures that were presented with initial model characterization in this authors publications since 2005, and computing algebraic topology, are receiving newer considerations for recursion on Kleene algebras, exploring only on how algebras are treated with a newer perspective there. Our model bases are called Fragment Consistency Models, where new techniques for creating generic models are defined that are infinitary correspondences to Robinson consistency. Infinitely positive language categories are defined and infinitary complements to Robinson consistency from the author's preceding papers are further developed to present new positive omitting types techniques and infinitary positive fragment higher stratified computing categories.

Chapter 3 becomes more intimate with admissible models for computability, defining admissible generic diagrams functors to admissible structures. Starting with infinite language category $L_{\omega 1, K}$ to start with infinitary language computability (Nourani, 1996), the author defines generic sets on for $L_{\omega 1, K}$. These infinitary computability areas are further developed in Chapters 7 and 8 with what is being accomplished by other authors for important enumerability and reducibility applications with manageable ordinals. Basic descriptive computing and admissible computable models are presented. Amongst the theorems is that there is a generic functor defining descriptive computable admissible models, whereby a specific descriptive computing is defined. Generic model computing diagrams and situations compatibility is characterized and Boolean model mappings are characterized. A brief on "description logic" brings us closer to where computer scientists have been engaged for the past decade. More practical algebraic computing deploying the recursion theorem on signatures algebraic trees is presented to with morphic tree computing program synthesizers.

Chapter 4 is a foundation for computation with algebraic subtree replacement systems for Initial or standard models. Algebraic tree computation theories are presented with generalized equational theories that with algebraic theories characterizing direct tree instantiation and realizations for initial models with free Skolemized proof trees. Methods for generating normal forms for subtree replacement systems are presented by defining model-theoretic forcing properties on algebraic trees. Positive forcing conditions are defined on algebraic trees such that canonical term initial models can be created. On that chapter there are further briefs on subtree replacement systems that might have normal forms with techniques based on algebraic closed groups. Wittgenstein's paradox is reviewed for insight on language term rewriting.

Chapter 5 in on Descriptive Sets and Infinitary Languages. The analytic hierarchy is briefed towards admissible sets and structures. Basic descriptive characterizations from the basics on set quantification to second order quantifications are examined. Admissible languages, Hanf number characterizations, and Boolean valued models are presented. The hyperarithmetical and the analytical hierarchy on subsets of Cantor and Baire spaces are briefed with the basic accomplishments there. The ordinary axiomatization of second-order arithmetic uses a set-based language in which the set quantifiers can naturally be viewed as quantifying over Cantor space. An equivalent definition of the analytical hierarchy on Baire space is given by defining the analytical hierarchy of formulas using a functional version of second-order arithmetic; then the analytical hierarchy on subsets of Cantor space can be defined from the hierarchy on Baire space. Set reducibility, admissible tree recursion, towards admissible set reducibility are presented. This area is further addressed on Chapters 7 and 8, for enumerability models.

Chapter 6 is based on this author's applications for positive forcing technique he developed since his dissertation and postdoctoral year at Boston-Cambridge, to computational complexity during the year 1984, at the Summer Logic Colloquium. The lower sections were developed some years later to present the relations to Admissible sets and Kripke-Platek Urelements and computational set genericity that he applied at 1984–1985 times to complexity. Sections 6.1–6.5 are briefs on what was written at GTE Research labs greater Boston area and with Harvard-MIT Universities associations, during 1984 times and circulated during 1985.

This chapter first presents positive forcing and the preceding chapter's applications to concrete and implicit complexity theory. It defines forcing conditions on r.e. sets to the polynomial time computability problems for recursive sets. It is a generalization of the earlier papers to recursive sets. It is shown that r.e. sets can be applied as positive forcing conditions. In the 80's, we showed that the P-NP problem is not independent of a sufficiently rich fragment of Universal Peano arithmetic and carries on the address what the exact question might be to resolve at models not proof-theory. The P-NP problem is considered as a problem of mutually reducing the two sets with Diophantine definability. The proof techniques combine major advances in recursion theory, number theory, model theory, and forcing defined on structures. A generic model of arithmetic is defined and a forcing companion for the arithmetic fragment sufficient to represent problems in P and NP as r.e. sets are presented. Sections 6.2 and 6.3 introduce a Hilbert like program for viewing computational complexity from the models view point, where the complexity questions are addressed as forcing and number theoretic problems that might have models without basing the techniques on a specific computation measure or system, thereby avoiding a direct computability comparison that has to have a recursive effective procedure or computability model to address.

Section 6.4 presents admissible sets and admissible recursion so far as basic computational complexity is concerned. There with the Kripke–Platek recursion on admissible sets complexity questions are addressed presenting important foundational areas the author addressed over the more recent years to revisit p-generic sets on admissible compact formulas. Further topological structures and more on complexity notions on the arithmetic hierarchy and Turing degrees are treated in a following chapter. Section 6.5 presents basic on the arithmetic hierarchy where applicable to concrete descriptive complexity. The following chapters address areas that are at time outside basic computability theory altogether. Such unsolvable problems arisen outside of computability theory are *computably enumerable* (c.e.). The c.e. sets are at times comparable to unbounded search problems, for example, formulas provable in some effective formal system. Degree theory is an important area of computability theory. The priority method introduced by Friedberg and Mučnik to solve Post's Problem is the first such technique.

Chapter 7 is an overview to Turing Degrees, arithmetic hierarchy, and enumerability degrees towards a structure for reducibility and abstract ordinal computability notions. Arithmetic hierarchy is presented based on decidable sets, language recognition, definability degrees, and Turing Jumps enumeration definability. Newer classical computability theory examines the degree structures that arise from reducibility on the power set $\wp(\omega)$ of the natural numbers: we say that a set A is "reducible" to a set B if there is a way to "compute" membership in A from membership information about B. There are several natural formalizations of this idea, giving different reducibilties. Such reducibility is always reflexive and transitive and thus induces a preorder on $\wp(\omega)$. The equivalence classes of sets reducible to each other are usually called "degrees," and the preorder on the sets induces a partial order on the degrees. Borel sets and the Borel hierarchy comparisons are presented and explored further on with admissible sets on Chapter 8.

The most commonly studied reducibility is Turing reducibility: a set A is Turing reducible to a set B if there is an algorithm that, on any input x, determines whether $x \in A$ in finitely many steps and making finitely many membership queries to B. A natural extension of Turing reducibility is enumeration reducibility: a set A is enumeration reducible to a set B if there is an algorithm to enumerate all elements of A from any enumeration of all the elements of B. Specific infinitary languages are studied followed by admissible set recursion and Kripke–Platek sets. Newer set reducibility based on model theory is presented, for example towards Muchnik reducibility and alternatives.

The isomorphism types for sets are examined for the Turing degree spectrums. A standard technique is to use countable ordinals for natural numbers, and countable sets play the role of finite sets. More structures on Turing degrees are examined based on Kallimulin pairs where we invoke KPU recursion urelements with isomorphism types towards lifts on K-pairs with direct product filters to examine degrees. This author's publications on generic set computability date back to the 1984 Summer logic colloquiums on pairing with generic sets to Peano arithmetic fragment models, when on new postdoc appointments at 1984 to KPU characterizations at the Summer logic colloquium 2003. Enumeration Definability and Turing Jumps are presented with new automorphisms and lifts defined based on compact admissible sets. Direct product filters are examined for K-paring, towards admissible enumeration model theory thereby comparing views to the Erschov hierarchy, recursion, and Sacks splitting, might areas to explore.

Chapter 8 presents Peano Arithmetic Models and Computability. Basic foundations from Gödel to Hilbert and von Neumann are briefed. Recursion on Arithmetic Fragments, Peano Arithmetic, Hyting, and Bounded Computations, are treated with respect to the computational reducibility and expressive nature. Reducibility based on product fields on fragments of Peano arithmetic is presented with this author and other's contributions since 2005. Next, infinitary theories and countable N models are studied as example areas to address reducibility and Turing degrees. Intuitionistic arithmetic fragments and the Intuitionist Topos are briefed with newer ac-

complishments that are a prelude to chapter on Reliability, indicating that
the intuitionistic arithmetic fragments are categorical. We have a glimpse
at the intuitionistic arithmetic hierarchy to have a feel for countable N
models where that section is followed by a categorical Enumeration Model Theory and computability on admissible sets are further developed with
computability on admissible sets, where we examine constructive and finite models for descriptive computability to further address reducibility
area on term algebra models on topological structures, admissible Hull, or
Hausdorf spaces.

Ordinal admissible computability models are examined towards ex-
amining the automata theoretic concrete computability areas. The initial
proofs on the areas rely on a class of finite automata with expressive power of the languages of ω-words and infinite trees. Generic computability
and filters on genericity and reducibility are stated that compare a newer
syntactic notions on c.e. genericity to the developments in the preceding
chapters, briefing on the newer authors. This area concludes with generic
Horn filter applications this author developed during the recent years to
the newer developments on degree pair definability for enumerability. Categoricity on degrees is reviewed with glimpses on degrees, determinacy,
projective sets, and saturation.

Chapter 9 presents realizability areas, in particular, on topos that in
hindsight the complexity structure might have been sited on. There are at
least five realizability areas that are not all mutually disjoint on thought
precepts:

   (i) realizing a type with closures omitting extras;
   (ii) realizing a model or type with morphisms and filter;
   (iii) realizability on topos, triposes, or sheaves;
   (iv) realizing universal arrows in categories;
   (v) Kleene and intuitionistic realizability.

From this authors perspective the topics were presented in part in his
publications including the 2014 book on functorial model theory during
contemporary times. Since the present volume is more on mathematics
from the computational perspective the author has minimized the exposure
to pure foundational areas from mathematics. The overview is as follows
and reconciled to a degree on the ending sections. Computability with
the intuitionistic schools this author we began to address on model computability on initial models during the past decade or more with a newer
look was at FMCS Vancouver 5–6 years ago, but dates 20 years. Newer

areas are treated in the author 2014 book on realizability on generic filters, ultrafiters and ultraproduct projective computability on saturated types, Gödel sets, and presheaves. From Longley (2012), we have the glimpse that amongst the preceding schools there is no general definition of what realizability is. Realizability interpretations for logics such as Heyting calculus as realizability applications, quintessential applicative programming models, and arithmetic type systems are the trends. Kleene first introduced the idea of reliability in a paper of 1945. This author's (Nourani, 2014) volume expounds on new realizability areas based on omitting types, forcing, and adjoin functors, with applications to Kleene structures. The treatments are on further on filters and ultraproduct models, beginning with Heyting and Kleene on intuitionistic logic. This chapter is a brief overview including the newer developments from authors during the past two years.

CHAPTER 2

COMPUTING CATEGORIES, LANGUAGE FRAGMENTS, AND MODELS

CONTENTS

2.1   INTRODUCTION

Chapter 2 presents the newer algebraic computing model areas starting with infinite language categories (Nourani, 1995) and specific computing models based on language fragments. Positive categories and horn categories are the new fragment categories defined and the applications to a positive process algebraic computing (Nourani, 2005c) is outlined. The areas pertaining to Kleene structures that were presented with initial model characterization in this authors publications since 2005 (Computing Algebraic Topology, or the newer book Nourani 2014) are receiving newer considerations on Leiße (2015) or RMCIS (2014) for recursion on Kleene algebras exploring only on algebras since Kozen (1994) but not far as categorical computing above. These areas are treated more with our newer perspective in this chapter.

For example, the author defined the category $\mathbf{L_{P,\omega}}$ to be the category with objects positive fragments and arrows the subformula preorder on formulas to present models. The model based on the Fragment Consistency Models where new techniques for creating generic models are defined. Infinitary positive language categories are defined and infinitary complements to Robinson consistency from the author's preceding papers are further developed to present new positive omitting types techniques and infinitary positive fragment higher stratified computing categories. Further, neoclassic model-theoretic consequences are presented in Nourani (2005a). Start with a well-behaved countable fragment of an infinitary language $L_{\omega1,K}$ defined by H.J. Keisler. The term fragment is not inconsistent with the terminology applied in categorical logic. A subclass F of class of all formulas of an infinitary language is called a fragment, if (a) for each formula φ in F all the subformulas of φ also belong to F; and (b) F is closed under substitution: if φ is in F, t is a term of L, x is a free variable in φ, the $\varphi(x/t)$ is in F.

Definition 2.1: By a fragment of $L\omega1,\omega$ we mean a set L<A> of formulas such that (1) Every formula of L belongs to L<A> (2) L<A> is closed under $\neg$, $\exists\, x$, and finite disjunction (3) if f (x) $\in$ L<A> and t is a term then f (t) $\in$ L<A> (4) If f $\in$ L<A> then every subformula of f $\in$ L<A>

Subformulas are defined by recursion from the infinite disjunction by sub $(V\Phi) = \cup$ sub $(\varphi) \cup \{V\Phi$ taken over the set of formulas in with the basis defined for atomic formulas} by sub(φ) = {φ}; and for compound formulas by taking the union of the subformulas quantified or logically connected with the subformula relation applied to the original formula as a singleton. Let C be a countable set of new symbols and form the first order language K by adding to L the constants c in C. We shall let K<A> be the set of all formulas obtained from formulas in L<A> by substituting finitley many free variables by constants c in C. The K<A> is the least fragment which contains L<A>. Each formulas in K<A> contains only faintly many c in C. We refer to the fragment by $L_{\omega1,K}$.

Let us define a small-complete category $\mathbf{L}_{\omega1,K}$ from $L_{\omega1,K}$. The category is the preorder category defined by the formula ordering defining the language fragment. The objects are small set fragments. There are three categories at play—the category $\mathbf{L}_{\omega1,K}$, the category **Set**, and the category **D<A, G>**. The D<A, G> category is the category for models definable with D<A, G> and their morphisms. It is applied towards the solution set theory presented in (Nourani, C.F. 1994), Solution Sets for Categories, ECCT, Tours France, and AMS 1994. The techniques we are presenting by the three categories save us from having to yet develop a categorical interpretation for $L_{\omega1,K}$, in categorical logic as in Lawvere's (1967) categorical formulation of logic or the current practice.

To state a brief glimpse on categories let us consider the following: A *category C* is comprised of:

1. A class **C**, with elements called *objects*;
2. A class denoted *M* with elements called *morphisms*. Each morphism *f* has a *source object a* and *target object b.* The expression *f*: $a \rightarrow b$, would be verbally stated as "f is a morphism from *a* to *b*." The expression **hom(*a, b*)**—alternatively expressed as **homC(*a, b*), mor(*a, b*)**, or *C*(*a, b*)—denotes the *hom-class* of all morphisms from *a* to *b*.
3. A binary operation °, called *composition of morphisms*, such that for any three objects *a, b,* and *c,* we have hom(*b, c*) × hom (*a, b*) $\rightarrow$ hom (*a, c*). The composition of *f*: $a \rightarrow b$ and *g*: $b \rightarrow c$ is written as *g°f* or *gf*, governed by two axioms:
   a. Associativity: If *f*: $a \rightarrow b$, *g*: $b \rightarrow c$ and *h* : $c \rightarrow d$ then *h*° (*g°f*) = (*h°g*) °*f*, and
   b. Identity: For every object *x*, there exists a morphism 1*x* : $x \rightarrow x$ called the identity *morphism* for *x*, such that for every morphism *f*: $a \rightarrow b$, we have 1*b°f* = *f* = *f°*1*a*.

From the axioms, it can be proved that there is exactly one identity morphism for every object. Some authors deviate from the definition just given by identifying each object with its identity morphism.

1. Objects will generally be denoted with capital letters A, B, C in C.
2. For morphisms we write $f: A \rightarrow B$ or $A \rightarrow f \rightarrow B$ to mean that f has source A and target B. We say that "f is a morphism from A to B."
3. $M(A, B)$ denotes the class of all morphisms from A to B, and is called the hom-set of M.

More succinct is the following:

A *category C* consists of the following three mathematical entities:

1. A class ob(C), whose elements are called *objects*;
2. A class hom(C), whose elements are called morphisms or maps; or *arrows*. Each morphism *f* has a *source object a* and *target object b.* The expression $f: a \rightarrow b$, would be verbally stated as "f is a morphism from *a* to *b*." The expression hom(a, b)—alternatively expressed as homC(a, b), mor(a, b), or C(a, b)—denotes the *hom-class* of all morphisms from *a* to *b*.
3. A binary operation $\circ$, called *composition of morphisms,* such that for any three objects *a, b,* and *c,* we have hom(b, c) $\times$ hom(a, b) $\rightarrow$ hom(a, c). The composition of $f: a \rightarrow b$ and $g: b \rightarrow c$ is written as $g \circ f$ or gf, governed by two axioms:
    a. Associativity: Iff: $a \rightarrow b, g: b \rightarrow c$ and $h: c \rightarrow d$ then $h \circ (g \circ f) = (h \circ g) \circ f$, and
    b. Identity: For every object *x*, there exists a morphism $1x : x \rightarrow x$ called the *identity morphism for x,* such that for every morphism $f: a \rightarrow b$, we have $1b \circ f = f = f \circ 1a$.

From the axioms, it can be proved that there is exactly one identity morphism for every object. Some authors deviate from the definition just given by identifying each object with its identity morphism.

2.2   LIMITS AND INFINITARY LANGUAGES

Start from the Op category formed by $\mathbf{L}_{\omega1,\kappa}$. Define a functor to utter the Logische Syntax der Sprache from the language category to the category **Set** until the model appears. A limit has to be created for the corresponding model to be defined from the language syntax. The following lemma is applied towards defining generic Functors for infinite language categories.

We define a small-complete category $\mathbf{L}_{\omega1,K}$ from $L_{\omega1,K}$. The category is the preorder category defined by the formula ordering defining the language fragment. Let us start with the basic definition:

Definition 2.2: A category is small if the set of its objects and its arrow set are small sets. Small sets are defined to be members S to a universe U, which has closure properties such that all standard set theoretic operations on S does not leave U. All ordinary mathematics is carried out at U, and known sets such as w, are in U. The model theory is all carried out with $L_{\omega1,K}$ positive forcing, hence there are limiting generic sets and small index sets.

Lemma 2.1: $\mathbf{L}_{\omega1,K}$ is a small-complete category.

Proof: To prove $\mathbf{L}_{\omega1,K}$ is small complete, that is, every small diagram has a limit, we can define a functor H: $\omega^{op} \to \mathbf{L}_{\omega1,K}$ and create a limit in $\mathbf{L}_{\omega1,K}$ as follows. The functor H is a list of sets H_n, consisting of the strings in $\mathbf{L}_{\omega1,K}$, and functions $f_i : H_i + 1 \to H_i$. Form the product set $\Pi_i H_i$. Let $l = \{l_0, l_1, l_2, ...\}$, with each $l_n \in H_n$, and the projections
　　$p_n : \Pi_i H_i \to H_n$, forming the following diagram.

$$
\begin{array}{ccccccc}
HO & \leftarrow & H1 & \leftarrow & & Hn & \leftarrow & Hn+1 & \leftarrow \\
\uparrow & & & & & & \uparrow \\
\Pi iHi & \leftarrow & & \leftarrow & include & & \leftarrow M = Lim\ H
\end{array}
$$

　　To form a limit at a vertex for commuting with the projections take the subset S of the strings in $\mathbf{L}_{\omega1,K}$, which match under f, that is, $f s_n + 1 = s_n$, for all n. The above A-diagram is well-defined because sets which are generic with respect to a base set, for example, the intended limit S are definable, and we only have to present limits for small A-diagrams.

　　The completeness proof, for example, that every functor from a small category to $L_{\omega1,K}$ has a limit, has bearings to the completeness proof for the category **Set**. The crux of the proof is there being limiting cones for arbitrary functors F : J $\to \mathbf{L}_{\omega1,K}$ for the K<A> fragment. We can define cones to base F from {}. The natural transformations are on functors sending arbitrary objects to sets on fragment string sets, arrowed by preorder functions. Since sets that are generic with respect to the base set are always definable by what the author had put forth in ((Nourani 1994, 1996) (also see Theorem 4.3, Section 4,), the limiting cones can always be defined.

Cones to base F from {} pick fragment sets. A limiting cone can always be defined with natural transformations on the preorder functors on $L_{\omega 1,K}$.

Define a functor $F: L_{\omega 1,K}{}^{op} \rightarrow$ **Set** by a list of sets M_n and functions f_n. The sets correspond to an initial structure on $L_{\omega 1,K}$. For example, to $f(t_1, t_2, ..., t_n)$ in $L_{\omega 1,K}$ there corresponds the equality relation $f(t_1, t_2, ..., t_n) = ft_1 ... t_n$. Let us refer to the above functor by the generic model functor since we can show it can defines generic sets from language strings to form limits and models. It suffices to define the functor upto an initial model without being specific as to what the model is, to have a generic functor. The specific model theoretic properties were explored over the past several years (see, Nourani, 2014). The generic model functor can define abstract models for computations defined with $L_{\omega 1,K}$. The proof for the following theorem follows from the sort of techniques applied by the present author to define generic sets with the fragment $L_{\omega 1,K}$ over a decade ago.

Theorem 2.1: The generic model functor has a limit.

Proof: Let us create a limit for the functor $F: L_{\omega 1,K}{}^{op} \rightarrow$ **Set** in **Set** as follows. Applying Lemma 1.1, the functor F is a list of sets F_n, consisting of (a) the sets corresponding to a free structure on $L_{\omega 1,K}$, for example, to $f(t_1, t_2, .. t_n)$ in $L_{\omega 1,K}$ there corresponds the equality relation $f(t_1, ..., t_n) = ft_1 ... t_n$ in Set; (b) the functions $f_i : F_i + 1 \rightarrow F_i$. Form the product set $\Pi_i F_i$. Let $l = \{l_0, l_1, l_2, ...\}$, with each $l_n \in F_n$, and its projections $p_n : \Pi_i F_i \rightarrow F_n$, forming the following diagram.

$$F0 \leftarrow F1 \leftarrow Fn \leftarrow Fn \quad 1 \leftarrow$$

$$\uparrow \uparrow$$

$$\Pi i F i \leftarrow \leftarrow include \qquad \leftarrow M = Lim \, F$$

To form a limit at a vertex for commuting with the projections take the subset S of the strings in S, which match under f, for example, $f \, s_n + 1 = s_n$, for all *n*. The above A-diagram is well-defined because by the above given definitions, theorems and sets which are generic with respect to a base set, for example, the intended limit M are definable.

Let us refer to the above functor by the *generic model functor* since it defines generic sets from language strings to form limits and models. The model theoretic properties are defined in Nourani (1997). The proof for the following theorem follows from the sort of techniques applied by the present author to define generic sets with the fragment $L_{\omega 1,K}$ by Nourani (1982, 1983).

2.3   GENERIC FUNCTORS AND LANGUAGE STRING MODELS

Creating limits amounts to defining generic sets on $L_{\omega 1,K}$ for $\mathbf{L_{\omega 1,K}}$. The present theory creates limits for infinitary languages to define generic models with language categories. Generic sets for infinitary languages have been applied by this author to define positive forcing over a decade and applied to categories and toposes.

Definition 2.3: Let M be a structure for a language L, call a subset X of M a generating set for M if no proper substructure of M contains X, for example, if M is the closure of X U {c(M): c is a constant symbol of L}. An assignment of constants to M is a pair <A, G>, where A is an infinite set of constant symbols in L and G: A → M such that {G(a): a in A} is a set of generators for M. Interpreting a by g(a), every element of M is denoted by at least one closed term of L(A). For a fixed assignment <A, G> of constants to M, the diagram of M, D<A, G>(M) is the set of basic (atomic and negated atomic) sentences of L(A) true in M. (Note that L(A) is L enriched with set A of constant symbols.)

Definition 2.4: A *generic diagram* for a structure M is a diagram D<A, G>, such that the G in definition above has a proper definition by a specific function set.

For example, Σ_1-Skolem functions are a specific function set. We abbreviate generic diagrams by G-diagrams.

For self-containment, we state the following theory T for the above language $L_{\omega 1,K}$, let T* be T augmented with induction schemas on the generic diagram functions in the authors' papers since the 80's.

Definition 2.5: Let (M, *a*)c in C be defined such that M is a structure for a language L and each constant *c* in C has the interpretation *a* in M. The mapping $c \rightarrow a_c$ is an assignment of C in M. We say that (M, *a*)c in C is canonical model for a presentation P on language L, iff the assignment *c a* maps C onto M, for example, M=(*a:c* in C).

2.4 FRAGMENT CONSISTENT ALGEBRAS

2.4.1 GENERIC PRODUCTS

The propositions in Section 2.1 and Theorem 2.1 are from basic model theory. Proposition 2.4 and Theorem 2.1 are from Nourani (2005).

Proposition 2.1: Let φ be a universal sentence. Then φ is a (finite) direct product sentence iff φ is equivalent to a universal Horn sentence.

Proposition 2.2: Let φ be a universal sentence. Then φ is a (finite) direct product sentence if and only if φ is equivalent to a universal Horn sentence.

Proposition 2.3: Let φ be an existential sentence. Then φ is a (finite) direct product sentence if φ is equivalent to an existential Horn sentence.

Let I be a nonempty set. Let S(I) be the set of all subsets of I. A filter D over I is defined to be a set $D < S(I)$ such that $I \in D$; if X, Y $\in$ D, $\cap Y \in$ D; if X$\in$ D an X<Z<I, then Z $\in$ D. Note that every filter D is a nonempty set since I $\in$ D. a filters are the trivial filter D = {I}. The improper filter D = S(I). For each Y < I, the filter D = {X <I; Y < X}; this filter is called the principal filter generated by Y. D is said to be a proper filter iff it is not the improper filter S(I).

Let E be a subset of S(I). By the filter generated by E we mean the intersection D of all filters over I which include E: D = $\cap$ {F: E < F and F is a filer over I}. E is said to have the finite intersection property iff the intersection of any finite number if elements of E is nonempty. The filter D generated by E, any subset E of S(I), is a filter over I.

Proposition 2.4: Let I be the set T*. Let $\varphi\,(x_1 \ldots x_n)$ be a Horn formula and let $\Re i \in$ I be models for language L. let $a_1 \ldots a_n \in \Pi_i \in$ I Ai. The $\Re i$ are fragment Horn models.

If $\{i \in I: \Re i \models \varphi[(a_1(i) \ldots a_n(i)]\}$ then the direct P_D on $\Re i \models \varphi[a_1 D \ldots a_n D]$, where D is the generic filter on T*.

Proof: Nourani (2007), for example, A Functorial Model Theory volume on Nourani (2012).

2.4.2. POSITIVE MORPHISMS AND MODELS

Definition 2.6: A formula is said to be positive iff it is built from atomic formulas using only the connectives '&', v and the quantifiers $\forall$, $\exists$.

Definition 2.7: A formula φ (x1, x2, ..., xn) is preserved under homorphisms iff for any homorphisms f of a model A onto a model B and all a1, ..., an in A if A $\models$ [a_1, ..., a_n] B $\models$ [fa_1, ..., fa_n].

Theorem 2.2: A consistent theory is preserved under homomorphisms iff T has a set of positive axioms.

Thus, if we apply positive formulas we can define a functorial model theory with initial models, where the arrows are homomorphisms. For what we have in mind we don't really need to apply nonpositive formulas. The functorial model theory, in its most abstract form, does not have to become specific to the formulas and what sort of morphisms.

Applying the definitions for positive formulas, we can state the following from Nourani (1997).

We can prove the following as a lemma and a consequent theorem.

Theorem 2.3: The embedding to form elementary chains on Fi's can be defined by a back and forth model design from the strings in $L_{\omega1,K}$ language fragments.

Theorem 2.4: By defining models corresponding to the Fi on the fragments as an w-chain from the elementary diagrams on the Th (A(Fi)), a generic model is defined by the limit.

2.5 POSITIVE CONSISTENCY AND OMITTING TYPES

Let us start from certain model-theoretic premises with propositions 2.5 and 2.6 known from basic model theory.

Proposition 2.5: Let A and B be models for L. Then A is isomorphically embedded in B iff B can be expanded to a model of the diagram of A.

Proposition 2.6: Let A and B be models for L. Then A is homomorphically embedded in B iff B can be expanded to a model of the positive diagram of A.

Further specific example shows the embedding from computability (Ehrig, 1977).

Let Σ be a set of formulas in the variables $x_1 \ldots x_n$. Let A be a model for L. We say that A realizes Σ iff some n-tuple of elements of A satisfies Σ in A. A *omits* Σ iff A does not realize Σ.

Definition 2.8: Let $\Sigma(x_1 \ldots x_n)$ be a set of formulas of L. A theory T in L is said to locally realize Σ iff there is a formula $\varphi (x_1 \ldots x_n)$ in L s.t.

1. φ is consistent with T;
2. for all $\sigma \in \Sigma$, $T \models \varphi \to \sigma$.

That is, every n-tuple of T which realizes φ satisfies Σ.
We say that T locally omits Σ iff T does not locally realize Σ.
For our purposes this author defined a positive realizability basis.

Definition 2.9: Let $\Sigma(x_1 \ldots x_n)$ be a set of formulas of L. Say that a positive theory T in L positively locally realize Σ iff there is a formula $\varphi (x_1 \ldots x_n)$ in L s.t.

1. φ is consistent with T;
2. for all $\sigma \in \Sigma$, $T \models \varphi$ or $T \cup \sigma$ is inconsistent.

Definition 2.10: Given models A and B, with generic diagrams D_A and D_B we say that D_A homomorphically extends D_B iff there is a homomorphic embedding f: $A \mapsto B$.

Consider a complete theory T in L. A formula $\varphi (x_1, \ldots, x_n)$ is said to be complete (in T) iff for every formula $\psi(x_1, \ldots, x_n)$, exactly one of $T \models \varphi \to \psi$ or $T \models \varphi \to \neg \psi$. A formula $\theta (x_1, \ldots, x_n)$ is said to be completable (in T) iff there is a complete formula $\varphi (x_1, \ldots, x_n)$ with T models $\varphi \quad \theta$. If that cannot be done θ is said to be incompletable.

Proposition 2.7: Let A and B be models for L. Then A is isomorphically embedded in B iff B can be expanded to a model of the diagram of A.

Proposition 2.8: Let A and B be models for L. Then A is homomorphically embedded in B iff B can be expanded to a model of the positive diagram of A.

Definition 2.11: Given models A and B, with generic diagrams D_A and D_B we say that D_A homomorphically extends D_B iff there is a homomorphic embedding f: $A \mapsto B$.

Consider a complete theory T in L. A formula $\varphi (x_1, \ldots, x_n)$ is said to be complete (in T) iff for every formula $\psi(x_1, \ldots, x_n)$, exactly one of $T \models \varphi \to \psi$ or $T \models \varphi \neg \psi$. A formula $\theta (x_1, \ldots, x_n)$ is said to be completable (in T) iff there is a complete formula φ $(x_1,..., xn)$ with $T \models \varphi \rightarrow \theta$. If that can't be done θ is said to be incompletable.

Theorem 2.5: (Nourani, 2005) Let L1, L2 be two positive languages. Let $L = L1 \cap L2$. Suppose T is a complete theory in L and $T1 \supset T$, $T2 \supset T$ are consistent in L1, L2, respectively. Suppose there is model M definable from a positive diagram in the language $L1 \cup L2$ such that there are models M1 and M2 for T1 and T2 where M can be homomorphically embedded in M1 and M2.

1. $T1 \cup T2$ is consistent.
2. There is model N for $T1 \cup T2$ definable from a positive diagram that homomorphically extends that of M1 and M2.

Let $L_{P, \omega}$ be the positive fragment obtained from the Kiesler fragment. Define the category $L_{P, \omega}$ to be the category with objects positive fragments and arrows the subfoumual preorder on formulas.

Define a functor F: $\mathbf{L}_{P, \omega}{}^{op} \rightarrow$ Set by a list of sets M_n and functions f_n. The functor F is a list of sets F_n, consisting of

1. the sets corresponding to an initial structure on $L_{P, \omega}$, for example, the free syntax tree structure, where to $f(t_1, t_2,.. t_n)$; in $L_{P, \omega}$ there corresponds the equality relation $f(t_1, ..., t_n)=ft_1...t_n$ in Set;
2. the functions $f_i : F_i+1 \rightarrow F_i$.

Proposition 2.7 (positive morphic extensions): The following are equivalent:

1. Every positive sentence holding on A also holds on B.
2. There are elementary extensions A < A,' B < B' such that B' is a homomorphic image of A'

2.6 POSITIVE FRAGMENT CONSISTENCY MODELS

Let $L_{P, \omega}$ be the positive fragment obtained from Kiesler fragment.

Define the category $\mathbf{L}_{P, \omega}$ to be the category with objects positive fragments and arrows the subfoumual preorder on formulas. Taking the Op category from that $\mathbf{L}_{P, \omega}{}^{op}$:

Define a functor $\mathbf{F}$: $\mathbf{L}_{P, \omega}{}^{op} \rightarrow$ **Set** by a list of sets M_n and functions f_n. The functor F is a list of sets F_n, consisting of

1. the sets corresponding to an initial structure on $L_{P,\omega}$, for example the free syntax tree structure, where to $f(t_1, t_2, .. t_n)$ in $L_{P,\omega}$ there corresponds the equality relation $f(t_1, ..., t_n) = ft_1...t_n$ in Set;
2. the functions $f_i : F_i+1 \to F_i$.

Theorem 2.6: Infinitary fragment consistency on algebras.

Let T<i, i+1> be complete theories for L<i, i+1> = Li intersect Li+1. Let Ti and Ti+1 be arbitrary consistent positive theories for the subfragments Li and Li+1, respectively, satisfying Ti contains T<i, i+1> and Ti+1 contains T<i, i+1>. Let Ai be A(Fi) and Bi be A(Fi+1).

1. There are iterated elementary extensions Bi+1> Bi and an embedding fi: Ai < Bi;
2. Slalom between the language pairs Li Li+1 gates to a limit model for L, a model M for $L_{P,\omega}$.

Proof: Starting with a basis model A<0,1> and B<0,1>, models of T_0 and T_1. A_0|L<0,1> and B_0|L<0,1> are models of a complete theory, therefore A_0|L<0,1> ≡ B_0|L<0,1>. It follows that the elementary diagram of A L<0,1> is consistent with the elementary diagram of B_0|L<0,1>. Let T<i, i+1> be the positive theory in L<i, i+1> = Li intersect Li+1 defined by T* (A(Fi)) intersect A(Fi+1)), where T* is the theory obtained from the positive theory T augment with inductive consequences. We shall construct a limit model M realizing (i) enroute. Every $\Sigma(x_1,...,x_n)$ a set of formulas of L<0,1> that is provable in T, T*(A_0|L<0,1>) positively locally realizes Σ, that is there is a formula φ $(x_1,...,x_n)$ in L<01> s.t. φ is consistent with T*(A_0|L<0,1>) and for all $\sigma \in \Sigma$, T* $\models \varphi$ or T* $\cup \{\sigma\}$ is not consistent. Every finite subset Σ of T*(A_0|L<0,1>) has a model, therefore T*(A0|L<0,1>) has a model M_0, where for every φ e L <0,1> T*(A0|L<0,1>) s.t. φ iff M0 $\models \varphi$. Starting with a basis models A_0 and B_0, we construct a positive tower. On the iteration we realize that by s preceding proposition at each stage there are elementary extensions to A_i and B_i there are elementary extensions A < A,' B< B' such that B' is a homomorphic image of A'. Therefore, there are elementary extensions B1> B0 and an embedding f1: A0 < B1 at L<0,1>. Passing to the expanded language L<0,1>A0, we have (A0, a)ae A0 ° L<0,1> A0 (B1, Ka) ae A0. g1 inverse is an extension of f1.

Iterating, we obtain the tower depicted sideways

$$A_0 < A_1 < A_2 < \text{----------}$$
$$\backslash \text{ f1 } | \text{ g1 } \backslash \text{f2 } | \text{ g2}$$
$$B_0 < B_1 < b2 < \text{-----}$$

Slalom between the language pairs Li Li+1 gates to a limit model for L. For each m, fm $\subset$ inverse(gm) $\subset$ fm+1, fm: Am−1 < Bm at L<m−1, m>. Let A = $\cup$ Am, m < ω, B= $\cup$ Bm, m < ω. B is isomorphic to a model B' such that A|L = B'|L. Piecing A and B' together we obtain a model M for $L_{P, \omega}$.

Define the category $\mathbf{L}_{H,\omega}$ to be the category with objects Horn fragments and arrows the subformula preorder on formulas.

Define a functor $\mathbf{F}: \mathbf{L}_{H,\omega}{}^{op} \rightarrow \mathbf{Set}$ by a list of sets Mn and functions fn. The functor F is a list of sets Fn, consisting of (a) the sets corresponding to an initial structure on $L_{P, \omega}$, for example the free syntax tree structure, where to f(t_1, t_2,.. t_n) in $L_{P, \omega}$ there corresponds the equality relation f(t_1, ..., t_n)=ft$_1$...t$_n$ in Set; (b) the functions f_i:F$_i$+1 $\rightarrow$ F$_i$.

Proposition 2.8: Infinitary Horn Fragment Consistency.

Let T<i, i+1> be the complete theory in L<i, i+1> = Li intersect Li+1, defined by Th (A(Fi) intersect A(Fi+1)), Let Ti and Ti+1 be arbitrary consistent positive Horn theories for the subfragments Li and Li+1, respectively, satisfying Ti contains T<i, i+1> and Ti+1 contains T<i, i+1>. Let Ai be A(Fi) and Bi be A(Fi+1). Starting with a basis model A0, A0|L<0,1> and B0|L<0,1> are models of a complete theory, hence, A0|L<0,1> $\equiv$ B0|l<0,1>. It follows that the elementary diagram of A0|l<0,1> is consistent with the elementary diagram of B0|L<0,1>.

(i) There are iterated elementary extensions Bi+1> Bi and an embedding fi: Ai < Bi;

(ii) Slalom between the language pairs Li, Li+1 gates to a limit model for L, a model M for $L_{H, \omega}$.

2.6.1 POSITIVE CATEGORIES AND HORN FRAGMENTS

Let $L_{P, \omega}$ be the positive fragment contained in the Kiesler fragment. Define the category

$L_{P,\omega}$ to be the category with objects positive fragments and arrows the subformula preorder on formulas.

Define a functor **F: $L_{P,\omega}{}^{op} \rightarrow$ Set** by a list of sets Mn and functions fn. The functor F is a list of sets Fn, consisting of (a) the sets corresponding to an initial structure on $L_{P,\omega}$, for example the free syntax tree structure, where to $f(t_1, t_2,.. t_n)$ in $L_{P,\omega}$ there corresponds the equality relation $f(t_1, \ldots, t_n)$=$ft_1 \ldots t_n$ in Set; (b) the functions $f_i:F_i+1 \rightarrow F_i$.

Theorem 2.7: Infinitary Fragment Consistency on Algebras Let T<i, i+1> be complete theories for L<i, i+1> = Li intersect Li+1. Let Ti and Ti+1 be arbitrary consistent positive theories for the subfragments Li and Li+1, respectively, satisfying Ti contains T<i, i+1> and Ti+1 contains T<i, i+1>. Let Ai be A(Fi) and Bi be A(Fi+1). (i) There are iterated elementary extensions Bi+1> Bi and an embedding fi: Ai < Bi; (ii) Slalom between the language pairs Li Li+1 gates to a limit model for L, a model M for L P, ω. Similar statement on Horn fragment consistency can be proved.

Theorem 2.8: Infinitary Horn Fragment Consistency Let T<i, i+1> be the complete theory in L<i, i+1> = Li intersect Li+1, defined by Th (A(Fi) intersect A(Fi+1)), Let Ti and Ti+1 be arbitrary consistent positive theories for the subfragments Li and Li+1, respectively, satisfying Ti contains T<i, i+1> and Ti+1 contains T<i, i+1>. Let Ai be A(Fi) and Bi be A(Fi+1). Starting with a basis model A0, A0|L<0,1> and B0|L<0,1> are models of a complete theory, hence, A0|L<0,1> ≡ B0|l<0,1>. It follows that the elementary diagram of A0|l<0,1> is consistent with the elementary diagram of B0|L<0,1>. (i) There are iterated elementary extensions Bi+1> Bi and an embedding fi: Ai < Bi; (ii) Slalom between the language pairs Li, Li+1 gates to a limit model for L, a model M for L P, ω.

2.6.2 FRAGMENT CONSISTENT KLEENE MODELS

Ordered computing structure have become a significant area for computing since Scott (1970). Generic diagrams have been applied to define model-theoretic computing in the author's papers since 1980's. Applications to Functor Computability is defined since 90's. Practical computing with fuunctors had been an important area in the authors' projects since 1978. ADJ initial categorical computing and ordered algebraic definitions are based on categories, algebraic theories, and quotient structures. Functorial model computing as it has also appeared in the author's publications applied generic models diagrams. ADJ define an ordered Σ-algebra and their homomorphisms with an ordering where the operations are monotonic and the homomorphisms which are monotonic for the ordering defined. The ordering's significance is its being operation preserving.

The author had over two decades, since a graduate student, developed a model-theoretic ordering for the initial ordered structures to reach for models for which operation preserving orderings are definable with a model theory on many-sorted categorical logic and its definability with finite similarity types. The ordering the author calls morphic might further be computationally appealing ever since computable functors the were defined by us. The author defines initial ordered structures with a slight change in terminology from ADJ-79, since the authors had not developed the relation between preorder algebras and admissible sets. Nor had they claimed a relation to admissible sets. That appears the author's ASL Leeds publications. Admissible sets are as Barwise-68. Functorial admissible models in the admissible set sense are due to Nourani 96–97. The following definition is a variation from ADJ (1978) and Nourani (1998).

Ordered computing structure have become a significant area for computing since Dana Scott's accomplishments during 1970's on. Generic diagrams have been applied to define model-theoretic computing in this author's papers since 1980's. Applications to Functor Computability is defined since 90's. Practical computing with fuunctors had been an important area in the authors' projects since 1978. ADJ initial categorical computing and ordered algebraic definitions are based on categories, algebraic theories, and quotient structures. Functorial model computing as it has also appeared in the author's publications applied generic models diagrams. ADJ define an ordered Σ-algebra and their homomorphisms with an ordering where the operations are monotonic and the homomorphism which are monotonic for the ordering defined. The ordering's significance is its being operation preserving. The author defined two calls morphic might further be computationally appealing ever since computable decades ago on model-theoretic ordering for the initial ordered structures to reach for models for which operation preserving orderings are definable for a model theory on many-sorted categorical logic and its definability with finite similarity type. The ordering the author functors the were defined by us.

Definition 2.14: A preorder $\ll$ on a Σ-algebra A is said to be morphic iff for every $\sigma \in \Sigma_s$, $s_1 \ldots s_n$ and a_i, $b_i \in$ Asi, and Bsi, respectively, if $a_1 \ll b_1$ for

$i \in [n]$ then, σA $(a_1, \ldots, a_n) << \sigma B$ $(b_1, \ldots, b_n)$. A Kleene algebra (Koz90) is an algebra $A = (A, +, 0,\ldots, 1, \_)$ such that $(A, +, 0)$ and $(A,\ldots, 1)$ are monoids, with $+$ commutative and idempotent, and satisfying

(1) $a(b + c) = ab + ac$ (4) $1 + aa^* \leq a^*$
(2) $a0 = 0$ (5) $1 + a^*a \leq a^*$
(3) $(a + b)c = ac + bc$ (6) $ab \leq b$ îZ $a^*b \leq b$
      $0a = 0$ (7) $ba \leq b$ îZ $ba^* \leq b$

Denote by KA the class of models of these axioms, and write Horn(KA) and Eq(KA) for the Horn and equational theories of KA respectively.

Proposition 2.9: Kleene structures can be granted with an initial model characterization with morphic preorders.

Proof: (Exercises).

We shall apply fragment consistent modeling techniques to generate Kleene models. A tree game degree is the game state a tree is at with respect a model truth assignment, for example, to the parameters on the Boolean functions on a game tree. Event algebras with monadic signature agents model consistent and complete computation logics.

Theorem 2.9: Let T be a ISL language theory. T is (a) a sound logical theory iff every axiom or proof rule in T preserves the tree game degree; (b) a complete logical theory iff there is a function-set pair <F, S> defining a canonical structure M such that M has a generic diagram definable with the functions F.

We'll apply fragment consistent model techniques to generate Kleene models. A tree game degree is the game state a tree is at with respect a model truth assignment, for example, to the parameters on the Boolean functions on a game tree.

Theorem 2.10: Let T be a ISL language theory. T is (a) a sound logical theory iff every axiom or proof rule in T preserves the tree game degree; (b) a complete logical theory iff there is a function-set pair <F, S> defining a canonical structure M such that M has a generic diagram definable with the functions F.

Proof: See the structural counterpart Theorem 3.2.

2.7 MORE ON KLEENE STRUCTURES

Here this author's interests in language fragment models since 1996 takes on a tangent to (Leiß, 2005) extend Kozen's theory *KA* of Kleene Algebra to axiomatize parts of the equational theory of context-free languages, using a least fixed-point operator μ instead of Kleene's iteration operator*. Although the equational theory of context-free languages is not recursively axiomatizable, there are natural axioms for subtheories KAF $\subseteq$ KAR $\subseteq$ KAG: respectively, these make μ a least fixed point operator, connect it with recursion, and express S. Greibach's method to replace left- by right-recursion and vice versa. Over KAF, there are different candidates to define * in terms of μ, such as tail-recursion and reflexive transitive closure. In KAR, these candidates collapse, whence KAR uniquely defines * and extends Kozen's theory KA.

Regular algebra is the equational theory that is definable by finitely many equations between regular expressions. Recently, two finite axiomatizations by theory of the algebra of regular languages, initiated by Kleene (1956). Redko (1996) showed that this theory is not axiomatic means have been given. Pratt (1992) is theory of Action Logic, ACT, enriches Kleene's set $\{+,0,\cdot,1,*\}$ of regular operations by left and right residuals $\leftarrow$ and $\rightarrow$, and is axiomatized by finitely many equations. Iteration * is characterized in ACT by its monotonicity properties and the equation $(x \rightarrow x)* = x \rightarrow x$ of 'pure induction.' Both ACT and KA are complete for regular algebra. Reasoning about context-free languages could also profit from algebraic and logical means based on axiomatic theories, for example, ADJ (1976) and Nourani (1997). Leiß (2015) applies least fix-point operators to treat the combinatorial complications of formal power series. Universal Horn axiomatizations for Kleene algebras is a newer trend. Recall that regular expressions are inductively defined by r: $= 0 \mid 1 \mid x \mid a \mid (r+r) \mid (r\cdot r) \mid r*$, where a ranges over a finite list or alphabet Σ of constants, and x over an infinite list of variables. With a least-fixed-point operator μ, we define μ-regular expressions by r:$= 0 \mid 1 \mid x \mid a \mid (r+r) \mid (r\cdot r) \mid r* \mid \mu xr$.

There are two standard interpretations of regular expressions: On the language tree assignments, L_Σ, variables range over the universe of all subsets (or formal languages) of the set $\Sigma*$ of finite sequences (or words) of elements of Σ, 0 denotes the empty set, 1 the singleton set containing the empty word ϕ only, the constant a denotes $\{a\}$, + is set union, $\cdot$ is element-wise concatenation, and * is the union of all finite concatenations of a language with itself. Let REG_Σ be the subclass of all regular languages over Σ, which are those elements of L_Σ that are the value of a closed regular expression, for example, one without free variables. In the relation interpretation, RK, variables range over the universe of all binary relations on a set K, with the empty relation as 0, the identity on K as 1, union and composition as $+$ and $\cdot$, and reflexive transitive closure as $*$.

As an interpretation for the constants a we can take any relations $Ra \subseteq K \times K$; however, as there is no canonical choice for these, it seems natural here to consider pure expressions only, for example, those containing no constants except 0 and 1. Since all operations are monotone with respect to set and relation inclusion, respectively, one can extend these standard interpretations to μ-regular expressions, with μ picking the least fixed point of monotone functions. The fundamental theorem of recursion theory on the natural numbers, saying that (i) a partial function f is definable by a system of Godel-Herbrand-Kleene-equations iff (ii) f is μ-recursively definable iff (iii) f is computable by a Turing-machine, has the following analog concerning the definability of formal languages:

Theorem 2.11: For every language $A \subseteq \Sigma*$, the following conditions are equivalent:

   (i)   A is definable by a system of regular equations,
   (ii) A is definable by a μ-regular expression,
   (iii) A is accepted by a pushdown-store automaton.

In characterizations of context-free languages in the literature, condition (ii) often is missing. This offers a nice and simple way to introduce pushdown-store automata and to explain why an implementation of recursion μ needs a stack. Here, definable by a system of regular equations is intended to be a component of the least solution of a system $x_1 = r_1(x_1, \ldots, x_m)$ $x_m = r_m(x_1, \ldots, x_m)$, where each r_i is a regular expression whose free variables are among the pairwise distinct recursion variables $x_1, \ldots, x_m$. Context-free grammars can be seen as systems $\{x_i = r_i\} 1 \leq i \leq m$ of regular equations where the r_i are $*$-free and in disjunctive normal form. One can apply μ-regular expressions as a naming system for context-free languages to axiomatize pieces of their equational theory.

That is done in three steps: First, add to the algebraic properties of $+$, 0, $\cdot$,1 the assumption that μ picks the least pre-fixed point of definable functions $\lambda x.r$, for each μ-regular expression $r(x, y1, \ldots, yn)$: $\forall y1 \ldots \forall yn$ $(r[\mu x.r/x] \leq \mu x.r \circ \forall x(r \leq x \rightarrow \mu x.r \leq x))$. To express minimality, add universal Horn-axioms, with conditional (in)equations for the quantifier free part. The resulting theory is a Kleene algebra with least fixed points, abbreviated KAF.

In a second step, extend KAF by equational axioms to obtain a theory of Kleene algebra with recursion, KAR, that can model iteration $*$ by recursion μ-a suitably restricted form of the least fixed-point operator. In KAR, we relate μ with the algebraic operations $+$ and $\cdot$ by $\forall a, b. (\mu x(b + ax) = \mu x(1 + xa) \cdot b \circ \mu x(b + xa) = b \cdot \mu x(1 + ax))$. (In the following, we often write a, b for free variables, while a_i will be a constant of Σ that may be added to the pure language, without adding any relations between these.) Using these equations and taking $\mu x(1 + ax)$ as $a*$, the least fixed point properties for $\mu x(b + ax)$ and $\mu x(b + xa)$ in KAF are just the properties of $*$ taken as axioms in Kozen's (1992) theory KA of Kleene algebra.

Hence, KAR is indeed a theory of Kleene algebras, with iteration generalized to recursion. While KAR proves all equations between regular expressions that are valid in the language interpretation, it cannot prove all valid equations between μ-regular expressions, nor can it be completed to that. In spite of this limitation, in a third step we look for larger fragments of the equational theory of context-free languages that have natural axioms. Greibach's way to eliminate left recursion in context-free grammars relies on an equivalence between grammars that can compactly be expressed as an equation schema between μ-regular expressions r, s, possibly containing x: $\mu x(s + rx) = \mu x(\mu y(1 + yr) \cdot s) \circ \mu x(s + xr) = \mu x(s \cdot \mu y(1 + ry))$.

Adding this schema to KAF gives a theory KAG, which extends KAR. It remains to be seen to what extent common reasoning about the equivalence between context-free grammars can be carried out within KAG. Equations can also be read as claiming continuity properties about $+$ and $\cdot$ continuous models of KAF and Conway's notion of Standard Kleene algebra satisfy the identities. Moreover, an equation between μ-regular expressions that is valid in the interpretation by context-free languages holds universally in the continuous models of KAF.

2.7.1 IFLCS AND EXAMPLE EVENT PROCESS ALGEBRAS

To prove Godel's completeness theorem, Henkin (1949) style proceeds by constructing a model directly from the syntax of the given theory. The structure is obtained by putting terms that are provably equal into equivalence classes, then defining a free structure on the equivalence classes. The computing and reasoning enterprise require more general techniques of model construction and extension, since it has to accommodate dynamically changing world descriptions and theories. Our techniques for model building as applied to agent computing allows us to build and extend models with diagrams. The minimal set of function symbols are those with which a model can be built inductively. The models presented and applied are called intelligent.

The models are initial, for example, unique upto isomorphism, and homomorphic to all models in the category of models definable with the specific generic diagrams. Specific agent term algebra models were presented and proved computable (Nourani, 1994). Canonical Models from models to set theory had been stated for arbitrary structures as follows. Generic diagrams allow us to define canonical models with specific functions. The are specific canonical models on agent term algebras presented at Nourani (1996). In the sections above we have shown how to define models with G-diagrams. Initial models are models definable by G-diagrams. A primitive set of operations of algebras of a class C is a set that implicitly defines the remaining operations. Formally, given any class C of algebras of a given signature S, a subset $T \subset S$ is called primitive for C when it determines the remaining operations of S, in the sense that any algebra with signature T has at most one expansion (same elements but additional operations) to an algebra in C. For example either $\cdot$ or $+$ is primitive for Boolean algebras because they each give away the underlying poset of the algebra, via $a = a \cdot b$ or $a + b = b$, which then determines all the Boolean operations. The author had published papers since 1980's (Nourani, 1979, 1980) on initial algebraic models that are obtained from a primitive set of operations, showing that the operations are definable with conservative extensions.

Theorem 2.12 (Pratt, 1992): The operations $+$ and; form a primitive set for ACT.

A Kleene algebra (Koz90) is an algebra $A = (A, +, 0, \ldots, 1, \_)$ such that $(A, +, 0)$ and $(A, \ldots, 1)$ are monoids, with $+$ commutative and idempotent, and satisfying:

(1) $a(b+c)=ab+ac$
(2) $a0=0$
(3) $(a+b)c=ac+bc \; 0a=0;$

(4) 1+aa* ≤a*
(5) 1+a*a ≤ a*
(6) ab ≤ ba*b ≤ b
(7) ba ≤ bba*≤ b

We denote by KA the class of models of these axioms, and write Horn(KA) and Eq(KA) for the Horn and equational theories of KA, respectively.

Theorem 2.13: ACT < KA.

Theorem 2.14: Eq(ACT) conservatively extends Eq(REG).

Theorem 2.15: Every finite Kleene algebra expands to an action algebra.

2.7.2 ISL ALGEBRAS

A String ISL algebra (Nourani, 2005) is a Σ-algebra with an additional property that the signature Σ's has a subsignature Λ is only on 1–1 functions.

Lemma 2.3: String ISL algebra extending a Kleene algebra A (A, +,0,...,1, *) such that (A, +,0) and (A,...,1) are monoids, with + commutative and idempotent, is Kleene.

Lemma 2.4 (author 2005): String ISL algebra homomorphically extending a algebra A (A, +, 0, ..., 1, *) such that (A, +, 0) and (A, ..., 1) are monoids, with + commutative and idempotent, is Kleene.

We can define specific ISL algebras based on specific signatures. For example, a Kleene ISL algebra is an algebra A (A, +, 0, ..., 1, *) such that (A, +, 0) and (A, ..., 1) are monoids, with + commutative and idempotent, and satisfying

(1) a(b+c)=ab+ac
(2) a0=0
(3) (a+b)c=ac+bc 0a=0
(4) 1+aa* ≤a*
(5) 1+a*a ≤ a*
(6) ab ≤ ba*b ≤ b
(7) ba ≤ bba*≤ b
    + and. associative respective and a+0=0+a= a a.1=1.a=a

Proposition 2.10 (author 2007) (Positive Process Event Fragments) Models for event expressions on Equational ACT are created by defining a language category on the signature, where the preorder category has language fragment sets as objects and is preordered with the preimplication ordering.

Proof (outline) applies the consistency on fragment models, lemmas 2.3 and 2.4, on Kleene structures.

The computational bases for REG on Kleene expression, parsing on REG languages and applications are being studied at Copenhagen Data-Logik group (e.g., Bjørn et al., 2014). ISL algebras might be applicable to avoid parsing ambiguity on regular expression trees.

2.7.3 PROCESS ALGEBRAS

In process algebra, a (potentially) infinite process is usually represented as the solution of a system of guarded recursive equations. Although specification and verification of concurrent processes defined in this way serve their purposes well, recursive operations give a more direct representation and are easier to comprehend. The Kleene star $*$ can be considered as the most fundamental recursive operation. In process algebra, the defining equation for the binary Kleene star reads $x*y = x \cdot (x*y) + y$, where $\cdot$ models sequential composition and $+$ models non-deterministic choice ($\cdot$ binds stronger than $+$). This section examines the Kleene star in the setting of process algebra, and considers some derived recursive operations, with a focus on axiomatisations and expressiveness in view of the above sections. Kleene introduced the binary operation $*$ for describing 'regular events.' He defined regular expressions, which correspond to finite automata, and gave algebraic transformation rules for these, notably $E*F = E(E*F) \vee F$ ($E*F$ being the iterate of E on F).

Kleene noted the correspondence with conventions of algebra, treating $E \vee F$ as the analog of $E + F$, and EF as the product of E and F. (Copi et al., 1958) introduced simpler and stronger nets and simpler operations. In particular the latter authors introduced a unary $*$ operation "[…] because the operation. Redko (1969) proves that there does not exist a sound and complete finite equational axiomatisation for regular expressions. Salomaa (1967) presented a sound and complete finite axiomatization for regular expressions, with a basic ingredient an implicational axiom dating back to Arden, namely (in process algebra notation): $x=(y \cdot x)+z \Rightarrow x=y*z$ if y does not have the so-called empty word property Kleene's beginning were:

Theorem 2.16: Finite automata model regular events.

Theorem 2.17: Each regular event can be described by a finite automaton.

Following Bergstra et al.'s notations, a labeled transition system is as follows:

aa√ A labeled transition system (LTS) is a tuple $\langle S, \{ a\to, a\to, \surd$

S is a set of states, $a \mid a \in A\}, s\rangle$, where $a'a\surd\ a'$ Expressions s $-\to$ s and s $-\to$ are called transitions. Intuitively, s $-\to$ s denotes

$-\to$ for $a \in A$ is a binary relation between states, a√

$-\to$ for $a \in A$ is a unary predicate on states, $s \in S$ is the initial state. that from state s one can evolve to state s' by the execution of action a, while a√

$-\to$ denotes that from state s one can terminate successfully by the execution of action a ($\surd$ is pronounced "tick").

Definition 2.15: (Strong bisimulation) A strong bisimulation is a binary, symmetric relation R over the set of states that satisfies

S1 R S2 and S $-a\to$S $\Rightarrow \exists$S' (S $\to$ S') & S' $-a\to$ R)
S1 R S2 & P $-a\to \surd \Rightarrow$ S2 $-a\to \surd$

Two states S1 and S2 are strongly bisimilar, notation S1 $\leftrightarrow$ S2, if there exists a strong bisimulation relation R with S1 R S2.

Milner (1984) was the first to consider the unary Kleene star in process algebra, modulo strong bisimulation equivalence. In contrast with regular expressions, this setting is not sufficiently expressive to describe all regular (i.e., finite-state) processes. Moreover, the merge operator $x\|y$ that executes its two arguments in parallel, and which is fundamental to process algebra, cannot always be eliminated from process terms in the presence of the Kleene star. Milner presented an axiomatisation for the unary The unary Kleene star naturally gives rise to the empty process which does not combine well with the merge operator. Therefore, Bergstra et al. (1993) applies the binary version $x*y$ of the Kleene star in process algebra, naming the operator after its discoverer. In order to increase the expressive power of the binary Kleene star in strong bisimulation semantics, Bergstra et al. (1993) proposed multi-exit iteration $(x_1,..., x_k)*(y_1,..., y_k)$ for positive integers k, with as defining equation $(x_1, ..., x_k)*(y_1, ..., y_k)=x_1 \cdot ((x_2,$

..., x_k, x_1)*(y_2, ..., y_k, y_1))+y_1. Aceto and Fokkink (1996) presented an axiomatic characterization of multiexit iteration in basic process algebra, modulo strong bisimulation equivalence. Partial models for data types are areas that address partial computation models adding more comprehension to these considerations are Broy-Wirsing (1985) and Nourani (2003).

Merging a glimpse from ISL algebras with the process algebras we have, for example, the following proposition. Considering message exchange as the only operation defined on processes. We can, for example, state the following.

Proposition 2.11: (Author, 2015) ISL Algebras are Kleene models for multiexit iterations module strong bisimulation equivalence for 1–1 process sequent exchanges, for example, if sequent processes S1 and S2 are string equivalent then an ISL algebra models S1≡S2. The equivalence is modeled with an ISL algebra.

Proof: Follows from lemma 2.3 and 2.4 and the definitions.

EXERCISES

1. Prove proposition 2.4.
2. Prove proposition 2.9: Kleene structures can be granted with an initial model characterization with morphic preorders.
3. Provide a detailed proof for Proposition 2.9.
4. Prove the following: (Theorem 2.12) For every language $A \subseteq \Sigma*$, the following conditions are equivalent:
(i) A is definable by a system of regular equations,
(ii) A is definable by a μ-regular expression,
(iii) A is accepted by a pushdown-store automaton.
5. Provide a complete proof for proposition 2.10.
6. Prove proposition 2.11.
7. Prove Theorem 2.16: Every finite Kleene algebra expands to an action algebra.

KEYWORDS

- automata languages
- event expressions
- fragment consistent algebras
- functorial models
- infinite language categories
- Kleene algebras

REFERENCES

ADJ 1973, Goguen, J. A., Thatcher, J. W., Wagner, E. G., Wright, J. B. "A Junction Between Computer Science and Category Theory," (parts I and II), IBM T. J. Watson Research Center, Yorktown Heights, NY Research Report, RC4526,1973.

ADJ 1979, Thather, J.W, Wagner, E. G., Wright, J. B. "Notes On Algebraic Fundamentals For Theoretical Computer Science," IBM T. J. Watson Research Center, Yorktown Heights, NY, Reprint from Foundations Computer Science III, part 2, Languages, Logic, Semantics, J. de-Bakker and J. van Leeuwen, editors, Mathematical Center Tract 109, 1979.

Arbib, M., Manes, E. "Arrows, Structures and Functors", The Categorical Imperative, Academic Press, 1977.

Barwise, J., "Syntax and Semantics of Infinitary Languages, "Springer-Verlag Lecture Notes in Mathematics, vol. 72, (1968). Berlin-Heidelberg-NY.

Bergstra, J. A., Bethke, I., Ponse, A. Process algebra with iteration and nesting. The Computer Journal, 37(4), 243–258, 1994.

Bergstra, J. A., Klop, J. W. Algebra of communicating processes with abstraction. Theoretical Computer Science, 37(1):77–121, (1985).

Bergstra, Wan Fokkink1 Alban Ponse, Process Algebra with Recursive Operations. The Netherlands. http://www.wins.uva.nl/research/prog/

Broy, M., Wirsing, M. Partial Models for Abstract Data Types, Acta Informatika 1985.

Copi, I. M., Elgot, C. C., Wright, J. B. Realization of events by logical nets. Journal of the ACM, 5, 181–196, (1958).

Ehrig, H. (1977). "Embedding Theorem in the Algebraic Theory of Graph Grammars." FCT 1977: 245–255.

Ehrig, H., Claudia Ermel, Gabriele Taentzer, A Formal Resolution Strategy for Operation-Based Conflicts in Model Versioning Using Graph Modifications. TU Berlin, Forschungsberichte 2011, 1.

Fokkink, W. J. 1996," On the completeness of the equations for the Kleene star in bisimulation." In M. Wirsing and M. Nivat, eds., Proceedings 5th Conference on Algebraic Methodology and Software Technology (AMAST'96), Munich, LNCS (1101). pp. 180–194. Springer-Verlag, 1996.

Fritz Henglein, Lasse Nielsen (2011). Regular Expression Containment: Conductive Axiomatization and Computational Interpretation. In *Proc. 38th ACM SIGACT-SIGPLAN Symposium on Principles of Programming Languages (POPL)*, January 2011.

Fritz Henglein, Lasse Nielsen. Bit-coded Regular Expression Parsing. (2011). In *Proc. 5th Int'l Conf. on Language and Automata Theory and Applications (LATA)*, Lecture Notes in Computer Science (LNCS). Springer, May 2011.

Gentzen, G. (1943). Beweisbarkeit und Unbewiesbarket von Anfangsfallen der trasnfininten Induktion in der reinen Zahlentheorie, Math Ann 119, 140–161.

Haas Leiß, (1992). "Towards Kleene Algebra with recursion, Computer Science Logic Lecture Notes in Computer Science Volume 626, 242–256.

Hartmut Ehrig. Embedding Theorem in the Algebraic Theory of Graph Grammars. FCT (1977). 245–255

Keisler, H. J. (1971). Model Theory for Infinitary Logic, North Holland, Amsterdam, 1971.

Kleene, S. C. (1956). Representation of events in nerve nets and finite automata. In C. Shannon and J. McCarthy, eds., Automata Studies, pp. 3–41. Princeton University Press, 1956.

Kozen, D. (1990). "On Kleene algebras and closed semirings." In B. Rovan, editor, Mathematical Foundations of Computer Science (1990). volume 452 of Lecture Notes in Computer Science, pages a Bytrsica 26–47, Bansk, (1990). Springer-Verlag.

Kozen, D. (1994). A completeness theorem for Kleene algebras and the algebra of regular events. Information and Computation, 110(2), 366–390.

Lawvere, F. W. (1963). "Functorial Semantics of Algebraic Theories," Proc. National Academy of Sciences, USA.

Mac Lane, S., Categories for The Working Mathematician, GTM Vol. 5, Springer-Verlag, NY Heidelberg Berlin, 1971.

Niels Bjørn Bugge Grathwohl, Dexter Kozen, Konstantinos Mamouras. KAT + B!. In *Proceedings of the Joint Meeting of the Twenty-Third EACSL Annual Conference on Computer Science Logic (CSL) and the Twenty-Ninth Annual ACM/IEEE Symposium on Logic in Computer Science (LICS)*, p. 44:1–44:10, (2014).

Niels Bjørn Bugge Grathwohl, Fritz Henglein, Dexter Kozen. (2013). Infinitary Axiomatization of the Equational Theory of Context-Free Languages. In *Proc. 9th Workshop Fixed Points in Computer Science* (FICS 2013), 44–55, September 2013.

Niels Bjørn Bugge Grathwohl, Fritz Henglein, Lasse Nielsen, Ulrik Terp Rasmussen. (2013). Two-Pass Greedy Regular Expression Parsing. In S. Konstantinidis, editor, *Implementation and Application of Automata*, volume (7982). of *Lecture Notes in Computer Science*, pages 60–71. Springer Berlin Heidelberg, 2013.

Niels Bjørn Bugge Grathwohl, Fritz Henglein, Ulrik Terp Rasmussen. Optimally Streaming Greedy Regular Expression Parsing. *To appear at ICTAC (2014)*. September (2014). Bucharest.

Nourani, C. F. (1982). "Two part paper on Types, Induction, and Inductive Completeness," Second Workshop on Theory and Applications of Data Types, University of Passau, Passau, West Germany.

Nourani, C. F. (1983). "Forcing With Universal Sentences," (1981). Proc. ASL, (1983). vol. 49, Boston. MA.

Nourani, C. F. (1991a). "Types, Induction, and Incompleteness," (1982). Technical Report (Revised 1991), EATCS Bulletin, Page 226, June 1992.

Nourani, C. F. (1994). "Functorial Model Theory and Infinite Language Categories," September (1994). Presented to the ASL, January (1995). San Francisco. (ASL Bulletins 1996).

Nourani, C. F. (1995). "Functorial Model Computing," FMCS, UBC Mathematics Department, Vancouver, Canada, June 2005.

Nourani, C. F. (1995). "Functorial Model Theory, Generic Functors and Sets," January 16, (1995). International Congress, Logic, Methodology, and Philosophy of Science, Florence, Italy, August 1995.

Nourani, C. F. (1995). The Term Rewriting Roller Coaster February 2nd, International Workshop on Term Rewriting, May 95, La Bresse, France.

Nourani, C. F. (1996). "Computable Functors and Generic Model Diagrams A Preview To The Foundations, December 1996.

Nourani, C. F. (1996). "Infinite Language Categories Limit Topology and Categorical Computing, Krief Overview," MFPS, Boulder Colorado, June (1996). Handbook Mathematical Logic, J. Barwise, editor, (1978). North-Holland.

Nourani, C. F. (1996a). "Slalom Tree Computing," (1994). AI Communications, Vol. 9. No.4, December (1996). IOS Press, Amsterdam

Nourani, C. F. (1997). "Functorial Consistency," May (1997). AMS 927, Milwaukee, Wisconsin, (1997). Abstract number 927–03–29.

Nourani, C. F. (1997). "Functorial Model Theory and Generic Fragment Consistency Models," October (1996). AMS-ASL, San Diego, January 1997.

Nourani, C. F. (1997). "Functorial Models, Admissible Sets, and Generic Rudimentary Fragments," Summer Logic Colloquium, (1997). Leeds.

Nourani, C. F. (1998). "Functorial Models, Admissible Sets, and Generic Rudimentary Fragments," March (1997). Summer Logic Colloquium, Leeds, July (1997). BSL, vol. 4, no.1, March (1998). www.amsta.leeds.ac.uk/events/logic97/con.html

Nourani, C. F. (2003). "A Sound and Complete Agent Logic Paradigm, 2000." Parts and abstract published at ASL, And FSS, AAAI Symposium.

Nourani, C. F. (2003). "Higher Stratified Consistency and Completeness Proofs," (2003). Helsinki, August 14–20.

Nourani, C. F. (2003). "KR, Predictive Model Discovery, and Schema Completion," Florida 2003 7th World Multiconference on Systemics, Cybernetics and Informatics (SCI 2002) to be Orlando, USA, in July 14–18, (2003). http://www.iiis.org/sci2002.

Nourani, C. F. (2005). "Functorial String Models," ERLOGOL-2005: Intermediate Problems of Model Theory and Universal Algebra, June 26–July 1, State Technical University/Mathematics Institute, Novosibirsk, Russia. www.nstu.ru/science/conf/erlogol-2005.phtml www.ams.org/mathcal/info/ 2005_jun26-jul1_novosibirsk.html

Nourani, C. F. (2005). Fragment Consistent Algebraic Models, July 2006 Abstract and presentation to the Categories Oktoberfest, U Ottawa, October 2005.

Nourani, C. F. (2005a). "Positive Categories and Process Algebras," Draft outline August 2005, Written to a perfunctory Stanford CS project.

Nourani, C. F. (2005b). "Functorial Model Computing," FMCS, UBC Mathematics Department, Vancouver, Canada, June 2005.

Nourani, C. F. (2005c). "Positive Categories and Process Algebras," Draft outline August (2005). Written to a perfunctory Stanford CS project. Abstract published at Topology and Computing, Paris Polytechnique, France Year?

Nourani, C. F. (2005d). "Fragment Consistent Algebras, July 2005.

Nourani, C. F. (2006). "Functorial Generic Filters," July (2005). ASL, Montreal, 2006.

Nourani, C. F., "Functorial Model Theory and Infinite Language Categories," September (1994). Presented to the ASL, January (1995). San Francisco. ASL Bulletins 1996.

Nourani, C. F., "Functorial String Models," ERLOGOL-2005: Intermediate Problems of Model Theory and Universal Algebra, June 26–July 1, State Technical University/Mathematics Institute, Novosibirsk, Russia. www.nstu.ru/science/conf/erlogol-2005.phtml; www.ams.org/mathcal/info/ 2005_jun26-jul1_novosibirsk.html

Nourani, C. F., "Higher Stratified Consistency and Completeness Proofs," April (2003). SLK (2003). Helsinki August 14–20. http://www.math.helsinki.fi/logic/LC2003/abstracts/

Nourani, C. F., Th. Hoppe (1994). "GF-Diagrams for Models and Free Proof Trees," Proceedings the Berlin Logic Colloquium, Homboldt University, May. Univesitat Potsdam, Germany.

Pratt, V. R. (1990). "Action Logic and Pure Induction," Invited paper, Logics in AI: European Workshop JELIA '90, ed. J. van Eijck, LNCS 478, 97–120, Springer-Verlag, Amsterdam, NL, Sep, (1990). Also Report No. STAN-CS-90–1343, CS Dept., Stanford, Nov. (1990). http://boole.stanford.edu/pub/jelia.pdf.

RAMiCS (2014). 14th International Conference on Relational and Algebraic Methods in Computer Science, 9 Jun (2005). Marienstatt im Westerwald, Germany, April 28 – Mai 1. http://mathcs.chapman.edu/ramics2014/index.html

Schroder, E. (1895). "Vorlesungenuber die Algebra der Logik (Exakte Logik). Dritter Band: Algebra und Logik der Relative." B. G. Teubner, Leipzig.

Scott, D. (1970). "Outline of a Mathematical Theory of Computation," Technical Monograph PRG-2, Oxford University Computing Lab., Proc. 4th Annual Princeton Conference on Information Sciences and Systems, pp.169–176.

Widerhold, G. (1992). "Mediation in the Architecture of Future Information Systems," IEEE Computer Magazine, Vol. 25, Number 5, 33–49.

CHAPTER 3

FUNCTORIAL ADMISSIBLE MODELS

CONTENTS

Infinite language categories were presented and their categorical properties are defined in author's 1994–95 papers. Computable functors for models is defined in December 1996 paper and a brief appeared at ASL (Nourani, 1997). By defining admissible generic diagrams, we define functors to admissible structures. Starting with infinite language category $L_{\omega 1, K}$ (Nourani, 1996) we define generic sets on $L_{\omega 1, \omega}$ Let L be the admissible language (Barwise, 1972) defined on the Keisler $L_{\omega 1, \omega}$ fragment. We have proved in Nourani (1997) the relations to descriptive computing and admissible computable models. Amongst the theorem is that there is a generic functor defining a descriptive computable admissible model. Let us start with a preliminary on infinitary languages. Section 3.2 presents an overview to infinitary languages, Section 3.3 addresses admissible sets and languages and examines the completeness and compactness areas for infinitary languages; Section 3.4 presents descriptive computing and a brief on the applications to admissible sets.

3.1   INFINITARY LANGUAGES BASICS

Based on the standard papers on syntax and semantics of infinitary languages (Barwise, 1967), handbook of mathematic logic and encyclopedia 2000, we have to paraphrase the obvious definitions. The definitions are stated here for self-containment. Given a pair κ, λ of infinite cardinals such that $\lambda \leq \kappa$, let us define a class of infinitary languages where we may form conjunctions and disjunctions of sets of formulas of cardinality $< \kappa$, and quantifications on variables of length $< \lambda$. Let L be a (finitely) base language with an arbitrary but fixed first-order language with any number of extralogical symbols. The infinitary language $L(\kappa, \lambda)$ has the following basic symbols:

- All symbols of L;
- A set Vr of individual variables, where the cardinality of Vr (written: |Vr|) is κ;
- A logical operator ∘ (infinitary conjunction).

The class of preformulas of $L(\kappa, \lambda)$ is defined recursively as follows:

- Each formula of L is a preformula;
- if φ and ψ are preformulas, so are $\varphi \circ \psi$ and $\neg \varphi$;
- if φ is a set of preformulas such that $|\Phi| < \kappa$, then $\circ \Phi$ is a preformula;

- if φ is a preformula and $X \subseteq Vr$ is such that $|X| < \lambda$, then $\exists X\varphi$ is a preformula; all preformulas are defined by the above clauses.

If Φ is a set of preformulas indexed by a set I, $\Phi = \{\varphi_i : i \in I\}$, the standard notation is to write $\Phi_{i \in I}$ for: $_i\Phi$, or, if I is the set of natural numbers, we write $\circ\Phi$ for: $\varphi_0 \circ \varphi_1 \circ \ldots$

If X is a set of individual variables indexed by an ordinal α, say $X = \{x_\xi : \xi < \alpha\}$, we write $(\exists x_\xi)_{\xi < \alpha} \varphi$ for $\exists X\varphi$. The logical operators $\vee$, $\rightarrow$, $\leftrightarrow$ are defined in the customary manner. We also introduce the operator's $\vee$ (infinitary disjunction) and $\forall$ (universal quantification) by

$$\vee\Phi =_{df} \neg\circ\{\neg\varphi : \varphi \in \Phi\}$$

$\forall X\varphi =_{df} \neg\exists X\neg\varphi$, and employ similar conventions as for $\circ$, $\exists$.

More specific fragments are presented on the following chapters.

Thus, $L(\kappa, \lambda)$ is the infinitary language obtained from L by permitting conjunctions and disjunctions of length $< \kappa$ and quantifications of length $< \lambda$. Languages $L(\kappa, \omega)$ are called finite-quantifier languages. The rest are infinite-quantifier languages. $L(\omega, \omega)$ is L itself: an anomaly can arise in an infinitary language but not in a finitary one. In the language $L_{\omega1, \omega}$ which allows countably infinite conjunctions but only finite quantifications, there are preformulas with so many free variables that they cannot be "closed" into sentences of $L_{\omega1, \omega}$ by prefixing quantifiers. Such is the case, for example, for the $L(\omega1,\omega)$-preformula $x_0 < x_1 \circ x_1 < x_2 \circ \ldots \circ x_n < x_{n+1} \ldots$, where L contains the binary relation symbol $<$. For this reason the following characterization of the standard model of arithmetic in $L_{1, \omega}$ is stated as follows: the standard model of arithmetic is the structure $N = \langle N, +, \cdot, s, 0\rangle$, where N is the set of natural numbers, $+$, $\cdot$, and 0 have their usual meanings, and s is the successor operation.

Let L be the first-order language appropriate for N. Characterization of the class of all finite sets in $L_{\omega1, \omega}$. Here the base language has no extralogical symbols. The class of all finite sets then coincides with the class of models of the $L_{\omega1, \omega}$ sentence $\vee_{n \in \omega} \exists v_0 \ldots \exists v_n \forall x(x = v_0 \vee \ldots \vee x = v_n)$. Let L be a countable first-order language (for example, the language of arithmetic or set theory) which contains a name n for each natural number n, and let $s_0, s_1, \ldots$ be an enumeration of its sentences. Well-orderings can be characterized as follows.

The base language L here includes a binary predicate symbol $\leq$. Let s_1 be the usual L-sentence characterizing linear orderings. Then the class of L-structures in which the interpretation of $\leq$ is a well-ordering coincides with the class of models of the $L_{\omega1,\omega}$ sentence $s = s_1 \circ s_2$, where $s s_2$ is $(\forall v_n)$ $\exists x_{n\in\omega} (V_{n\in\omega} (x = v_n) \circ {}_{\circ n\in\omega} (x \leq v_n))$.

Notice that the sentence s_2 contains an infinite quantifier: it expresses the essentially second-order assertion that every countable subset has a least member. It can in fact be shown that the presence of this infinite quantifier is essential: the class of well-ordered structures cannot be characterized in any finite-quantifier language. This example indicates that infinite-quantifier languages such as $L(\omega_1,\omega_1)$ behave similar to second-order languages, with similar defects, however, similar expressive power.

Many extensions of first-order languages can be translated into infinitary languages. Another language translatable into $L_{\omega1,\omega}$ in this sense is the weak second-order language obtained by adding a countable set of monadic predicate variables to L which are then interpreted as ranging over all finite sets of individuals. Languages with arbitrarily long conjunctions, disjunctions and (possibly) quantifications may also be introduced. For a fixed infinite cardinal λ, the language $L(\infty, \lambda)$ is defined specifying its by class of formulas, $Frm(\infty, \lambda)$, to be the union, over all $\kappa \geq \lambda$, of the sets $Frm(\kappa, \lambda)$. Thus, $L(\infty, \lambda)$ allows arbitrarily long conjunctions and disjunctions, in the sense that if Φ is an arbitrary subset of $Frm(\infty, \lambda)$, then both $\circ\Phi$ and $V\Phi$ are members of $Frm(\infty, \lambda)$. But $L(\infty, \lambda)$ admits only quantifications of length $< \lambda$: all its Formulas have $< \lambda$ free variables. The language $L(\infty, \infty)$ is defined in turn by specifying its class of formulas, $Frm(\infty, \infty)$, to be the union, over all infinite cardinals λ, of the classes $Frm(\infty, \lambda)$. So $L(\infty, \infty)$ allows arbitrarily long quantifications in addition to arbitrarily long conjunctions and disjunctions.

3.2   ADMISSIBLE LANGUAGES

To reach well-behaved infinitary language fragments that have the important compactness property let us examine hereditary finite set codings on $H(\omega)$. There is the usual definability and Gödel coding for logic that is well-known. An equivalent coding structure for the first order language of arithmetic is the structure $\langle H(\omega), \in H(\omega)\rangle$ of *hereditarily finite sets*, where a set x is *hereditarily finite* if x, its members, its members of members, etc., are all finite. This coding structure takes account of the fact that first order formulas are naturally regarded as finite sets.

Each finite set of codes of L-sentences is a member of $H(\omega)$. The compactness theorem for L might be stated as follows:

Theorem 3.1: If $\Delta \subseteq \text{Sent}(L)$ is such that each subset $\Delta_0 \subseteq \Delta$, $\Delta_0 \in H(\omega)$ has a model, so does Δ.

Proof consequence of the Hampf coding completeness theorem.

Theorem 3.2: If $\Delta \subseteq \text{Sent}(L)$ and $\Sigma \in \text{Sent}(L)$ satisfy $\Delta \vDash \sigma$, then there is a deduction D of Σ from Δ such that $D \in H(\omega)$.

Proof (Exercises).

The compactness theorem for ordinary $L_{\omega1,\omega}$ fails since there are sentences a set $\Gamma \subseteq \text{Sent}(\omega1,\omega)$ such that Each countable subset of Γ has a model but Γ does not. Example: Let L be the language of arithmetic augmented by ω_1 new constant symbols $\{c_\xi : \xi < \omega_1\}$ and let Γ be the set of $L(\omega1,\omega)$-sentences $\{\sigma\} \cup \{c_\xi \neq c_\eta : \xi \neq \eta\}$, where Σ is the $L_{\omega1,\omega}$ sentence characterizing the standard model of arithmetic. Each countable subset of Γ has a model but Γ does not. However, there are well-behaved fragments and admissible infinitary languages with the compactness property upheld. Recall also that we introduced the notion of deduction in $L_{\omega1,\omega}$ since such deductions are of countable length it follows from the above that there is a sentence $\Sigma \in \text{Sent}(\omega_1,\omega)$ such that $\Gamma \vDash \sigma$, but there is no deduction of Σ in $L_{\omega1,\omega}$ from Γ.

Now the formulas of $L_{\omega1,\omega}$ can be coded as members of $H(\omega_1)$. Now $H(\omega_1)$ is closed under the formation of countable subsets and sequences. So we can restate compactness as follows: Each $\Gamma_0 \subseteq \Gamma$ such that $\Gamma_0 \in H(\omega_1)$ has a model, but Γ does not. There is a sentence $\Sigma \in \text{Sent}(\omega1,\omega)$ such that $\Gamma \vDash \sigma$, but there is no deduction $D \in H(\omega_1)$ of Σ from Γ. It follows that theorem 3.1 and 3.2 fail the when "L" is replaced by "$L(\omega1,\omega)$" and "$H(\omega)$"by "$H(\omega_1)$." It can be further shown that the set $\Gamma \subseteq \text{Sent}(\omega1,\omega)$ in the Theorems 3.1 and 3.2 may be taken to be Σ_1 on $H(\omega_1)$. Thus, the compactness and generalized completeness theorems fail even for Σ_{1-} sets of "L" is replaced by "$L_{\omega1,\omega}$ and "$H(\omega)$" by "$H(\omega_1)$."

From Theorem 3.1, we observer that thee reason why the generalized completeness theorem fails for Σ_{1-}sets in $L_{\omega1,\omega}$ is that $H(\omega_1)$ is not really "closed" under the formation of deductions from Σ_{1-} sets of sentences in $H(\omega_1)$. So in order to remedy this it would seem natural to replace $H(\omega_1)$ by sets A which are, in some sense, closed under the formation of such deductions, and then to consider just those formulas whose codes are in

A. An overview on how this can be done is as follows. First, we identify the symbols and formulas of $L_{\omega1, \omega}$ with their encoding in $H(\omega_1)$, For each countable transitive set A, let $L_A = (Frm(L_{\omega1, \omega})) \cap A$.

Now, let us consider sublanguage fragments for L_A for $L_{1, \omega}$ if the following conditions are satisfied:

i. $L \subseteq L_A$
ii. if φ, $\psi \in L_A$, then $\varphi \circ \psi \in L_A$ and $\neg\varphi \in L_A$
iii. if $\varphi \in L_A$ and $x \in A$, then $\exists x\varphi \in L_A$
iv. if $\varphi(x) \in L_A$ and $y \in A$, then $\varphi(y) \in L_A$
v. if $\varphi \in L_A$, every subformula of φ is in L_A
vi. if $\Phi \subseteq L_A$ and $\Phi \in A$, then $\circ\Phi \in L_A$.

The notion of deduction in L_A is defined in the customary way; if Δ is a set of sentences of L_A and $\varphi \in L_A$, then a deduction of φ from Δ in L_A is a deduction of φ from Δ in size disproportionate cosmetics: $L_{\omega1,\omega}$ every formula of which is in L_A. We say that φ is deducible from Δ in L_A if there is a deduction D of φ from Δ in L_A; under these conditions we write $\Delta \vdash_A \varphi$. In general, D will not be a member of A; in order to ensure that such a deduction can be found in A it will be necessary to impose further conditions on A. We define the Hanf number $H(L)$ of a language L to be the least cardinal κ such that, if an L-sentence has a model of cardinality κ, it has models of arbitrarily large cardinality.

Let A be a countable transitive set such that L_A is a sublanguage of $L_{\omega1,\omega}$ and let Δ be a set of sentences of L_A. We say that A (or, by abuse of terminology, L_A) is Δ-closed if, for any formula φ of L_A such that $\Delta \vdash_A \varphi$, there is a deduction D of φ from Δ such that $D \in A$. It can be shown that the only countable language which is Δ-closed for arbitrary Δ is the first-order language L, for example, when $A = H(\omega)$. However, Barwise discovered that there are countable sets $A \subseteq H(\omega_1)$ whose corresponding languages L_A differ from L and yet are Δ-closed for all Σ_1 sets of sentences Δ. Such sets A are called admissible sets. Informally, they are extensions of the hereditarily finite sets in which recursion theory, and hence proof theory, are possible.

From these definitions there is the following compactness theorem.

Theorem 3.3: (Barwise 1967) Let A be a countable admissible set and let Δ be a set of sentences of L_A which is Σ_1 on A. If each $\Delta' \subseteq \Delta$ such that $\Delta' \in A$ has a model, then so does Δ.

Proof: (Exercises).

The presence of "Σ_1" here indicates that this theorem is a generalization of the compactness theorem for recursively enumerable sets of sentences.

There is an alternate version of the Barwise compactness theorem, useful for constructing models of set theory applies countable transitive sets.

Definition 3.1: $L(\kappa, \lambda)$ is a preformula which contains $< \lambda$ free variables. The set of all formulas of $L(\kappa, \lambda)$ will be denoted by $\mathrm{Frm}(L(\kappa, \lambda))$ or simply $\mathrm{Frm}(\kappa, \lambda)$ and the set of all sentences by $\mathrm{Snt}(L(\kappa, \lambda))$ or abbreviated as $\mathrm{Snt}(\kappa, \lambda)$.

Having defined the syntax of $L(\kappa, \lambda)$, we next brief on the semantics. Since the extralogical symbols of $L(\kappa, \lambda)$ are just those of L, and it is these symbols which determine the form of the structures in which a given first-order language is to be interpreted, it is natural to define an $L(\kappa, \lambda)$-structure to be simply an L-structure. The notion of a formula of $L(\kappa, \lambda)$ being satisfied in an L-structure A (by a sequence of elements from the domain of A) is defined in the same inductive manner as for formulas of L except that we must add two extra clauses corresponding to the clauses for $\circ\Phi$ and $\exists X\varphi$ in the definition of preformula. In these two cases we naturally define: $\circ\Phi$ is satisfied in A (by a given sequence) $\Leftrightarrow$ for all $\varphi \in \Phi$, φ is satisfied in A (by the sequence); $\exists X\varphi$ is satisfied in A $\Leftrightarrow$ there is a sequence of elements from the domain of A in objective correspondence with X which satisfies φ in A.

These informal definitions need to be tightened up in a rigorous development, but their meaning should be clear to the reader. Now the usual notions of truth, validity, satisfiability, and model for formulas and sentences of $L(\kappa, \lambda)$ become available. In particular, if A is an L-structure and $\Sigma \in \mathrm{Sent}(\kappa, \lambda)$, we shall write $A \vDash \Sigma$ for A is a model of σ, and $\vDash \Sigma$ for Σ is valid, that is, for all A, $A \vDash \sigma$. If $\Delta \subseteq \mathrm{Snt}(\kappa, \lambda)$, we shall write $\Delta \vDash \Sigma$ for Σ is a logical consequence of Δ, that is, each model of Δ is a model of σ.

Similar arguments, applied with further augmentations of the axioms and rules of inference, reaches completeness theorems for many other finite-quantifier languages. If just deductions of countable length are admitted, then no deductive apparatus for $L(\omega 1, \omega)$ can be set up which is adequate for deductions from arbitrary sets of premises, that is, for which the theorems would hold for every set $\Delta \subseteq \mathrm{Snt}(\omega 1, \omega)$, regardless of cardinality. This is because there is a first-order language L and an uncountable set Γ of $L_{\omega 1, \omega}$ sentences such that Γ has no model but every countable subset of Γ does. For example, let L be the language of arithmetic augmented by ω_1 new constant symbols $\{c_\xi: \xi < \omega_1\}$ and let Γ be the set of $L(\omega1,\omega)$-sentences $\{\sigma\} \cup \{c_\xi \neq c_\eta: \xi \neq \eta\}$, where Σ is the $L(\omega1,\omega)$-sentence characterizing the standard model of arithmetic.

This example also shows that the compactness theorem fails for $L(\omega1,\omega)$ and so also for any $L(\kappa, \lambda)$ with $\kappa \geq \omega_1$. On the other hand, the upward Löwenheim-Skolem theorem in its usual general form does not hold for all arbitrary infinitary languages. For example, the $L_{\omega1,\omega}$-sentence characterizing the standard model of arithmetic has a model of cardinality $\aleph_0$ but no models of any other cardinality. However, that has a transcending alternative for infinitary languages. We define the Hanf number H(L) of a language L to be the least cardinal κ such that, if an L-sentence has a model of cardinality κ, it has models of arbitrarily large cardinality. The existence of H(L) is readily established. For each L-sentence Σ not possessing models of arbitrarily large cardinality let $\kappa(\sigma)$ be the least cardinal κ such that Σ does not have a model of cardinality κ. If λ is the supremum of all the $\kappa(\sigma)$, then, if a sentence of L has a model of cardinality λ, it has models of arbitrarily large cardinality.

The values of the Hanf numbers of infinite-quantifier languages such as $L(\omega1,\omega_1)$ are sensitive to the presence or otherwise of large cardinals, but must in any case greatly exceed that of $L(\omega1,\omega)$. A result for L which generalizes to $L(\omega1,\omega)$ but to no other infinitary language is the Craig Interpolation Theorem: If $\sigma, \tau \in Snt(\omega1,\omega)$ are such that $\vDash \Sigma \to \tau$, then there is $\theta \in Snt(\omega1,\omega)$ such that $\vDash \Sigma \to \theta$ and $\vDash \theta \to \tau$, and each extralogical symbol occurring in θ occurs in both Σ and τ.

The proof is an extension of the first-order case. Finally, we mention one further result which gene realizes nicely to $L(\omega1,\omega)$ but to no other infinitary language. It is well known that, if A is any finite L-structure with only finitely many relations, there is an L-sentence Σ characterizing A up to isomorphism. For $L(\omega1, \omega)$ we have the following generalization known as the Scott's Isomorphism Theorem.

Theorem 3.4 (Scott): If A is a countable L-structure with only countably many relations, then there is an $L_{\omega1,\omega}$ sentence whose class of countable models coincides with the class of L-structures isomorphic with A.

The restriction to countable structures is essential since countability cannot in general be expressed by an $L_{\omega1,\omega}$ sentence.

3.3   ADMISSIBLE MODELS

We restate a brief on example admissible models to start with. We define initial ordered structures with a slight change in terminology, since we have not yet defined the relation between preorder algebras and admissible sets. Admissible sets are as Barwise (1968). Functorial admissible models in the admissible set sense are due to the author's (Nourani 1996–1997).

Definition 3.2: A preorder $<<$ on a $\sum$-algebra A is said to be morphic iff for every s in s, s1...sn and ai, Ki in Asi, Ksi, respectively, if ai, bi for i in [n] then, sA (a1, ..., an) $<<$ sB (b1, ..., bn).

To save confusion with admissible sets, what we a call in definition 3.5 a morphic preorder here is what was called an admissible order by Thatcher Wagner, et al. (1979). The ordering's relation to admissible sets and functors are explored at Nourani (1997).

Proposition 3.1: If $<<$ is a morphic preorder on a $\sum$-algebra A, then the equivalence relation $\sim$ determined by $<<$ is a congruence relation.

Proposition 3.2: The morphic quotient structure for an ordered $\sum$—algebra A, is an ordered $\sum$-algebra.

We apply function and relation definably in the following theorem. To have a glimpse as to what it implies let us see an example. Let A be a model with built-in Skolem functions. A function $f \in A$, for A the model's universe, is said to be definable iff there exits a formula φ (xyz1...zn) in the language and elements a1, ..., an $\in$ A such that for all a, K $\in$ A, A $\vDash$ φ [aba1, ..., an] iff f(a)= b. We have defined generic Functors on the category $\mathbf{L}_{\omega 1, K}$ with computable hom sets (Nourani, 1996).

Definition 3.3: For functions f and g in a structure for a language L, define the morphic preorder $f << g$ iff there are formulas φ and γ from L such that the formulas define f and g, respectively, and φ is a subformula of γ in the sense of definition 3.1.

3.4   INFINITE LANGUAGE CATEGORIES

Define a functor F: $\mathbf{L}_{\omega 1, K} \rightarrow$ Set by a list of sets Mn and functions fn. The functor F is a list of sets Fn, consisting of (a) the sets corresponding to a morphic preorder on $\mathbf{L}_{\omega 1, K}$ defined by the deniability order from the subformula Preorder on $\mathbf{L}_{\omega 1, K}$. Starting with he free syntax tree structure, where to f(t1, t2,.. tn) in $L_{\omega1,K}$ there corresponds the equality relation f(t1, …, tn)=ft1…tn in Set; (b) the functions fi:Fi+1 → Fi.

Theorem 3.5 (author): The morphic preorder functor defines initial models for $L_{\omega1,K}$.

To prove Theorem 3.5 there are two routes. Route A applies (Keisler, 1971; Nourani, 1982, 1997) with new techniques similar to Robinson's consistency theorem, however, with varying languages and fragments called functorial consistency. Route B is outlined on the following paragraph from this author's ASL publications and Nourani (2014), where functorial morphic models are defined. One consequence of Nourani (1998) is a theorem, which can get us Theorem 3.6's proof. It itself is obtained by techniques, which are by no means obvious applying admissible sets and Urelements.

Theorem 3.6 (author): The morphic preorders on initial structures are definable by formulas, which are preserved by end extensions on fragment consistent models.

Towards the proof: Since (Nourani, 1997) European Research council brief as follows. At the Summer 1996 European logic colloquium (Nourani, 1996) had put forth descriptive computing principles, defining descriptive computable functions. Amongst the theorems at is that for A an admissible computable set, A is descriptive computable.

Generic diagrams were applied to define admissible computable sets and models.

Definition 3.4: A model is admissible iff its universe, functions, and relations are defined with or on admissible sets.

Theorem 3.7 (author 1997): Admissible models are obtained by taking a reduct from the admissible hull to the Skolem hull definable by a generic diagram.

ADJ (1979) defines ordered initial MacLane (1971) categories important for computing. The ordering's significance is its being operation preserving. We define a model-theoretic ordering for the initial ordered structures to reach for models for which operation preserving orderings are definable. The author's dissertation was in part on a model theory on many-sorted categorical logic and its definability with finite similarity type. That proved to be a newer look at intuitionistic forcing a year or two later and announced at ASL, Boston. The ordering we call morphic might further be computationally appealing ever since computable functors were defined by Nourani (1996, 1998).

Definition 3.5: For functions f and g in a structure for a language L, define the morphic preorder f $\ll$ g iff there are formulas j and y from L such that the formulas define f and g respectively, and j is a subformula of y in the sense of definition of the fragments for infinitary logic (Keisler, 1973).

We can apply our generic diagrams and sets from + forcing Nourani (1982) to the S formulas (Barwsie) 1972; Kripke (1964, 1965) Platek (1966) by applying a theorem form (Feferman-Kreisel, 1966.) It might also put our mind at serenity with the preorder functors and initial structures on the forthcoming chapters.

Theorem 3.8 (Ferferman-Kriesel): If a formula q is preserved under end extensions then it is equivalent to a S formula.

We apply the Ferferman-Kriesel theorem to check for conditions on admissible sets starting with what we call generic rudimentary sets. With positive forcing (Nourani, 1983) and fragment consistency models (Nourani, 1997), we can check if the formula belongs to a condition for fragment consistent Functorial models.

Kiesler (1971) and Barwise (1969, 1972) admissible fragments are also pertaining. We had defined functors from the language category $\mathbf{L}_{\omega 1,K}$ defined on the (Keisler, 1973) $L_{\omega 1,K}$ fragment to the category Set. Its arrows are the preorder arrows and its objects the fragments. Its properties and further areas called functorial model theory, admissible models, and ordered structures are defined by the author 1997 on. We conclude with the morphic order functor theorem. The functor creates limits on the category Set applying the morphic preorder to $L_{\omega 1,K}$

Theorem 3.9 (Author): The morphic preorder functor defines initial models for $L_{\omega 1,K}$.

The model has the morphic preorder property.

Proof (e.g., Nourani, 2014).

3.5   A DESCRIPTIVE COMPUTING

A computational epistemology is presented by defining computation on G-diagrams for models. Deterministic and probabilistic epistemology are defined on the specific description logic based on infinitary logics and model theory. An inherent computability theorem is proved, comparing the two computational epistemics. By applying a G-diagram formulation of situations and possible worlds a compatibility theorem is proved for situations in terms of Generic diagram (G-diagram). Further epistemics for computation is defined by introducing der Vielleicht Vorhandenen, and defining an epistemic for computing. Techniques and formulations of computation on generalized diagrams are presented for computations applications to planning as an example for computations with ordinal epistemic states, based on world models definable from logical syntax.

By defining descriptive computing on generic diagrams we can relate admissible sets to descriptive computing and prove certain admissible sets are descriptive computable. A correspondence between models definable by G-diagrams for models, their computations, and Boolean algebras are presented for descriptive computing. The correspondence of modalities (Nournai, 1998) to Possible Worlds via Model Sets and the containment of the possible worlds approach by the generic diagrams, implies that we can present a model-theoretic formulation for the dynamics of the possible worlds computing and for world descriptions for artificial intelligence applications.

The interest of Artificial Intelligence for the past 25 years in concepts and theories of logic, computational linguistics, cognition, and vision was motivated by the urge to bridge the human-computing theories for intelligence and cognition gap, with logic, metaphysics, philosophy and linguistics, to design practical Artificial Intelligence systems. For the present author such interests were in automatic programming with types and rules, computational logic, and at more recent times, directed towards research in AI reasoning, and planning (Nourani, 1998, 1991, 1984) and the meta-mathematics of such concepts. The present section presents a brief overview basis for a logic and model theory towards defining a descriptive computational epistemology with, for example, admissible sets.

Thus, with computational epistemology we can formally address epistemology, robot planning and theorem proving, and knowledge representation, and logical cognition. It presents computation with generalized diagrams, allowing us to build models of AI theories from the syntax of the logical language. Expressing knowledge about concepts and concept hierarchies fro planning on G-Diagrams. There is a Tarski style declarative semantics, which allows them to be seen as sub-languages of predicate logic. One starts with primitive concepts and roles, and can use the language constructs (such as intersection, union, role quantification, etc.) to define new concepts and roles. Concepts can be considered as unary predicates which are interpreted as sets of individuals whereas roles are binary predicates which are interpreted as binary relations between individuals.

The computational issues are pursued for artificial intelligence planning applications by the author elsewhere. We present formulations of computations, possible worlds, and situations by generalized diagrams and conclude with some technical results for such formulation that is of computational significance. We present some formulations of symbolic computation and knowledge representation on generalized diagrams, following our earlier research papers (Nourani 1998, 1991). Thus, by symbolic computation we can address issues of knowledge representation, robot planning, and theorem proving. The latter computational issues are pursued for planning applications in Nourani (1993).

We present formulations of computations, possible world's, and situations by generalized diagrams and conclude with some technical results for such formulation which are of computational significance. The specific areas that I will address will be outlined with two goals in mind: to formulate computation on diagrams encoding possible worlds and epistemic states, and to shed light on areas that have relations to what we define as computational epistemology. The concepts applied by the present direction for research is that of possible worlds and epistemic states, applied to computation and Artificial Intelligence by this author starting from 1986 and the subsequent publication of some of the ideas and further research in Nourani (1998, 1991).

3.5.1   COMPUTING MODEL DIAGRAMS

The generic diagram defined by this author in the 1980's is a diagram in which the elements of the structure are all represented by a minimal family of function symbols and constants, such that it is sufficient to define the truth of formulas only for the terms generated by the minimal family of functions and constant symbols. Such assignment implicitly defines the diagram. This allows us to define a canonical model of a theory in terms of a minimal family of function symbols. Generalized diagrams are precisely what allow us to build models from the syntax of a theory, thus allow for symbolic computation of models and theories. The correspondence of modalities to Possible Worlds (Nourani 1994, 1998) and the containment of the possible worlds approach by the generalized diagrams approach of this author implies that we can present a model-theoretic formulation of the concept of modal objects with varying properties, with cross product of modes formed from various generalized diagrams corresponding to each mode.

Now, let us examine the definition of situation and view it in the present formulation.

Definition 3.6: A situation consists of a nonempty set D, the domain of the situation, and two mappings: g and h. (i) g is a mapping of function letters into functions over the domain as in standard model theory. (ii) h maps each predicate letter, pn, to a function from Dn to a subset of $\{t, f\}$, to determine the truth value of atomic formulas as defined below. The logic has four truth values: the set of subsets of $\{t, f\}.\{\{t\}, \{f\}, \{t, f\},0\}$. The latter two is corresponding to inconsistency, and lack of knowledge of whether it is true or false.

Due to the above truth-values, the number of situations exceeds the number of possible worlds. The possible worlds being those situations with no missing information and no contradictions. From the above definitions the mapping of terms and predicate models extend as in standard model theory. Next, a compatible set of situations is a set of situations with the same domain and the same mapping of function letters to functions. In other worlds, the situations in a compatible set of situations differ only on the truth conditions they assign to predicate letters.

Restating definitions from Chapter 2:

Definition 3.7: Let M be a structure for a language L, call a subset X of M a generating set for M if no proper substructure of M contains X, for example, if M is the closure of X U $\{c[M]$: c is a constant symbol of L$\}$. An assignment of constants to M is a pair <A, G>, where A is an infinite set of constant symbols in L and G: A $\rightarrow$ M, such that $\{G[a]$: a in A$\}$ is a set of generators for M. Interpreting a by g[a], every element of M is denoted by at least one closed term of L[A]. For a fixed assignment <A, G> of constants to M, the diagram of M, D<A, G>[M] is the set of basic [atomic and negated atomic] sentences of L[A] true in M. (Note that L[A] is L enriched with set A of constant symbols.)

Generic diagrams, denoted by G-diagrams, were what we defined since 1980's to be diagrams for models defined by a specific function set, for example Σ_1 Skolem functions.

Definition 3.8: A Generic diagram for a structure M is a diagram D<A, G>, such that the G in definition 3.6 has a proper definition by specific function symbols.

Remark: The functions above are those by which a standard model could be defined by inductive definitions.

To decide compatibility of two situations we compare their generalized diagrams. Thus, we have the following Theorem.

Theorem 3.10: Two situations are compatible iff their corresponding generalized diagrams are compatible with respect to the Boolean structure of the set to which formulas are mapped (by the function h above, defining situations).

Proof: The G-diagrams encode possible worlds and since we can define a one-one correspondence between possible worlds and truth sets for situations. Thus, computability is definable by the generic diagrams.

Now, let us present a Descriptive Computing (Nourani, 2003) applying generalized diagrams. We define descriptive computation to be computing with G-diagrams for the model and techniques for defining models with G-diagrams from the syntax of a logical language. G-diagrams are diagrams definable with a known function set. Thus, the computing model is definable by G-diagrams with a function set. An example function set might be Σ1 Skolem functions. The corresponding terminology (Martin, 1977) in set theory refers to sets or topological structure definable in a simple way. Thus, by descriptive computation we can address artificial intelligence planning and theorem proving, for example. The latter computational issues are pursued by the author in Nourani (1994, 1997).

The logical representation for reaching the object might be infinitary only. We have seen above that the artificial intelligence problem from the robot's standpoint is to acquire a decidable descriptive computation for the problem domain. We had proved (Nourani, 1994) two specific theorems for descriptive computing on diagrams. A compatibility theorem applies descriptive computing to characterize situation compatibility. Further, a computational epistemic reducibility theorem is proved by the descriptive computing techniques on infinitary languages in the preceding sections.

A deterministic epistemics is defined and it is proved not reducible to known epistemics. We further define H[κ] from Barwise (1968) for any cardinal κ, let H[κ] be the set of sets x such that the transitive closure of x, TC[x], has cardinality less than κ. For κ regular, L<H[κ]> is the language usually denoted by Lκ,ω.

The following are this author's extrapolations based on the new definitions.

Definition 3.9: A set is Descriptive Computable iff it is definable by a G-diagram with computable functions.

Proposition 3.3: For descriptive computable sets the set H[κ] is definable from a G-diagram by recursion.

Theorem 3.11: For A an admissible computable set, A is descriptive computable.

Proof: Since A is admissible, by a theorem stated in Barwise (1968) applying Gentzen system completeness for Lω1,ω and Beth definability, A ⊆ H[ω1] can be proved. Thus, by the proposition A is descriptive.

3.5.2 BOOLEAN COMPUTING DIAGRAMS

Applying generic diagrams to define knowledge with Skolem functions the computation the Skolemized formulas are defined and expanded to assign truth values to the atomic formulas. Boolean algebras can be defined corresponding to the formulas with Skolemized trees. A truth valuation can be defined with respect to a class of Boolean algebras $\mathfrak{R}$ as a homomorphism from a closure of formulas to an algebra in $\mathfrak{R}$.

For example we can define B = <B, min, max, −, 0, 1> where min and max are defined by induction on depth of Skolemized trees. By the theorem below from Richter (1973) logical implication for formulas and models for classes of Boolean algebras can be reduced to only the Boolean algebra over {0,1}.

Theorem 3.12: (Richter 1973). If $\mathfrak{R}$ is the class of Boolean algebras and $\mathfrak{R}$' contains only the Boolean algebra over {0,1} then |= for $\mathfrak{R}$ is equivalent to |= for $\mathfrak{R}$'.

As this author defined in Nourani (1995) a "plan" deploying standard artificial intelligence techniques, is a sequence of operations in the universe that could result in terms that instantiate the truth of specific plan goal formulas in a modeled world with a discourse language. Thus, a plan can be represented by a tree. In planning with model diagrams the plan trees involving free Skolemized trees are carried along with the proof tree for a plan goal. We call such trees Free Proof Trees. The idea is that if the free proof tree is constructed then the plan has a model in which the goals are satisfied. The model is the standard model definable by the generic diagram for which the free Skolemized trees were constructed. Thus, we the following Free Proof Tree Sound Computing Theorem.

Theorem 3.13: (author) For the free proof trees defined for a goal formula from the G-diagram there is a standard model satisfying the goal formulas. It is the model definable by the G-diagram.

Proof: outline A free proof tree is a proof tree that is constructed with free Skolem functions from a G-diagram. The idea is that if the free proof tree is constructed for a goal formula, then from the G-diagram there is a model satisfying the goal formula satisfied. The model is the standard model for which the free Skolemized trees were constructed. Thus, we have transformed the model theoretic problems of computing to that of descriptive computing with G-Diagrams.

What it implies for Skolemized trees and Boolean algebras is the computation by the preceding theorems.

Theorem 3.14: (author) There is a Boolean algebra $\mathfrak{R}'$ in the class $\mathfrak{R}$, $\mathfrak{R}'$ containing only $\{0,1\}$, such that for the standard model A defined by the G-diagram theorem. A $\models$ is equivalent to $\models$ for $\mathfrak{R}'$.

Proof: follows from Theorems 3.12 and 3.13 and Exercise 3.6.

The term computational epistemology was put forth as a key word in the abstract to define possible new direction for research in AI and computational logic. Thus, we have a theoretical basis for computational epistemology. The world models, objects, symbols, languages, and epistemic computations for their representation, are important and at times imply the application of new conceptual theories that have to be developed in the forthcoming research and practice of computational epistemology. Amongst new areas defined is descriptive Boolean computing on diagrams and descriptive computing with admissible sets. The cardinality concepts have started to be applied to computational knowledge representation.

3.6 DESCRIPTION LOGIC

This section is an overview to description logics that are newer versatile encompassing knowledge representation formalism unifying to a logical basis to more traditional systems. Our developments come closer to constructive description logic. There for we state a brief overview based on Paiva (2002). However, description logics are much more logic-based than previous formalisms but somewhat "unconventional" as logics. Logics usually have syntax and a semantics (description logics are no exception) and they usually can be placed in the hierarchy of other more traditional logics without any problems. From this perspective description logics are quite different. Before discussing the problem of placing description logics into traditional hierarchies of logics, we must examine the basic syntax of description logics.

Different description logics have different description languages. One of the basic description languages in the handbook of description logics is ALC. Concepts descriptions in ALC are formed according to the following syntax rule:

$$C, D \rightarrow A \mid T \mid \bot \mid \neg C \mid C \sqcap D \mid C \sqcup D \mid \forall R.C \mid \exists R.C$$

where C, D range over concepts, R ranges over names of roles and A stands for the atomic concepts.

Recall that the semantics of classical ALC is given by interpretations $I = \langle \Delta I, \cdot I \rangle$ where ΔI is a non-empty set, every concept A is mapped by the interpretation function $\cdot I$ to a set $AI \subseteq \Delta I$ and every role name R is mapped to a bi.

Now if you think that reasoning constructively is a useful paradigm and you want to describe constructive or intuitionistic description logics, say CDL, you have nary relation RI on ΔI. This function is extended to arbitrary concepts as expected to, in principle. One can use a translation like t1 and you can think of the system of constructive description logic you're describing as a subsystem of the well-known intuitionistic first-order logic system IFOL. Or you can use the modal translation t2 *and you can consider your target system of intuitionistic description logic as embedded into a constructive version of the modal system Km. There are several proposals for "the" constructive version of K in the literature.*

It is somewhat remarkable that these two routes can lead to the same place. Getting to the same place in the classical case is no coincidence: As van Benthem has shown (van Bentham, 1998), the fact that classical modal logic can be seen as a fragment of first-order logic, as well as an extension of it, is a consequence of the standard translation. The "Modal Invariance Theorem," says that a first-order formula is definable by a modal formula if and only if it is invariant under bisimulation. One would like an analogous result to be true about the constructive systems for DL. Paive (2002) adapts classical ALC to a constructive system using the two routes outlined above. The syntax of such constructive system is the same in both cases. Concept descriptions in this constructive description logic CDL language obey the following syntax rule

$$C, D \rightarrow A \mid T \mid \perp \mid C \sqcap D \mid C \sqcup D \mid C \rightarrow D \mid \forall R.C \mid \exists R.C$$

where C, D range over concepts, A is an atomic concept and R ranges over names of roles, as before. As usual in constructive logics, since $\neg C$ is simply an abbreviation for $C \rightarrow \perp$ we do not need to consider it. In compensation we must add in the constructive implication of concepts, which in classical description logic is a derived concept. Also it is just a convenience to have the true concept T, as it could be defined as $\neg \perp$.

The aim of this section is to describe the system of constructive description logic IALC obtained from ALC by considering its translation into (a subset of) traditional intuitionistic first order logic IFOL.

The disadvantage of IALC, when compared to classical ALC or even Wansing and Odintsov's inconsistency-tolerant system CALC is that we cannot use the easy semantics of classical first-order logic to give "meaning" to IALC constructors. Instead we must use the translation t1 of IALC into IFOL, which is quite standard. This translation is actually parameterized on two new intuitionistic variables, say x and y.

Thus so tx1: IALC $\rightarrow$ IFOL

On the one hand translation of the basic notions of the description logic in question, (respectively concepts and roles) into FOL unary predicates (for concepts) and into FOL binary predicates (for roles) as well as translation of the chosen required connectives into themselves. Note that this translation (let us call it t1:ALC → FOL) takes a description logic concept C into a predicate C(x), with a free variable x, where we think of C(x) as the collection of the x's that happen to be instances of the concept C.

This translation is the basis of several of the computational complexity results that underlay most of the theoretical work on description logics. On the other hand, very early on, it was realized that description logics can also be considered as a notational variant of the multimodal K logic, if one thinks of roles in the description logic as accessibility relations in multimodal Km. In this case the translation (call it t2:ALC → Km) from the description logic to the modal logic Km transforms concepts C into atomic modal propositions C, intersections of concepts into propositional conjunctions, unions of concepts into propositional disjunctions and most importantly, taking the existential and universal role restrictions into modal operators as follows: t2(∃Ri.C) = Qi(t2(C)) and t2(∀Ri.C) = Pi(t2(C)). Thus when we try to place description logics into traditional hierarchies of logics, we have to consider translations and there are in principle, two very different translations to work with. For instance one translation uses predicate logic, while the other is propositional. It remains to develop an admissible description logic based on the preceding sections towards characterizing a constructive description logic the applies generic and modal diagrams. That is a topic outside the scopes of this chapter.

3.7   HIGH FIDELITY TREE COMPUTING

The present chapter defines new tree computing techniques and an algebraic computing theory defining recursion tree amplification applied to algebraic synthesis as an illustration. Structures are designated by signatures similar to the way the key signature in music designates a structure. Signatures define the way the functions interact with the carriers for objects, scenes, and images. Signatures can designate music tones with icons and objects and signatured structures on which morphings can be defined. Precise mathematical morphings can be defined and be computed with agents carrying on the morphisms. A canonical initial algebraic synthesizer is defined with categorical equalizes.

The Tree Amplification Principle (TAP) is put forth based relating a minimal function set to a recursion theoretic gain concept, defining computing efficiency by a gain minimization function. The present computing defined is towards a high fidelity computing theory, with amplification, equalizers, and filters. The preliminary categorical properties and defines a HIFI computing logic based on functorial model theory. The techniques we have put forth are not the same as a categorical interpretation for logic as in Categorical Logic. Thus, it has it own theoretical properties and application areas. We define a generic functor with infinitary language topology, for example, the functor on an Infinitary language $_{L\omega1,K}$. The present computing defined is towards a High Fidelity computing theory, with amplification, equalizers, and filters.

3.7.1   FUNCTORIAL MODEL THEORY AND HIFI COMPUTING

For the computing theories defined for HIFI logic the functorial model theory and the limit definitions are essential to get models defined. Functorial computing models are definable based on what the author presented in Nourani (1994,1996). The computing applications are at preliminary stages. It is apparent we can define a computing theory based on a functorial model theory for INFLC and HIFI computing. INFLCS can be viewed as an abstract language for defining computation. There is a categorical recursion theory defined by HIFI computing by which computing models can be defined for INFLCS. Let us start with the preliminaries to Functorial Model Theory. We present some definitions from this author's publications that allow us to define standard models of theories that are CTA's. The standard models are significant for tree computational theories, for example, Chapter 4.

Definition 3.10: Let (M, a)c in C be defined such that M is a structure for a language L and each constant c in C has the interpretation a in M. The mapping c → a_c is an assignment of C in M. We say that (M, a)c in C is canonical model for a presentation P on language L, iff the assignment c→ a maps C onto M, for example, M=(a:c in C).

Definition 3.11: A standard model M is a canonical model inductively definable with a generic diagram.

Definition 3.12: Let M be a set and F a finite family of functions on M. (F, M) is called a monomorphic pair, provided for every f in F, f is injective, and the sets {Image(f):f in F} are pairwise disjoint.

The inductive definitions of the ordinal ω and the class of formulas of a first order language are monomorphic. The definition is basic in defining induction for abstract recursion-theoretic hierarchies and inductive definitions. We had put forth variants of it with axiomatizations in [7]. What is added in the author's papers is the definition of generalized standard models with monomorphic pairs. This definition was put forth by the present author around 1982 for the computability problems of initial models. Note that a standard model M, of base M and functionality F, is a standard model inductively defined by <F, M> provided the <F, M> forms a monomorphic pair. The definitions are applied in the sections to follow.

3.7.2   GENERIC FUNCTOR INITIAL MODELS

To present our model theory for categorical computing we present model theoretic diagrams referred to by G-diagrams. These diagrams are to be distinguished form the arrow diagrams for categories. There is a two-diagram terminology in the present paper. Model-theoretic diagrams, we refer to them by M- Diagrams; and arrow diagrams, as in commutative diagrams. We refer to these by A-Diagrams. The G-diagrams are model-theoretic diagrams, defined by this author around 1980's, thus G- diagrams are algebraic M-Diagrams. The generalized diagram (G-diagram) is a diagram in which the elements of the structure are all represented by a minimal set of function symbols and constants, such that it is sufficient to define the truth of formulas only for the terms generated by the minimal set of functions and constant symbols.

Define a functor F: $L_{\omega1,K}{}^{op} \to$ Set by a list of sets Mn and functions fn. The sets correspond to an initial structure on $L_{\omega1,K}$. For example, to f(t1, t2,.. tn) in $L_{\omega1,K}$ there corresponds the equality relation f(t1, t2, ..., tn) = ft1...tn. Let us refer to the above functor by the generic model functor since we can show it can defines generic sets from language strings to form limits and models. It suffices to define the functor upto an initial model without being specific as to what the model is, to have a generic functor. The specific model theoretic properties are not defined in the present chapter and are presented in alternate chapters here. The generic model functor can define abstract models for computations defined with $L_{\omega1,K}$. The proof for the following theorem follows from the sort of techniques applied by the present author to define generic sets with the fragment $L_{\omega 1, K}$ in 1982.

From the preceding chapters a basic theorem is given below.

Theorem 3.15: The generic model functor has a limit.

Proof: (e.g., Nourani, 2014).

The following are the basics to functorial models on $L_{\omega 1, K}$. Define a small-complete category $L_{\omega 1, K}$ from $L_{\omega 1, K}$. The category is the preorder category defined by the formula ordering defining the language fragment. Present a functor from this category onto Set. Define a category for models definable by D<A, G>, call it D<A, G>. This is the category with objects being models definable form the diagram D<A, G> and arrows their morphisms. There are three categories at play—the category $L_{\omega 1, K}$, the category Set, and the category D<A, G>. The D<A, G> category is the category for models definable with D<A, G>. The techniques we are presenting by the three categories save us from having to yet develop a categorical interpretation for $L_{\omega 1, K}$, in categorical logic as in Lawvere's 1967 categorical formulation of logic or the current practice. Taking limits at $L_{\omega 1, K}$ we could show by a universal embedding from the D<A, G> model in the limit, and by the product object properties, solution sets defining initial objects.

3.7.3   INITIAL TREE ALGEBRAS AND AMPLIFICATION

To define TAP, we start with defining gain functions. Let F be a minimal set of functions, for example type constructors, from which a CTA and a standard initial model with signature is definable, where standard in the sense of the definitions above, and definable is in the model-theoretic sense, put forth by this author.

Definition 3.13: A function f is tree-computable iff f is definable with and f is a computable function.

Let {f0, f1, f2, …} be the set of functions -tree-computable. Further let d<Fi>(fi) be the recursion depth for Fi-trees appearing in fi, with Fi in F. Then a gain function for fi is defined from d<Fi>(fi). Let G<F>(fi) = Sum {d<Fi>(fi) x WSum(Fi), for all Fi's in F}, where WSum is a weighted sum function defined from the G<F> for the subtrees formed from Fi at each recursion level. The weighted sum function might be defined from the weight of the subtrees rooted by fi at each recursion level. The weight W of a subtree tj, denoted W(tj) = Sum { G<Fm>, Fm in tj, Fm in F}. G<Fi> = 1 for a nullary function Fi. For Fi of arity (t1, …tn), G<Fi> = Sum {W(t1), …, W(tn)}.

The WSum for Fi is defined by induction as follow.

For a depth 0 term Fi(t), WSum(0)(Fi) = W(tj) For a depth n term, WSum(n)(Fi) = n x WSum(n–1)(Fi)

G<F> for the computable functions {f0, f1, f2, …} on a signature might be defined by the Maximum {G<F>(fi)}. This is not the only way to define gain, but it is a way to start.

Brief Overview – For a signature and minimal functions F for defining initial models on signature we have defined gain as follows.

For a signature and a choice F, a gain set is defined for the functions on the signature by {G<F> (f1), G<F>(f2), ….}, where each G<F>(fi) defines gain on F for a function fi definable from a signature. We have further defined gain for all computable functions on a signature, to be G<F> = U {G<F>(fi): fi a computable function definable on }, where U is the least upper bound operation.

Thus, we have defined gain functions for functions computed by a minimal functions set F. For a particular signature, and a set F of minimal functions, we have defined gain on F for all computable functions definable on the signature. This is only a starting formulation, based on what the author we presented in (Nourani 1994), from which new tree computing results could be obtained, shedding light on how the choice for F and a signature, leads to complex or simpler computation. These are topic for future research and outside the intended goals of the present paper.

The TAP Principle: The trees welcomed into the computations by the choice F, are amplified by recursion with gain G<F>. Computational efficiency could be defined by gain minimization.

Measures for computational complexity based on TAP could be defined, thus allowing us to base the selection on the computing model and the computing that is intended to minimize gain, thus have efficient computation.

As an example, the function f(x, y) = f(g(x), h(x, y)) defined in terms of functions g and h is computed as follows. For terms t1, and t2 taken, f(t1, t2) sends out f(g(t1), f(g(t1), h(t1, t2)))

```
∧ ∧ ∧ ∧
/ t'1 \ / t'2 \
```

terms with g's and h's are put out. The t1' and t2' are then taken on, putting out the trees with g's and h's terms that are the output.

$f(\wedge, \wedge) \Rightarrow g(/ \backslash), h(/ \backslash, / \backslash) \Rightarrow \wedge \wedge$

$/ t'1 \backslash / t'2 \backslash$

The peak are modulated by the specific signature functions.

3.7.4 THE RECURSION THEOREM

To get a start for what we can define with the basis for an abstract recursion theory based on the concepts defined by the present paper. The recursion theorem is the following.

Definition 3.14: Let X be any fixed set. A function $\Gamma: \wp(X) \to \wp(X)$ is called an operator over X. Γ is said to be inclusive iff for all $Y \subseteq X$, $Y \subseteq \Gamma(Y)$, monotone iff all $Y \subseteq Z \subseteq X$, $\Gamma(Y) \subseteq \Gamma(Z)$, and inductive iff Γ, for example, either inclusive or monotone.

We define by transfinite recursions the sequence $\Gamma[\sigma]$ by $\Gamma[\sigma] = \Gamma(\cup \{\Gamma\tau: \tau < \sigma\})$ and set $\Gamma^\wedge = \cup \{\Gamma \sigma: \Sigma\epsilon \text{ Or}\}$. We write $\Gamma<\sigma>$ for $\cup \{\Gamma \tau: \tau < \sigma\}$ so that $\Gamma[\sigma] = \Gamma(\Gamma<\sigma>)$. Think of the set $\Gamma^\wedge$ as being "built up" is stages. Starting with the empty set, $\Gamma(\varnothing)$, $\Gamma(\Gamma(\varnothing))$, ...are obtained. $\Gamma[\sigma]$ is called the σ-th stage.

Inductive definitions are closures in the following sense. Let $Y \subseteq X$, F a family of finitary functions on X.

Let $S \subseteq X$, F a set of finitary functions on X. Let $\alpha(f)$ be the arity f in F. For each pair (S, F) define an inductive operator $\Gamma <S, F>(Y)$, for Y a subset of S by $\Gamma <S, F>(Y) = S \cup \{f(Y'): f \text{ in } F \ \& \ Y' \text{ in } Y\alpha(f)\}$

$\Gamma<S, F>$ is called the closure of S under F. Since $\Gamma<S, F>$ is monotone, it is the smallest set inductively S- closed under F.

Inductive definitions are closures in the following sense. Let $Y \subsetneq X$, F a family of finitary functions on X.

To get a start for what we can define with the basis for an abstract recursion theory (based on the concepts defined by the present paper). The recursion theorem is the following:

Theorem 3.16: For any monomorphic pair (S, F) and any set X*, suppose G: X*→ X* and for each f in F, f*: X*α(f) →X*, then there exists a unique function $\Theta: \Gamma <S, F> \to$ X* such that (i) for all s in S, Θ (s) = G(s), (ii) for all f in F and all x in $\Gamma <S, F>$, $\Theta(f(x)) = f^* (\Theta(x0), ..., \Theta(x \alpha(f)-1))$ []

Proof: Let $\Gamma = \Gamma<S, F>$, X^*, G and f be given as in the hypothesis. We define functions Θr: $\Gamma<r> \rightarrow X^*$ by ordinary recursion as follows. $\Theta i = G$, for $i = 0$. Suppose Θr is defined and $X \varepsilon \Gamma< r+1>$. If $x \varepsilon \Gamma<r>$ we set $\Theta r+1(x) = \Theta r(x)$. If $x \varepsilon \Gamma <r+1>\sim \Gamma<r>$, then by the assumption that (Y, F) is monomorphic there exits a unique $f \varepsilon F$ and $z \varepsilon k(f) \Gamma<r>$ such that $x = f(z)$, where $k(f)$ selects terms from $\Gamma<r>$ for f's arity. We then set $\Theta r+1(x) = f^*(\Theta r(z0), ..., \Theta r(zk(f)-1))$. Finally, $\Theta = \cup \{\Theta r: r \varepsilon \omega\}$. Θ satisfies conditions (i) and (ii).

Defining an abstract recursion for algebraic theories is another enterprise altogether. HIFI computing is combining abstract recursion with recursion on a functorial algebraic theory to define a computing theory. It is not defining semantics for an existing computing theory. Hence it has to define its foundations anew. Let us start with a gain recursion theorem.

Theorem 3.17: For any monomorphic pair (S, F), and a computation sequence defined on the signature, there is a unique gain function G<F> defined.

Proof: For a signature Σ and a choice F, a gain set is defined for the functions on the signature by $\{G<F> (f1), G<F>(f2), ...\}$, where each $G<F>(fi)$ defines gain on F for a function fi definable from a signature Σ. We have further defined gain for all computable functions on a signature Σ, to be $G<F>(\Sigma) = \cup \{G<F>(fi): fi$ a computable function definable on $\Sigma\}$, where U is the least upper bound operation.

For the monomorphic pair S, F, take $\Gamma(S, F)$ to be the tree inductive closure for the functions S defined by the signature Σ. Since gain is defined on Σ-computable functions by inductive definitions, by definition 4.1, for each function f in S, it is sufficient to view the tree rooted with f, with maximum height. Let $G:X \rightarrow X$ be a function on the f-trees, which for a computation sequence on a signature Σ, for each f in Σ, returns the f-tree with maximum height. By a well-known basic recursion theorem for algebraic theories, the function G, extends to a unique Σ-homomorphism from the set of f-trees in the computation sequence to the Σ-algebra on X with the Σ-definable functions. Let us refer to it by morphism recursion. By Theorem 4.3, there is a unique function $\theta: \Gamma (S, F) \rightarrow X$, where X is the set of f-trees defined by the computation sequence, assigning terms from the inductive set to X such that (i) $\Theta (s) = G(s)$ for all s in S appearing in the computation sequence;

(ii) for all f in F and all x in Γ <S, F>, Θ(f(x)) = f* (Θ(x0), …, Θ(x α(f)–1)). Composing Theorem 5.3 with morphism recursion for algebraic theories, we can obtain gain direct from G. Hence a unique gain is defined.

3.7.5 ADMISSIBLE GAIN SYNTHESIZER

A formal specification is based on a formal language and makes use of some defining axioms and possibly mathematical structures, in characterizing modules or programs. A formal specification in our approach consists of a signature specifying the type names, and the operations on types, including the rank and the arity of the operations. It also includes a set of axioms which recursively define the operations. Such specifications capture the intended meaning of the computation sequences possible using a given set of objects and operations, thus specifying a program. Synthesis homomorphism is defined by setting a correspondence between source syntax trees in the specification language and a target language. The target language might be executable syntax trees, obtained by user-defined metaprograms. Algebraic specification languages (e.g., Wirsing, 1995) are characterized by their underlying logic, their constructs supporting a particular programming paradigm and their structuring mechanisms.

The homomorphic mappings preserve the operation on trees in the image algebra of trees for the target language. The algebraic synthesizer, illustrated as an example here, is designed based on the what we refer to by the TAP. The practical applications for TAP are the synthesizer this author has put forth in Nourani (2010) and various tree computing projects that the author and others have launched during the recent years. Informally TAP is what allows us to filter the trees that are principally involved in computing to be able to achieve synthesis. TAP itself is not the filtering process, but it is what happens to computing trees as a result of the programming constructs and selections that are applied at source programming time. Mistakes made at programming time are amplified at synthesis and compute time.

Design with TAP allows us to drive the synthesizer with the amplification 'gain' on the trees that are admissible for a particular computing, to achieve a synthesizer suitable for selected computing criteria. Let us make at least an informal statement of how TAP can be defined and how 'gain' on amplification is defined and applied by the present project. Modules specified or programming language abstract syntax trees are defined by recursion from a set of functions. The functions are amplified by recursion. Thus, there is a 'gain' function on trees rooted by the defining function that might be defined from the recursions defining the set of all trees. The synthesizer makes use of a minimal set of Skolem functions and type constructors, explicit in the syntax of the specification language, abstract syntax definition, and the definition of the homomorphisms that implement the synthesis. The minimal set of functions are amplified to define the trees that drive the synthesizer.

Example application areas are a class of algebraic theories called "recursion-closed," which generalize the rational theories studied by ADJ (1978). This work is motivated by the problem of providing the semantics of arbitrary polyadic recursion schemes in the framework of algebraic theories (Gallier, 1981) that begins with a premise that "rational theories" are insufficient and that it is necessary to introduce a new class of "recursion-closed" algebraic theories. This new algebraic theories, is defined and studied, and "free recursion-closed algebraic class of theories" are proved to exist.

EXERCISES

1. Provide a detailed proof for Theorem 3.1
2. Prove Theorem 3.2.
3. Show that if we add to the usual first-order axioms and rules of inference the new axiom scheme

    (a) $\circ \Phi \rightarrow \varphi$

    (b) for any countable set $\Phi \subseteq \mathrm{Frm}(\omega 1, \omega)$ and any $\varphi \in \Phi$, together with the new rule of inference

    $$\frac{\varphi_0, \varphi_1, \ldots, \varphi_n, \ldots}{\circ_{n \in \omega} \varphi_n}$$

    and allow deductions to be of countable length. Writing $\vdash^*$ for deducibility in this sense, we then have the $L(\omega 1, \omega)$-Completeness Theorem.
    Hint: This deductive system is adequate for deductions from countable sets of premises in $L(\omega 1, \omega)$. That is, for any countable set $\Delta \subseteq \mathrm{Sent}(\omega 1, \omega)$, $\Delta \models \varphi \Leftrightarrow \Delta \vdash^* \varphi$. This completeness theorem can be proved by modifying

the usual Henkin completeness proof for first-order logic, or by employing Boolean-algebraic methods.

4. Prove that the class of L-structures isomorphic to N coincides with the class of models of the conjunction of $L(\omega1,\omega)$ sentences:
(where 0 names 0):

°$m \in \omega$ °$n \in \omega$ $sm + sn0 = sm+n0$
°$m \in \omega$ °$n \in \omega$ $sm0 \cdot sn0 = sm \cdot n0$
°$m \in \omega$ °$n \in \omega-\{m\}$ $sm0 \neq sn0$
$\forall x \forall m \in \omega$ $x = sm0$

The terms $s_n x$ are defined recursively by

$$
\begin{aligned}
S_0 &= x \\
S_{n+1} x &= s(s_n x)
\end{aligned}
$$

5. Prove Theorem 3.3.
6. Prove Theorem 4.5: There is a Boolean algebra $\Re'$ in the class $\Re$, $\Re'$ containing only $\{0,1\}$, such that for the standard model A defined by the G-diagram Theorem 4.2. A $\models$ is equivalent to $\models$ for $\Re'$.

KEYWORDS

- **admissible languages**
- **admissible models**
- **descriptive computing and sets**
- **infinite language categories**
- **recursion on infinite trees**

REFERENCES

ADJ, Goguen, J. A., Thatcher, J. W., Wagner, E. G., Wright, J. B. (1978). in "Proceedings, 17th IEEE Symposium on Foundations of Computer Science, Houston, Texas, October 1976," pp. 147–158; "Mathematical Foundations of Computer Science, 1978," Lecture Notes in Computer Science, Vol. 64, Springer-Verlag, New York/Berlin, 1978.

Alechina, N., Mendler, M., de Paiva, V., Ritter, E. (2001). Categorical and Kripke Semantics for Constructive Modal Logics. In Computer Science Logic (CSL'01), Paris, September 2001.

Baader, F., Calvanese, D., McGuiness, D., Nardi, D., Patel-Schneider, P. (2003). The Description Logic Handbook: theory, implementations and applications. Cambridge University Press, 2003.

Barwise, J. (1967). Infinitary Logic and Admissible Sets. Ph.D. Thesis, Stanford Holland.

Barwise, J. (1968). "The Syntax and Semantics of Infinitary Languages," Springer-Verlag Lecture Notes on Mathematics, vol. 72, 1968.

Barwise, J. (1972). "Implicit Definability and Compactness in Infinitary Languages," in The Syntax and Semitics of Infinitary Languages, Edited by J. Barwise, Springer-Verlag LNM, vol.72, Berlin-Heidelberg, NY.

Barwise, J. (1975). Admissible Sets and Structures, Berlin: Springer-Verlag.

Barwise, J. (ed.), (1975). Mathematical Logic, Amsterdam: North-Holland, 233–282.

Barwise, J., Feferman, S. (eds.), (1985). Handbook of Model-Theoretic Logics, New York: Springer-Verlag.

Baumgartner, J. (1974). "The Hanf number for complete $L_{\omega_1, \omega}$ sentences (without GCH)," Journal of Symbolic Logic, 39, 575–578.

Brauner, T., de Paiva, V. Towards Constructive Hybrid Logics. In Methods for the Modalities 3, Nancy, France, 2003.

Gallier, J. H. (1981). Recursion Closed Alebraic theories, Journal of Computer and System Sciences. Journal of Computer and System Sciences, Volume 23, Issue 1, August (1981). pp. 69–105.

Hanf, W. P. (1964). Incompactness in Languages with Infinitely Long Expressions, Amsterdam: North-Holland.

Heidegger, M. (1962). "Die Frage nach dem Ding," Max Niemeyer Verlag, Tubingen, 1962. Hofmann, M. (2005). Proof-theoretic Approach to Description-Logic. In Proc. of Logic 5, 229–237.

Karp, C. (1964). Languages with Expressions of Infinite Length, Amsterdam: North-Holland.

Keisler, H. J. (1974). Model Theory for Infinitary Logic, Amsterdam: North-Holland.

Keisler, H. J., Julia F. Knight, (2004). "Barwise: Infinitary Logic And Admissible Sets," Journal of Symbolic Logic, 10(1), 4–36

Kripke, S.(1964) "Transfinite recursion on admissible ordinals," I, II, JSL, 1964, 29, 161–162, abstract.

Kripke, S. (1964), "Transfinite recursion on admissible ordinals", J. Symbolic logic 29, 161–162, JSTOR 2271646.

Kripke, S. A. (1965) 'Semantical analysis of intuitionistic logic I'. In: Crossley J. N.; Dummett M. A. E.; (eds.): Formal Systems and Recursive Functions. Amsterdam: North- Holland, 1965, 92–130.

Makkai, M. (1977). "Admissible Sets and Infinitary Logic," Handbook of University.

Martin, D. A. (1977). "Descriptive Set Theory: Projective Sets," Handbook Mathematical Logic, 783–815.

McCarthy, J. Notes on Formalizing Context. In Proc. of the 13th Joint Conference on Artificial Intelligence (IJCAI-93), 555–560, 1993.

Mendler, M., Valeria de Paiva, (2005). Constructive CK for Contexts. In Proceedings of the First Workshop on Context Representation and Reasoning, CONTEXT'05, Paris, France.

Nadel, M. (1985). "$L_{\omega_1, \omega}$ and Admissible Fragments," in J. Barwise and S. Feferman (eds.) (1985). 271–287.

Nourani, C. F. (1984). "Equational Intensity, Initial Models, and Reasoning in AI: A Conceptual Overview," Proc. Sixth European AI Conference, Pisa, Italy, (1984). North-Holland.

Nourani, C. F. (1988). "Diagrams, Possible Worlds, and The Problem of Reasoning in Artificial Intelligence," Proc. Logic Colloquium, (1988). Padova, Italy, Journal of Symbolic Logic. http://www.univie.ac.at on author's name.

Nourani, C. F. (1991). "Planning and Plausible Reasoning in Artificial Intelligence, Diagrams, Planning, and Reasoning," Proc. Scandinavian Conference on Artificial Intelligence, Denmark, May (1991). IOS Press.

Nourani, C. F. (1993). "Computation on Diagrams, and Free Solemnization," Summer Logic Colloquium, England, July 1993.

Nourani, C. F. (1994). "Functorial Model Theory and Infinitary Language Categories," September (1994). ASL, January (1995). San Francisco, (see ASL Quarterly, Summer (1996). for the ASL Wisocnsin update).

Nourani, C. F. (1994). "Towards Computational Epistemology—A Forward," Proc. Summer Logic Colloquium, Clermont-Ferrand, France, July 1994.

Nourani, C. F. (1994). High Fidelity Computing With Abstract Models, Categories and Recursion, March 15, (1994). ASL-Spring, Kansas City.

Nourani, C. F. (1997). "Descriptive Computing-The Preliminary Definition," Summer Logic Colloquium, July (1996). San Sebastian Spain. See AMS April (1997). Memphis.

Nourani, C. F. (1997). "Functorial Admissible Models," ASL April1997, MIT, Cambridge.

Nourani, C. F. (1998). "Computability, KR and Reducibility, March (1998). ASL, Toronto, May 1998.

Nourani, C. F. (1998). Admissible Models and Peano Arithmetic, ASL, March (1998). Los Angeles, CA. BSL, vol.4, no.2, June 1998.

Nourani, C. F. (1998). Syntax Trees, Intentional Models, and Modal Diagrams For Natural Language Models, Revised July 1997. Uppsala Logic Colloquium, August (1998). Uppsala University, Sweden.

Nourani, C. F. (1998a). "Functorial Models, Admissible Sets, and Generic Rudimentary Fragments, March (1997). Summer Logic Colloquium, Leeds, July 1997. BSL, vol. 4, no.1, March 1998. www.amsta.leeds.ac.uk/events/logic97/con.html

Nourani, C. F. 1995" Free Proof Trees and Model-Theoretic Planning, February 23,1995, Automated Reasoning AISB, England, April 1995.

Odintsov, S., Wansing, H. Inconsistent-tolerant Description Logic: motivation and basic systems. In Trends in Logic: 50 Years of Studio Logic, eds. V. Hendriks and J. Malinowski, J. van Benthem. Modal Logic in Two Gestalts. In Advances in Modal Logic, vol. II, eds. M. de Rijke, H. Wansing and M. Zakharyashev, CSLI Publications, Stanford, 73–100, 1998.

Platek, Richard Alan (1966). Foundations of recursion theory, Thesis (Ph.D.)–Stanford University, *MR 2615453.*

Richter, M. M. (1978). Logikkalkule, Teubner Studeinbucher, Tubner, Leibzig, 1978.

Schild, K. A correspondence theory for terminological logics: preliminary report. In Proc. of the 11th Joint Conference on Artificial Intelligence (IJCAI-91), 466–471, 1991.

Scott, D. (1961). "Measurable Cardinals and Constructible Sets," Bulletin of the Academy of Polish Sciences, 9: 521–524.

Scott, D. (1965). "Logic with Denumerably Long Formulas and Finite Strings of Quantifiers," The Theory of Models, J. Addison, L. Henkin, and A. Tarski (eds.), Amsterdam: North-Holland, 329–341.

Scott, D., Tarski, A. (1958). "The sentential calculus with infinitely long expressions," Colloquium Mathematicum, 16, 166–170.

Spohn, W. (1986). Ordinal Conditional Functions: A Dynamic Theory of Epistemic States, Universitat Munchen, Report, 1986.

Valeria de Paiva, (2002). Constructive Description Logics: what, why and how, Xerox PARC, 3333. Coyote Hill Road, Palo Alto CA 94304, USA.

Wijesekera, D. 1990. Constructive modal logic I. Annals of Pure and Applied Logic, 50, 271–301.

Wirsing, M. (1995). Recent Trends in Data Type Specification, Lecture Notes in Computer Science, 906, 81–115.

FURTHER READING

Ackermann, W., Solvable Cases of the Decision Problem. North-Holland, Amsterdam, 1954.

ADJ-Goguen, J., Thatcher, J., Wagner, E., Wright, J., Abstract Data Types as Initial Algebras and Correctness of Data Representations. Conference on Computer Graphics, Pattern Recognition and Data Structure, May 1975, 89–93.

ADJ-Goguen, J.A., Thatcher, J.W., Wagner, E.G., An Initial Algebra Approach to the Specification, Correctness, and Implementation of Abstract Data Types. "Current Trends in Programming Methodology," Vol.4, Ed. Yeh, R., Prentice-Hall.

Aho, A., Sethi, R., Ullman, J. (1972). Code Optimization and Finite Church-Rosser Systems. in Proceedings of Courant Computer Science Symposium 5, Ed. Rustin, R., Prentice Hall.

Backus, J. (1978). Can Programming be Liberated from the von Neumann Style!' A Functional Style and Its Algebra of Programs. CACM 21, 8, 613–641.

Birkhoff, G., Lipson, J.D. (1970). Heterogeneous Algebras. Journal of Combinatorial Theory 8, 115–133.

Boone, W. (1959). The Word problem. Ann. of Math. 2, 70, 207–265.

Boyer, R., Moore J, (1979). A Computational Logic. Academic Press.

Boyer, R., Moore, J. (1975). Proving Theorems About LISP Functions. JACM 22, 129–144.

Burstall, R. M. (1969). Proving properties of Programs by Structural Induction. Computer, J. 12, 41–48.

Church, A., Rosser, J.B. (1936). Some Properties of Conversion. Transactions of AMS 39, 472–482. Computer Science Department Report No. STAN-CS-80–785.

Curry, H. B., Feys, R. (1958). Combinatory Logic Vol. Z. North-Holland, Amsterdam.

Davis, M. (1973). Hilbert's Tenth Problem is Unsolvable. Amer. Math. Monthly 80, 3, 233–269.

Evans, T. (1951). The Word Problem for Abstract Algebras. J. London Math. Sot. 26, 6471.

Gerard Huet, Derek C. Oppen, Equations and Rewrite Rules: A Survey, Stanford Verification Group, Report No. 15, 1980.

Guttag, J. V., Horning, J. J. (1978). The Algebraic Specification of Abstract Data Types. Acta Informatica 10, 27–52.

Hermann, G. (1926). Die Frage der endlich vielen Schritte in der Theorie der PoZynomideale, Math. Ann., Vol 95, 736–738.

Hindley, R. (1969). An Abstract Form of the Church-Rosser Theorem Journal of Symbolic Logic 34, 4, 545–560.

Huet, G. (1973). The Undecidability of Unification in Third Order Logic. Information and Control 22, 257–267.

Koffmann, M., O'Donnell, M. (1979.) Interpreter Generation Using Tree Pattern Matching. 6th ACM Conference on Principles of Programming Languages.

Kozen, D. (1977). Complexity of Finitely Presented Algebras. Ninth ACM Symposium on Theory of Computing, May, 164–177.

Lankford, D.S., Ballantyne, A.M. (1977). Decision Procedures for Simple Equational Theories With Commutative Axioms: Complete Sets of Commutative Reductions. Report ATP-35, Departments of Mathematics and Computer Sciences, University of Texas at Austin, March 1977.

Lipton, R. and Snyder, L., On the Halting of Tree Replacement Systems. Conference on Theoretical Computer Science, U. of Waterloo, Aug. 1977, 43–46.

Makanin, G. S. (1977). The Problem of Solvability of Equations in a Free Semigroup. Akad. Nauk. SSSR, TOM 233, 2.

Malcev, A. I. (1958). On Homomorphisms of Finite Groups. Ivano Gosudarstvenni Pedagogicheski Institut Uchenye Zapiski, vol. 18, pp. 49–60.

Matiyasevich, Y. (1970). Diophantine Representation of Recursively Enumerable Predicates. Proceedings of the Second Scandinavian Logic Symposium, North-Holland.

McCarthy, J. (1960). Recursive Functions of Symbolic Expressions and Their Computations by Machine, Part, I. CACM 3, 4, 184–195.

McNulty, G. (1976). The Decision Problem for Equational Bases of Algebras. Annals of Mathematical Logic, 11, 193–259.

Meyer, A. R., Stockmeyer, L. (1973). Word Problems Requiring Exponential Time. Fifth ACM Symposium on Theory of Computing, April, 1–9.

Musser, D. L., A Data Type Verification System Based on Rewrite Rules. 6th Texas Conf. on Computing Systems, Austin, Nov. 1978.

Nash-Williams, C. St. J. A. (1963). On Well-quasi-ordering Finite Trees. Proc. Cambridge Phil. Sot. 59, 833–835.

O'Donnell, M. (1977). Computing in Systems Described by Equations. Lecture Notes in Computer Science 58, Springer Verlag.

Plaisted, D. (1978). Well-Founded Orderings for Proving Termination of Systems of Rewrite Rules. Dept. of Computer Science Report 78–932, University of Illinois at Urbana- Champaign, July.

Post, E. (1947). Recursive Unsolvability of a Problem of Thue. J. Symbolic Logic 12, 1–11.

Presburger, M., Uber die Vollstandigkeit eines gewissen Systems der Arithmetik ganzer Zahlen, in welchem die Addition als einzjge Operation hervortritt. Comptes-Rendus du ler CongrCs des Mathematician's des Pays Slaves, 1929.

Rosen, B.K. (1973). Tree-Manipulation Systems and Church-Rosser Theorems. JACM 20, 160–187.

Scott, D. (1970). Outline of a Mathematical Theory of Computation. Monograph PRG-2, Oxford University Press.

Seidenberg, A. (1954). A New Decision Method For Elementary Algebra and Geometry. Ann. of Math., Ser. 2, 60, 365–374.

Sethi, R. (1974). Testing for the Church-Rosser Property. JACM 21, 671–679. Erratum JACM 22 (1975), 424.

Staples, J. (1975). Church-Rosser Theorems for Replacement Systems. Algebra and Logic, ed. Crossley, J., Lecture Notes in Math., Springer Verlag, 291–307.

Stickel, M. E. (1975). A Complete Unification Algorithm for Associative-Commutative Functions. 4th International Joint Conference on Artificial Intelligence, Tbilisi.

Tarski, A. (1951). A Decision Method for Elementary Algebra and Geometry. University of California Press, Berkeley.

Tarski, A. (1968). Equational Logic. Contributions to Mathematical Logic, ed. Shiite et al., North-Holland.

Thatcher, J., Wagner, E., Wright, J. (1978). Data Type Specifications: Parameterization and the Power of Specification Techniques. Tenth ACM Symposium on Theory of Computing.

Tserunyan, A. Introduction to Descriptive Set Theory. www.math.uiuc.edu/~anush/Notes/dst_lectures.pdf

Yiannis, N. Moschovakis (2009). Descriptive Set Theory, Professor of Mathematics. University of California, Los Angeles and Emeritus.

CHAPTER 4

INITIAL TREE COMPUTING AND LANGUAGES

CONTENTS

4.1   INTRODUCTION

This chapter presents the foundations for a theory and application to computation with algebraic trees based on initial algebras developed by Nourani and colleagues since early-1977 at IBM Research Yorktown Heights (ADJ, 1976), and on University of California, TU Berlin, SRI International, and Munich. We start by defining standard initial models of computation theories. A tight formulation of generalized educational theories is presented by algebraic theories and subtree replacement systems. This is perhaps the only direction in computing research where algebraic theories are presented along with their model-theoretic formulations. There are a few disciplines of computing and mathematics at play, abstract model theory, abstract recursion, algebraic theories, computability with trees and computational logics for AI.

Here, the methods for constructing canonical initial models for equational trees are presented applying the properties of algebraic sub tree replacement systems, for example, the Church-Rosser property and the initial canonical modeling techniques. It is shown that subtree replacement systems rewrite to initial algebras by providing the mathematical foundations for a theory of term rewriting. Some practical development of this author in subtree replacement systems and some of our recent mathematical results put together in the present chapter to provide new foundations for some of the theory and practice of computing with subtree replacement systems, and to present new practical techniques. Thus, we have a computational characterization of initial standard models. Then, a bridge from initial algebras and model theory is made onto the theory of tree rewriting. The formulations shows how there are free proof trees and free Skolemized trees with generalized standard models.

The generic model diagram techniques are further applied to put together formulations of initial standard models and free proof trees that are generated from free Skolemized diagrams. Next, we present methods of generating normal forms subtree replacement systems by defining model-theoretic forcing properties on algebraic trees. Positive forcing conditions are defined on algebraic trees such that infinitary models could be generated for algebraic subtree replacement systems. Models are essential for solving term rewriting problems and we present methods that show how to define them. The sections show how some subtree replacement systems could have normal forms by presenting methods of generating algebraic closed structures for their presentations. Thus, we solve some important computability problems by supporting our methods of implementing term rewriting systems with some theoretical foundations. Applications of the *method of free proof trees* to some problems in artificial intelligence is also alluded to in brief here and presented by this author in for example (Nourani, 1996).

The paradox *"No course of action could be determined by a rule because every course of action can be made to accord with the rule"* is what is always faced by subtree replacement systems at "rule for interpreting the rule" problem emphasized by Wittgenstein. To interchange Wittgenstie's "what is on the mind" with "what is encoded onto a machine's mind, the meaning attached to functions is only what a partial model for what has gone on thus far can imply. It is precisely the Wittgenstein's private language problem. The section before last culminates where the thoughts had been applied and are applicable to ASRS. There are new applications to tree computing for AI (Nourani, 1996) and computational linguistics. We present foundations for guiding tree rewriting by model fragments as conditions. The private language (Wittgenstein, 1959) problem faced at tree computing can be remedied by keeping in mind condition models as partial models (Broy-Wirsing, 1982).

4.2 INITIAL MODELS AND THEIR ALGEBRAIC FORMULATION

4.2.1 THE BASICS

To prove Godel's completeness theorem, Henkin style, one proceeds by constructing a model directly from the syntax of the given theory. This structure is obtained by putting terms that are provably equal into equivalence classes, then defining a free structure on the equivalence classes. For example, G(ui, tar) is defined to be the formal string 'Guitar.' Constant symbols are added to the language to instantiate existential formulas. The computing and reasoning enterprise require more general techniques of model construction and extension, since it has to accommodate dynamically changing world descriptions and theories. This author's techniques for model building since 1980's applied to computing and the problem of reasoning allows us to build and extend models through model diagrams. This requires us to focus attention on the notion of generalized diagram. Such diagrams were developed by this author to build models with a minimal set of generalized Skolem functions (see, Nourani, 1991). The minimal set of function symbols are those with which a model can be built inductively. We focus our attention on such models, since they are Initial and computable.

4.2.2   CANONICAL MODELS

A technical example of algebraic models defined from syntax had appeared in defining initial algebras for equational theories of data types. Computing models of equational theories of computing problems are presented by a pair (Σ, E), where Σ is a *signature* (of many sorts, for a sort set S) and E a set of Σ-equations. Let $T<\Sigma>$ be the free tree word algebra of signature Σ. The quotient of $T<\Sigma>$, the word algebra of signature Σ, with respect to the Σ-congruence relation generated by E, will be denoted by $T<\Sigma, E>$, or $T<P>$ for presentation P. $T<P>$ is the "initial" model of the presentation P. The Σ-congruence relation will be denoted by $\equiv P$. One representation of $T(P)$ which is nice in practice consists of an algebra of the canonical representations of the congruence classes. This is a special case of generalized standard models defined here.

In what follows $gt1\ldots tn$ denotes the formal string obtained by applying the operation symbol g in Σ to an n-tuple t of arity corresponding to the signature of g.

Furthermore, g_C denotes the function corresponding to the symbol g in the algebra C.

Definition 4.1: Let Σ be an S-sorted signature. Then a canonical term Σ-algebra (Σ-CTA) is a Σ-algebra C such that
   (1) $|C|$ is a subset of $T<\Sigma>$ as S-indexed families;
   (2) $gt1\ldots tn$ in C implies t is in C and; $g_C (t1, \ldots, tn) = gt1\ldots tn$.
         For constant symbols
   (3) Special case of constants must also hold as well, with $g_C = g$.

The following theorem is from ADJ (1978) with generalized versions for slalom trees (Nourani, 1996).

Theorem 4.1: Let C be a Σ-algebra. Let $P = (\Sigma, E)$ be a presentation. Then C is Σ-C is isomorphic to $T<P>$, iff

(i) C satisfies E;
(ii) g_C (t1, ..., tn) $\equiv$ gt1...tn.

Note: (ii) must also hold for constants with g_C = g; $\equiv$ refers to the Σ-congruence generated by E;.

Proof (e.g., Thatcher et.al., 1978; Nourani, 1996).

Example of group Strings: for example, a canonical term algebra constructed on representatives for a group structure on $\{x, y\}^*$, such that x2=e, y3=e, and xy=yx, where e denotes the empty string for the Kleene * operation. Thus, these constitute a set of representatives for the above. Let G= {e, x, y, xy, y2, xy2}.The canonical term model is defined by putting the obvious group structure on the six-element set. x.y = xy; e.z =z for all z in G; y.xy = xy2; y.y2 = e; xy. xy2 = e

x.y2 = xy2; y.xy2 = x. This is a canonical model of the cyclic group of order six. The model satisfies the conditions of the CTA-theorem. This model is an example of what we call a standard model for an equational theory. we have formal definitions for the models. Let us define how these models are to be put forth.

Definition 4.2: Let M be a set and F a finite family of functions on M. We say that (F, M) is a monomorphic pair, provided for every f in F, f is injective, and the sets {Image(f):f in F} and M are pair-wise disjoint.

This above definition is basic in defining induction for abstract recursion-theoretic hierarchies (Hinman, 1979) and inductive definitions. What was new in our papers is that of the definition for generalized standard models with monomorphic pairs. This definition was put forth by Nourani in 1982 for the computability problems of initial models. Instances are state were state in Chapter 3

Definition 4.3: A standard model M, of base M and functionality F, is a structure inductively defined by <F, M> provided the <F, M> forms a monomorphic pair.

We will review these definitions in the sections to follow.

4.2.3   GENERIC DIAGRAMS OF INITIAL MODELS

The generic diagrams (G-diagram) c.f. Chapters 2 and 3, are diagrams in which the elements of the structure are all represented by a minimal set of function symbols and constants, such that it is sufficient to define the truth of formulae only for the terms generated by the minimal set of functions and constant symbols. Such assignment implicitly defines the diagram. This allows us to define a canonical model of a theory in terms of a minimal family of function symbols. The minimal set of functions that define a G-diagram are those with which a standard model could be defined by a monomorphic pair. Formal definition of diagrams are stated here, generalized to G-diagrams, and applied in the sections to follow.

Definition 4.4: Let M be a structure for a language L, call a subset X of M a generating set for M if no proper substructure of M contains X, for example, if M is the closure of X U {c(M): c is a constant symbol of L}. An assignment of and G: A → M, such that {G(a): a in A} is a set of generators for M. Interpreting a by g(a), every element of M is denoted by at least one closed term of L(A). For a fixed assignment <A, G> of constants to M, the diagram of M, D<A, G>(M) is the set of basic (atomic and negated atomic) sentences of L(A) true in M. (Note that L(A) is L enriched with set A of constant symbols.)

Definition 4.5: A G-diagram for a structure M is a diagram D<A, G>, such that the G in definition 4.4 has a proper definition by a minimal set of function symbols, for example, Σ_1 Skolem functions.

Remark: The minimal set of functions above are those by which a standard model could be defined by a monomorphic pair for the structure M. Thus, initial models could be characterized by their G-diagrams. Further practical and the theoretical characterization of models by their G-diagrams are presented by this author, for example, Chapter 3. This builds the basis for some sections that follow where we show how initial models could appear out of thin air within our formulation.

4.3 TREE COMPUTING ON ALGEBRAIC THEORIES

In this section, we present methods of constructing initial models by algebraic tree rewriting. Thus, we show how initial algebras can be defined by subtree replacement and tree rewriting. Subtree replacement systems in computing for equational theories and theorem have been for quite some time a subject of research by this author and many others at the present time. Let us present a brief overview and formulation of algebraic subtree replacement systems and connect it to our recent computational views with G-diagrams. We formulate subtree replacement systems with algebraic theorems in the present chapter from Nourani (1996). First, we define algebraic subtree replacement systems.

Consider a set A of Σ-terms for an S-sorted signature Σ. Let E be a set of Σ-equations. Let S* denote the set of all strings u =u1...un. For u in S, including the empty string; let S+ denote the set of all non-empty strings of sorts. For s in S and u=u1...un in S*, let A(u, s) denote the set of all Σ-expressions of sort s, with variables (usually x1, ..., xn) of sorts u = u1...un. For u in S* and v in S+, with v = v1...vm, let A(u, v) denote the set of all tuples e = (e1, ..., em) with e in A(u, vi); say that e is a tuple expression, of arity u and coarity v, or a Σ v-tuple expression in u-tuple variables; wrote e:u → v. Then A is the union over all s in S of A s = A (1, s), where 1 denotes the empty string. Define a set of Σ-rewrite rules (or Σ-equations) for an algebra A of signature Σ, to be a set of pairs (f, g) with f, g in some A(u, s), u in S* and s in S.

Let E(u, s) be the set of all rules (f, g) with (f, g) in A(u, s). To define the set R[E] of applications of rules in E to constant expressions, for example, expressions without variables, we need to define the notion of substitution. Given f: u →v and h: w →u, let foh denote the result of substituting the ui for the corresponding variable xi in f, yielding f o h: w →v. Then R[E] is the set of all pairs (q o foh, q o g oh) such that (f, g) in E(u, s), q in A(s, s'), and h in A(1, u). This is all pairs resulting from substituting a tuple h of constant expressions (without variables) into (f, g), then substituting the result for a variable in some expression one instance at a time. We refer to (A, R[E]) as an *algebraic subtree replacement system*, and say that q o g oh results from rewriting q o f oh.. We say that <q, h> is a substitution.

This formulation is following the usual practice in algebraic theories of defining substitution as tree composition. As an example, let f = x1+(x1+x2)+x3. Then f =g is the associative law. An application of this rule rewrites (2+((3+5)+1)*8) to ((2+(3+5))+1)*8. Here q = x1*8 and h =(2, (3+5),1). The best way to visualize this process is to look at the tree representation of the above expressions. Given an algebraic subtree replacement system (A, R[E]), the relation defined by R[E] on A will be denoted by "=>." We write t => t,' for t and t' in A, iff t can be rewritten to t' by a finite number of applications of the replacement rules in (A, R[E]). Let => * denote the reflexive transitive closure of =>. Then we say that (A, R[E]) is *Church-Rosser* iff for all a, b, c in A such that a =>*b and a =>*c, there is a d in A such that b => *d and c =>*d.

An algebraic replacement system has the (FTP) iff there is no infinite sequence of rewrites x1 => x2 => … A term t in A is said to be in normal form with respect to (A, R[E]) if it cannot be rewritten by the rules of (A, R[E]). Such term t is said to be *irreducible*. If reduced forms are unique, we say that (A, R[E]) has *the unique termination property* (UTP). The significance of the UTP and FTP is that under such circumstances unique normal forms exist.

Definition 4.6: Let Σ be an S-sorted signature, E a set of Σ-equations. An equational theory with signature Σ and axioms E is a formal system with Σ-equations as its only formulas, with axioms that are those in E, and rules of inference that are reflexivity, symmetry, transitivity, and substitution properties of equality, for example, 1. t =t; 2. t =t' => t'=t; 3. t =t' and t'=t' => t=t" 4. t =t' => g o t oh = g o t' o h, where <q, h> is a substitution. Here t, t,' and t' are Σ-terms (including variables).

Definition 4.7: Let T be an equational theory with signature and axioms E, with an S-sorted signature. A proof of t=t' in T is a finite sequence b of Σ-equations ending in t=t' such that if q=q' is in b, then either q=q' in E, or q=q' is derived from 0 or more previous equations in E by one application of the rules of inference. Write T <ST>|— t=t' for "T proves t=t' by algebraic subtree replacement system." When the set of axioms E is not specific, and is only an implicit set, we write R instead of R[E]. Deduction with equational theories is closely related to rewrite rule simplifications. Given a set R of Σ-equations, for a signature Σ, let Ro be the union of R with its converse. Furthermore, let R be Ro viewed as a set of rewrite rules, for example, l =r in Ro is viewed as l => r in R. Let T(R) be the equational theory of signature Σ and axioms R. The following lemma expresses a relationship between T(R) and R that clarifies the theoretical development and is useful for our applications.

Lemma 4.1: Let R be a set of Σ-equations. Let R be the set of algebraic Σ-rewrite rules obtained by considering each equation l =r in Ro as a rule l => r, then for t, t' in T<Σ>, t => * t' iff T(R) <ST>|— t = t.'

Proof (e.g., Nourani 1996).

Recall that a presentation (Σ, E) defined an equational theory of signature Σ and axioms E. Next we show how canonical models can be constructed by algebraic subtree replacement system. A definition and what we have done thus far gets us to where we want to go: the canonical algebraic term rewriting theorems. $\Sigma <s1, sn, s>$ denotes the part of the signature with operations of arity $(s1, ..., sn)$ and coarity s, with Csi the carrier of algebra C of sort si.

Definition 4.8: Let R be a convergent set of Σ-rewrite rules, for example, $T <\Sigma, R>$ has the FTP and UTP properties, let $[t]$ denote the R-reduced form of t in $T<>$. Σ Let $|C|$ be a subset of $|T<\Sigma>|$, for g in $\Sigma <s1...sn, s>$ and ti in C si, define $gC (t1, ..., tn) = [g(t1, ..., tn)]$. If this always lies in C, then C becomes a Σ— algebra, and we say that (C, R) represents a Σ-algebra A iff the Σ-algebra so defined by (C, R) is Σ-isomorphic to A.

The following intermediate theorem gives sufficient conditions for constructability of an initial model for an equational presentation. It is the mathematical justification for the proposition that initial models can be automatically implemented (constructed) algebraic subtree replacement systems with normal forms defined by a minimal set of functions that are Skolem functions or type constructors.

Theorem 4.2: Let Σ be an S-sorted signature, R a convergent set of Σ-rewrite rules. Let $|C|$ be a subset of $|T<\Sigma>|$. Define $gC(t1, ..., tn) = [g(t1, ..., tn)]$. Furthermore, assume that $[f] = f$ for all f in $\Sigma(1, s)$ If there exists a subset CF of Σ such that $|C| = |T<CF>|$ and the following conditions are satisfied

1. $gC(t1, ..., tn)$ in C whenever ti in C, where g has nontrivial arity $(s1, ..., sn)$, for ti of sort si;
2. for all g of nontrivial arity $(s1,..., sn)$ with ti in C, g in CF, $gC(t1, ..., tn) = gt1, ...tn$; in particular for a constant g, $gC = g$;
3. for g in $\Sigma - CF$ of arity $(s1,..., sn)$, $gC(t1, ..., tn)=t$, for some t in T<CF>;

Then: (i) C is a canonical term algebra; and
(ii) (C, R) represents $T <\Sigma, R>$, R is R viewed as a set of Σ-equations.

Proof (Author, for example, Nourani, 1996).

Theorem 4.3: Let Σ be an S-sorted signature, and R a convergent set of rewrite rules such that $[g] = g$. Define a Σ-algebra structure C on $T<\Sigma>$ by $gC(t1, ..., tn) = [g(t1, ..., tn)]$. Let C* be the smallest sub Σ-algebra of C. Then C is a canonical term algebra consisting of R normal forms and (C, R) represents $T <\Sigma, R>$.

Proof (Exercises, for example, Nourani 1996)

Corollary 4.1: Let Σ be an S-sorted signature, R a convergent set of Σ-rewrite rules then. Let |C| be subset of |T<Σ>|. Define gC (t1, …, tn) = [g(t1, ….tn)].

   If there exists a subset CF of Σ such that |C| is subset of |T<Σ>|, and
   (1) g_C(t1, …, tn) in C whenever ti in C, for g of nontrivial arity (s1, …, sn);
   (2) for all g in Σ of nontrivial arity (s1, …, sn) with ti in C, gC (t1, …, tn)=gt1…tn; in particular g_C = g for g a constant symbol;
   (3) for g in Σ – CF, gC(t1, …, tn) = t for t in C; then:
      (i) C is a canonical term algebra; and
      (ii) C with R represents T <Σ, R>.

Proof (Nourani, 1996)

   Note that CF does not appear in the statement of the above theorem, but it is implicit in claiming a C*. CF is how we could get a C*.

   The above theorems and the corollary point out the importance of the constructor signatures in computational characterization of initial models. These are the minimal set of functions that by forming a monomorphic pair with the base set, bring forth an initial model by forming the free trees that define it. Thus, an initial free model is formed. The model than can be obtained by algebraic subtree replacement systems. The G-diagram for the model is also defined from the same free trees. The conditions of the theorems are what you expect them to be that canonical subset be closed under constructor operations, and operations outside the constructor signature on canonical terms yield canonical terms.

   The group example, cyclic group of order six on the set {x, y}, is revisited here again: The normal forms are the six element of the set {e, x, y, xy, y2, xy2}. Now consider the following three basic group axioms: e.x = x; x-1.x = e; (x.y). z = x. (y.z). Its completion thorough the algorithm (Knuth-Bendix, 1970) is attained by the following additional equations.

$$e-1 = e; (y-1)-1 = y; (y.y")-1 = y"1.y; y-1.(y.z) = z; z.e= z; y. (y-1.z) = z$$

$$y.y-1 = e.$$

   Using the above equations as a set of left-right rewrite rules R, the group, the group structure on {x, y}* with R defines a canonical group structure C isomorphic to the cyclic group of order six. The set of type constructors for the given group along with the set R or rewrite rules satisfies the conditions of theorems above. Thus, (C, R) is isomorphic to the cyclic group structure and represents the initial tree algebra T<Σ, R>. More recent incarnations for such algebraic tree rewriting is Hausmann (2010).

4.4   TREE REWRITING, ALGEBRAS, AND INFINITARY MODELS

4.4.1   POSITIVE FORCING INFINITARY MODELS

In the section above, we showed how to design initial models on algebraic free trees. We further showed how they could be implemented by automatic methods from algebraic subtree replacement systems In the present section the author develops the intricate mathematics that formed the basis for the model-theoretic forcing basis for algebraic tree computing and algebraic subtree replacement systems. The basic results proved in this section is that the partially ordered set defined by the power set of T*(W), under the natural subset ordering, defines a forcing property for which T*(W) generates a generic set.

Thus, the closure construction contains all one needs to conclude the truth of formulas in the particular computing world defined. The formulation of forcing that this author presented since 1981 and put forth in brief here is based on model-theory and set-theory within a well-behaved fragment of infinitary logic $L_{\omega 1, \omega}$. The language and some of the formulation, for example the notion of a forcing property is the same as that of Keisler (1973), except for its algebraic formulation here.

The Keisler formulation is quite elegant and encodes all the usual notions of forcing put forth. But from that point on things are not the same. We formulate theories, models, and forcing properties and conditions on algebraic theories and initial models. First, we take a countable language L and then enrich it with a countable supply of constants to name individuals. A well-behaved fragment of L1,ω is applied (see appendix for further definitions). Call the enriched language L[C]. We shall use the symbol M for a model and M for the set of individuals in its universe. Throughout this paper our theories do not admit trivial one-point models. Now we put forth some definitions.

Definition 4.9: A forcing property for a language L is a triple $\Im = (S, <, f)$ such that

(i) (S, $\leq$) is a partially ordered structure with a least element, for ex-
     ample, 0;
(ii) f is a function which associates with each p in S a set f(p) of atomic
     sentences of L[C];
(iii) whenever p $\leq$ q, f(p) $\subseteq$ f(q);
(iv) let l and t be terms of L[C] without free variables and p in S; let $\gamma\varphi$
     be a formula of L[C] with one free variable. Then if (l$\eta\neg$=t) is in
     f(p) then (t=l) is in f(q) for some q $\geq$ p. φ(t) in f(p) implies φ(l) is
     in f(q) for some q $\geq$ p. For some c in C and q $\geq$ p, (c=l) is in f(q).
     The elements of S are called conditions for $\Im$.

Now, consider a presentation P, its initial standard model M, obtained
by taking the quotient of its free tree algebra (word algebra) with respect
to the congruence relation generated by the equations of the presentations
on the trees. Enrich the model with the names of the function symbols that
can define the G-diagram for the model. In other words, add on Skolem
function names that can complete the set and define a G-diagram corresponding to what amounts to the set of functions that can define the model
by a monomorphic pair.

Lemma 4.2: ($\wp$(T*(P), $\subseteq$, f), where P is the powered operation, and f an
assignment of free variables in T*(P) into M, is a forcing property.

Proof: First we note that such f indeed does associate a set f(p) of atomic
sentences of Lc with each p in S; for f defines a function g, g:(T*(P)) $\rightarrow$
Γ, whereΓ is the set of atomic sentences of Lc. Given any r in $\wp$(T*(P)),
r is a subset of T*(P), therefore, g assigns to each free variable in r, from
M, which is assumed to be built out of Lc. Thus, f in turn maps to a set of
atomic sentences of Lc by virtue of T*(P) being strictly a set of equational
formulas. Thus, f implicitly defines the mapping. Furthermore, ($\wp$(T*(P)),
$\subseteq$) is clearly a partially ordered structure, with the least element being the
empty set. To show that (iii) of definition 3.1 is satisfied: just note that
since for any p, q in p, q T*(P), therefore f(p) $\subseteq$ f(q). Checking (iv) takes
a bit more work: if (l=t) is in f(p) the (t=l) is in f(q) for some q > p, by the
symmetric property of equality in T(P).

Next, if (l=t) and φ(l) in f(p) then by substitution of equals in T(P), φ(t)
is in f(q) for some q > p, q a subset of T*(P). Last, to show that for some c
in C and q>p, (c=l) in f(q). This follows because Lc has an infinite supply
of names for constants, and by assigning to each representative term in M,

equations of the form c=l, for l a term, would be consistent with M and T*(P). The assignment of names to terms can be done in a predetermined manner within the countable fragment such that c=l is in some $q > p$. This completes the proof of the lemma.

Next, we define positive forcing.

Definition 4.10: The relation p ||— + φ, read "p positively forcesφ" is defined for conditions p an q in S as follows: for an atomic sentence φ, p ||—+ φiff φ is in p; for an open formula $\varphi$of the form f(X) = g(X), p ||—+ φ, iff for all c in L[C], where c is an n-tuple of constants form L[C] and all q, $p < q$ implies there is an r such that $r > q$ and r ||—+ f(c) = g(c).

Definition 4.11: Let $\mathfrak{J} = (S, <, f)$ be a positive forcing property, we say that a subset G of S is positively generic iff

(i) p in G and q <p implies q in G;

(ii) p, q in G, implies there is an r in G such that $p < r$ and q <r;

(iii) for each sentence φ, there exists p in G, such that either p ||—+ φ or there is no q in S, q >p such that q ||—+ φ

A special case of the above definition is when S consists of sets of formulas. For such S, we can make a substitution: subset relation for <. Clause (iii) can then be stated:

(iii)' for each sentence φ, there exists p in G such that either p ||—+ φ or p U {φ} is not a condition for F.

Next, we state the positive forcing theorem that forms the basis for model building with infinitary logic on algebraic trees. We should note that the notion of condition in the present formulation is always relative to a standard model M, in the sense defined by the present paper. That is the model with respect to which we take a UIC. Thus, conditions are always formulae consistent with M and UIC.

Conditions are infinitary in the present formulation, therefore, we are at a universe where the conditions are infinitary.

Theorem 4.4: Let T(P) be an equational theory, M its standard model obtained by a monomorphic pair. Then $\wp$ (T*(P)) is positively generic for T(P).

Proof: We show that S = $\wp$(T*(P)) satisfies the conditions put forth in definition 4.4, with the subset relation as the ordering on S. To check (i) note that T*(P) is the maximal element of $\wp$ (T*(P)) and every condition p is a subset of T*(P). If p is in S, and $q > p$, then clearly there is a condition r such that r > q and r > p, namely p U q. this follows form the fact that both p and q must be consistent with T*(P) and further they must be modeled by M. Finally, to show that the third requirement of 4.4 is satisfied: suppose not and consider the condition T*(P), where T*(P) does not force φ, but for some q > T*(P), q ‖—+ φ. One such q is T*(P) U {φ}, but then by definition of ‖—+, T*(P) ‖—+ φ.

Thus, we have exhibited an explicit construction for a generic set. We have further shown that UIC has the genericity property. It is important to note that within the present formulation of forcing, the set to which we force is known. Thus, there is an infinitary set, in principle, that could contain all that various conditions could imply for the forcing property. That is what makes this method and results applicable for computing. That does not mean that one could get his hands on it, however, to trivialize the computational implications. Thus, we have exhibited an explicit construction for a generic set.

We have further shown that UIC has the genericity property. It is important to note that within the present formulation of forcing, the set to which we force is known. This is in part a consequence of us having shown that positive forcing has a forcing companion for the forcing property. Thus, there is an infinitary set, in principle, that could contain all that various conditions could imply for the forcing property. That is what makes these techniques and results applicable for computing. That does not mean that one could get his hands on it, however, to trivialize the computational implications.

Thus, to state an mathematical result and construction of an infinite set, the forcing companion, does not imply that it could be invoked in computing practice to force a solution to everyday problem, then and there. But it helps to generate models and to figure out what sets are not conditions in proofs. Hence, it has computational significance. Most other notions of forcing do not have forcing companions there by free construction. Thus, in those formulations it is not known to what set things are forced to. We want to show the reader that by the present approach we can generate infinitary models of computational theories. The methods could then be applied to a variety of problems from subtree replacement completion problems to proof trees. These application areas are presented in the subsequent two sections.

4.5   FREE PROOF TREES AND COMPUTING MODELS

4.5.1   GENERATING MODELS BY POSITIVE FORCING

In this section, we present the applications of the methods of positive forcing to the construction of infinitary models of equational theories and thus for algebraic subtree replacement systems. We apply the methods to get completion for algebraic subtree replacement systems and solve word problems for algebra and computing. We start with a definition or two. First, the notion of an algebraic closed group: Let C be a class of algebraic structures of the same signature Σ. Let A be an algebra in C. Consider a set S of sentences made from the elements of A; the variables x1, x2, ...y, z, the operations of Σ, equality and the Boolean operations. Call S consistent over A if A can be embedded in an algebra A' in C such that S is satisfied.

Definition 4.12: Let I be the class of all set S of sentences as defined above. A is called algebraically closed if every set S in I that is consistent over A is satisfied in A.

From this definition, we get as special cases the usual notions of algebraically closed fields, and similar notions for algebraically closed groups. For example, if we let C be the class of (commutative) fields, and I the class of singletons S = $\{\varphi\}$, where each φ is a polynomial equation in a single variable quantified existentially; we get algebraically closed fields. The notion of an algebraically closed group is technically difficult to realize because if g in G for a group G, g not e, for e the identity, then there is no solution of x-1x = g in any group H containing G, with G a subgroup. The above abstract definition (4.10) gets around the problem by arriving at a similar notion for groups as well as for fields. Now, let us define what generic models are and then show how to get generic models to solve computing problems.

Definition 4.13: A generic set is said to generate M iff M is a canonical model and every sentence O of L[C] which is forced by some p in G holds in M.

Let us revisit our canonical term models. Such models are useful for computing algebraic closed groups as we shall show. Recall that canonical term models were obtained from some canonical term algebras that were defined on representatives of congruence classes. If we have enough function symbols in the signature or we know how to add Skolem functions to get enough functions to define a G-diagram for, then it is easy to show that we can define a canonical model, which is also a canonical term initial model. The following theorem is what is required to show that.

Theorem 4.5: $\wp\,(T^*(P))$ generates M the canonical initial model for $T^*(P)$.

Proof: The following lemma was applied to prove the theorem.

Lemma 4.1: The signature enriched canonical term algebra M, modeling $T(P)$, the equational theory presented by P, is the canonical model of $T(P)$, in the sense of definition of canonical for positive forcing.

Proof: Each constant c in C is interpreted by either a constant m or by a term built by constants and function symbols in M.

Now, the proof of the theorem goes as follows

Proof of theorem By the above lemma, M is a canonical model, and for every sentence φ forced by some p in $\tilde{A}\,(T^*(P))$, if φ is atomic, then be definition it must be in p to be forced, so it is modeled by M. For an arbitrary equation of the form $f(X) = g(X)$, if it is forced by some p, then there is an r >=p such that r forces e =(f(c)= g(c)). By induction e in r for all c, an n-tuple of constants from M, and M models e for all c. Therefore, M models $f(X) = g(X)$.

From the above definitions and theorems we conclude how generic set generates a model. This allows us to construct algebraically closed groups as follows.

4.5.2 ALGEBRAICALLY CLOSED GROUPS

Consider groups with signature consisting of -1, multiplication, and the identity e plus constants, with group axioms

$$e.x = e; x{-}1.x = e; (x.y).z = x. (y.z)$$

Given the above presentation, the UIC yields a generic set G. The generic set G generates a model in which the original group, for example A, is embedded. Every universal group sentence consistent over A can be embedded in some condition p in G; and every existential sentence has an instance in a condition p in G; therefore, they are satisfied by the generic model M generated by UIC. We can get specific and define an algebraically closed group by forcing. For example, let the signature of groups be defined as above.

Let R be the set of terms over the group signature. define t1 ≡t2 iff there is a condition p in Ω, where Ω is the ℑ-generic filter UIC (e.g., Nourani, 2014), such that p ‖–+ t1=t2. ≡ is easily seen to be an equivalence relation by the definition of the forcing property. It is also a congruence relation with respect to the operations of the signature −1 and.. That we have an instance of a generic filter (see Chapter B.6, Handbook Mathematical Logic; Barwise, 1983). The congruence condition can be readily checked with respect to ≡. Now, let B be the canonical term algebra isomorphic to the quotient of R with respect to ≡.

Theorem 4.6: B = (B,..., −1, e) is an algebraically closed group.

Proof: (Exercises).

The reader should not be dismayed by all that is found in the mathematics literature on unsolvability of such problems. For here, models are generated by infinitary conditions and sets, to model a theory. We are not dealing with finitely generated models. The are models generated by infinitary forcing conditions. Neither are we solving word problems in the proof theoretic sense. We present and generate models, that could have many new Skolem functions added on such that a set of axioms is modeled. Thus, word problems are solved in an indirect manner and all that talk about unsolvability is irrelevant to it. Of course, that is not to say that such area of research is irrelevant. It is not always easy to do the Skolemization, especially if we are dealing with problems in mathematics. In computing, however, there is some intuition on what functions generate the computing model.

4.6 WORD PROBLEMS AND THE SRS ROLLER COASTER

4.6.1 THE ROLLER COASTER

In the following, we show how by defining positive conditions on trees ASRS coasts by tree substitution to normal forms. The enclosed figure is from what had been presented at Nourani (1995). The figure indicates the conditions by light blue rectangles. The tree terms are on a roller coaster ride where the condition peaks modulate the ride. The normal forms emerge on the plateau for the specific tree normal form. The process of term rewriting to normal forms is a roller coaster ride from the pint of view of the intended normal form. The analogy is to what goes on at quantum physics frames of reference. The enclosed is the view at the tree normal form frame of reference.

Some methods are a longer ride, some methods could appear as if the computer time is never enough, and some could not terminate and the normal form lost in the forest of trees. Those who have tried to do completion and automatic tree rewriting could well imagine that the process feels like that, if you could think of yourself as one of the algebraic terms. Thus, the methods we have put forth are really good for the term rewriting theories. We show that by an example below. The result of the last two sections have some practical applications for solving word problems and completion of algebraic subtree replacement systems. The implications are that if one could construct a canonical term initial group for a group presentation, then one in effect has solved the word problem for groups, or any algebraic structure for that matter. Solving such problems is of great importance to computing.

The group problem has received considerable attention since the paper of Knuth-Bendix. Of course, that's not where things end. The problem of tree completion for algebraic subtree replacement systems is really what computing is interested in. Those are the sorts of problems we had formulated in the beginning sections. We are interested in completing subtree replacement systems such that we can tree rewrite to implementations of computing models, thus carrying out symbolic computation for AI and other applications.. The completion Is what gets us the Church-Rosser, UTP, and FTP properties. Newer applications with categorical computing is Beke (1990).

Here, we take the cyclic group example, again, and show how to form an algebraically closed group with it. Let $\{x, y\}*$ be the set of strings on the alphabet $\{x, y\}$ such that $x2=e$, $y3=3$, and $xy=yx$, where e denotes the empty string (the identity of the structure). Every string in $\{a, b\}*$ is equivalent to one of the six strings e, a, b, ab, b2, ab2. Thus, these constitute a set of representatives for the above. Let (e, a, b, ab, b2, ab2} = G. The canonical term algebra is defined by putting the obvious group structure on the six-element set (i.e., $a.b = ab$; $e.x = x$ for all z in G; b. $b2= e$; $ab.b2 = x$; $ab.ab2 = e$; $a.b2 = ab2$; $b.ab2= a$; etc.).

In this case, by picking a generating set for the diagram we can define a canonical initial model, which is by the results of the above sections an algebraically closed group, the word problem is solved indirectly, because the normal forms are forced to the canonical representatives of congruence classes. That is to say once, you are on a roller coaster ride, you bet your sweet life that you are more than happy that the cars are forced to stay on their track and the passenger is forced not to fall off and to stay put until the darn ride is over.

4.6.2   LANGUAGES AND WITTGENSTEIN'S PARADOX

There is insight from Wittgenstein which we can be brought onto the SRS to apply the techniques we have presented in the preceding section. Starting with Wittgenstein's Paradox No course of action could be determined by a rule because every course of action can be made to accord with the rule. A brief overview to the Paradox (Kripke, 1980).

The example of the + operation on natural numbers and an arbitrary quus function is what Kripke's review applies to the Paradox.

Define quus by x + y = x+y, if x, y < 37; + 5, otherwise

At a momentary laps of reason when our mind is computing + might be computing quus. Ordinary time our mind might have learned to internalize a set of instructions for computing ordinary addition and always under normal mind computing do the addition the way it is meant by natural number ordinary addition. The skeptic argues there can be any new answer to the add-on function since there can always be an interpretation for the previous intentions, as appeared by a set of rewrite rules, in the SRS case. At SRS we are always facing the "rule for interpreting the rule" problem emphasized by Wittgenstein. To interchange Wittgenstein's "what is on the mind" with "what is encoded onto a machine's mind, the meaning attached to functions is only what a partial model for what has gone on thus far.

It is precisely the Wittgenstein's private language problem. The models defined by the preceding sections are relativizing truth such that the rules are in "the mind" only as far as the model completion has allowed thus far. The operation + defined by a set of rewrite rules only computes a quus function while the tree rewriting model is being completed. The mind gets off the roller coaster at quus for some time until the model is completed. The instructions for computing normal forms are in a private language in the mind of the ASRS Coaster on a slate. It is directions not a finite list of particular operations redefined in the past to compute a function defined by a set of ASRS equations.

We have had examples in planning with GF-diagrams that part of the plan that involves free Skolemized trees is carried along with the proof tree for a plan goal. The idea is that if the free proof tree is constructed then the plan has a model in which the goals are satisfied. The model is the initial model of the AI world for which the free Skolemized trees were constructed. Nondeterminism (Floyd, 1967) and free proof trees by GF-diagrams is useful in realizing robot plans since there usually are many indeterminate formulas in the world diagram, whose actual values are inconsequential in an immediate plan of action, thus may be assigned a free Skolem function to satisfy a goal.

4.7   CONCLUDING COMMENTS

We have presented the foundations and methods of computation by algebraic subtree replacement systems. It is shown how to implement algebraic tree rewriting such that free initial models are obtained by automatic techniques. Term rewriting systems are formalized by algebraic theories and a model theory for such computation is developed with abstract recursion to have computable standard models. Computation is then formulated with Generalized diagrams and proof trees on algebraic subtree replacement systems such that free proofs could be obtained by tree rewriting.

The model theory and the proof system is put together by a formulation of positive forcing on a infinitary defined for algebraic theories such that a foundation for models and proofs could be presented. It is further shown how algebraically closed initial models could be generated for subtree replacement systems such that the methods of normal form tree rewriting could have solutions. Theoretical results and practical computational techniques are presented. The mathematics of algebraically closed structures for computing were circulated by the author since 1990's. Applicability computing problems are pointed out and some of our new directions for research in AI appears in Nourani (1996). Private languages and Wittgenstein's paradox are reviewed for insight into term rewriting. Term rewriting private language computing is presented in brief as a new area for term rewriting applications.

EXERCISES

1. Prove Lemma 4.1.
2. Prove Theorem 4.2.
3. Prove Theorem 4.3.
4. Prove Theorem 4.6.

KEYWORDS

- algebraic tree rewrite computing
- equational theories
- initial models
- model computing

REFERENCES

Albert Burroni, (1993). *Higher-dimensional word problems with applications to equational logic,* Theoret. Comput. Sci. 115, no. 1, 43–62.

Beke, Tibor, (2010). Categorification, Term Rewriting and the Knuth-Bendix Procedure, March UML.

Bishop, E. (1967). Foundations of Constructive Analysis, McGraw Hill, New York.

Boone, W. (1959). "The Word Problem," Annals of Mathematics, 207–265.

Brouwer, L. E. J., Intuitionitische Zerlegung mathematischer Grundbegriffe, Jber. Deutsch. Math. Verein, 33, 251–256.

Broy, M. (1988). Equational specification of partial higher-order algebras, Theoretical Computer Science, Volume 57, Issue 1, April, 3–45.

Broy, M., Wirsing, M. (1982). "Partial Abstract Types," Acta Informatica, Springer-Verlag, 18, pp. 47–64.

Donald E. Knuth, Peter B. Bendix, (1970). *Simple word problems in universal algebras,* Computational Problems for Abstract Algebra (Proc. Conf., Oxford, 1967), Pergamon Press.

Fikes, R. E., Nilsson, N. J. (1971). "Strips: A New Approach to the Application of Theorem Proving to Problem Solving", AI 2, pp. 189–208.

Floyd, R. M. (1967).Nondeterminstic Algorithms, JACM 14, 636–644.

Genesereth, M. R., Nilsson, N. J. (1987). Logical Foundations of Artificial Intelligence, Morgan-Kaufmann.

Goguen, J. A. Thatcher, J. W., Wagner, E. G., Wright, J. B. "An Introduction to Categories, Algebraic Theories and Algebras," IBM Research Report, RC5369, Yorktown Heights, N. Y., April 1975.

Goguen, J. A., Thatcher, J. W., Wagner, E. G., Wright, J. B. (1973). "A Junction Between Computer Science and Category Theory," (parts I and II), IBM T. J. Watson Research Center, Yorktown Heights, N. Y. Research Report, RC4526.

Handbook of Mathematical Logic, North-Holland, second edition, (edited by J. Barwise), 1978.

Heinrich, H., Nondeterminism in Algebraic Specifications and Algebraic Programs, Birkhauser, Boston, 1993.

Henkin, L. "On Mathematical Induction," American Mathematical Monthly, 67, 1960.

Henkin, L., "The Completeness of First Order Functional Calculus", Journal of Symbolic Logic," vol. 14, 1949.

Hinman, P. G. Recursion Theoretic Hierarchies, Spring-Verlag, 1980.

Hyting, A. Intuitionism An Introduction, North Holland, 1956.

Keisler, H. J., "Forcing and the Omitting Types Theorem, in Studies in Model Theory, Mathematical Association of America, New York, 1967.

Kelly, G. M. On MacLane's conditions for coherence of natural associativity's, commutabilities, etc., J. Algebra 1 (1964), 397–402.

Knuth, D. E., Bendix, P. B. "Simple Word Problems in Universal Algebra," in Computational Problems in Abstract Algebra, edited by J. Leech, Pergamon Press, Oxford and New York, 1969.

Kripke, S. A., Wittgenstein- On Rules and Private Language, Harvard University Press, 1980.

Ludwig Wittgenstein, Philosophical Investigation, Third Edition, Prentice Hall.

Neumann, B. H., "The Isomorphism Problem for Algebraically Closed Groups," in Word Problems, Boone et.al. editors, North-Holland, (1973). 553–562.

Nourani, C. F. "A Model Theoretic Approach To Specification and Implementation of Abstract Data Types," (1978). revised (1979). Proc. Programming Symposium, April (1979). Paris, Springer-Verlag Lecture Notes in Computer Science, Vol. 83, Berlin.

Nourani, C. F. "Diagrams, Possible Worlds, and the Problem of Reasoning in Artificial Intelligence," Logic Colloquium, Padova, Italy, (1987). Proc.in Journal of Symbolic Logic.

Nourani, C. F. "Forcing with Universal Sentences and Initial Models," 1981–82, Annual Meeting of the Association for Symbolic Logic, December (1983). Proc. in Journal of Symbolic Logic.

Nourani, C. F. "The Term Rewriting Roller Coaster," 2nd Intl. Workshop on Tree Rewrite Termination, May (1995). LaBress, France, Sponsored by INRIA Lorraine and Center de Rechreche en Informatique de Nancy.

Nourani, C. F. (1983). "Abstract Implementations and Their Correctness Proofs", JACM, vol.3, no. 2.

Nourani, C. F. (1983). "Equational Intensity, Initial Models, and AI Reasoning, Technical Report", A: Conceptual Overview, in Proc. Sixth European Conference in Artificial Intelligence, Pisa, Italy, September 1984. North-Holland.(Extended version to be published)

Nourani, C. F. (1985). "On the Power of Positive Thinking On Trees", Abstract Brief LICS (1985). Stanford.

Nourani, C. F. (1991). "Planning and Plausible Reasoning in AI, "Proc. Scandinavian Conference in AI, May (1991). Rocklike, Denmark, 150–157, IOS Press.

Nourani, C. F. (1993). "Computation on Diagrams and Free Skolemization", Summer Logic Colloquium, England.

Nourani, C. F. (1996). "Slalom Tree Computing, A Tree Computing Theory For AI," AI Communications, December (1996). IOS Press, Amsterdam.

Nourani, C. F. (1995). How Could There Be Models For Nothing and Proofs For Free, Brief Abstract IDPL, May (1995). Datmstadt.

Nourani, C. F. (2003). Versatile Abstract Syntax Meta-Contextual Logic and VR Computing, 36th Lingustische Kolloquium, Austria Proceedings 36th Lingustische Kolloquium, Austria Proceedings of the 35th Colloquium of Linguistics, Innsbruck. Europa Der Sprachen: Sprachkopetenz-Mehrsprachigeit-Translation, TIEL II: Sprache und Kognition, Sonderdruc (2003). Lew N. Zybatow (HRSG.).

Nourani, C. F. (1980). "Abstract Implementations and Complete and Consistent Extensions: Correctness and Construction," University of Michigan, Ann Arbor, March (1980). (Sent to FOCS-80). Also see UCLA TCS Report, UMI, Ann Arbor.

Nourani, C. F. (1981). "On Induction for Types: Models and Proofs," Revised June (1981). Presented at the 2nd Workshop on Theory and Applications of Abstract Data Types," University of Passau, Germany, 1982.

Nourani, C. F. (1982). "Positive Forcing, (1982).

Nourani, C. F. (1982). "The Connection Between Positive Forcing and Tree Rewriting," Proc. Logics In Computer Science Conference (LICS) and ASL, Stanford University, July 1985. Proc. Journal of Symbolic Logic.

O'Donnell, M. J., Computing in Systems Described By Equations, Lecture Notes in Computer Science, 58, Springer-Verlag, New York, 1977.

Tarski, A. Mostowski and R. M. Robinson, Undecidable Theories, North-Holland, 3rd edition, 1971.

Thatcher, J. W., Wagner, E. G., Wright, J. B. (1978). "Data Type Specification: Parameterization and the Power of Specification Techniques, "Proc. 10th Ann. Symposium on Theory of Computing (STOC), San Diego, CA. ACM, New York, 119–132..

Tibor Beke, *Operands from the viewpoint of categorical algebra,* Higher homotopy structures in topology and mathematical physics, Contemp. Math., no. 227, American Mathematical Society, (1999). pp. 29–47.

Further Reading

Ackermann, W., Solvable Cases of the Decision Problem. North-Holland, Amsterdam, 1954.

ADJ-Goguen, J., Thatcher, J., Wagner, E., Wright, J., Abstract Data Types as Initial Algebras and Correctness of Data Representations. Conference on Computer Graphics, Pattern Recognition and Data Structure, May 1975, 89–93.

ADJ-Goguen, J.A., Thatcher, J.W., Wagner, E.G., An Initial Algebra Approach to the Specification, Correctness, and Implementation of Abstract Data Types. "Current Trends in Programming Methodology," Vol.4, Ed. Yeh, R., Prentice-Hall.

Aho, A., Sethi, R., Ullman, J. (1972). Code Optimization and Finite Church-Rosser Systems. in Proceedings of Courant Computer Science Symposium 5, Ed. Rustin, R., Prentice Hall.

Backus, J. (1978). Can Programming be Liberated from the von Neumann Style!' A Functional Style and Its Algebra of Programs. CACM 21, 8, 613–641.

Birkhoff, G., Lipson, J.D. (1970). Heterogeneous Algebras. Journal of Combinatorial Theory 8, 115–133.

Boone, W. (1959). The Word problem. Ann. of Math. 2, 70, 207–265.

Boyer, R., Moore J, (1979). A Computational Logic. Academic Press.

Boyer, R., Moore, J. (1975). Proving Theorems About LISP Functions. JACM 22, 129–144.

Burstall, R.M. (1969). Proving properties of Programs by Structural Induction. Computer, J. 12, 41–48.

Church, A., Rosser, J.B. (1936). Some Properties of Conversion. Transactions of AMS 39, 472–482.

Curry, H.B., Feys, R. (1958). Combinatory Logic Vol. Z. North-Holland, Amsterdam.

Davis, M. (1973). Hilbert's Tenth Problem is Unsolvable. Amer. Math. Monthly 80, 3, 233–269.

Evans, T. (1951). The Word Problem for Abstract Algebras. J. London Math. Sot. 26, 6471.

Gerard Huet, Derek, C. Oppen, Equations and Rewrite Rules: A Survey, Stanford Verification Group, Report No. 15, 1980.

Guttag, J.V., Horning, J.J. (1978). The Algebraic Specification of Abstract Data Types. Acta Informatica 10, 27–52.

Hermann, G. (1926). Die Frage der endlich vielen Schritte in der Theorie der PoZynomideale, Math. Ann., Vol 95, 736–738.

Hindley, R. (1969). An Abstract Form of the Church-Rosser Theorem Journal of Symbolic Logic 34, 4, 545–560.

Huet, G. (1973). The Undecidability of Unification in Third Order Logic. Information and Control 22, 257–267.

Koffmann, M., O'Donnell, M. (1979.) Interpreter Generation Using Tree Pattern Matching. 6th ACM Conference on Principles of Programming Languages.

Kozen, D. (1977). Complexity of Finitely Presented Algebras. Ninth ACM Symposium on Theory of Computing, May, 164–177.

Lankford, D.S., Ballantyne, A.M. (1977). Decision Procedures for Simple Equational Theories With Commutative Axioms: Complete Sets of Commutative Reductions. Report ATP-35, Departments of Mathematics and Computer Sciences, University of Texas at Austin, March 1977.

Lipton, R. and Snyder, L., On the Halting of Tree Replacement Systems. Conference on Theoretical Computer Science, U. of Waterloo, Aug. 1977, 43–46.

Makanin, G. S. (1977). The Problem of Solvability of Equations in a Free Semigroup. Akad. Nauk. SSSR, TOM 233, 2.

Malcev, A. I. (1958). On Homomorphisms of Finite Groups. Ivano Gosudarstvenni Pedagogicheski Institut Uchenye Zapiski, vol. 18, pp. 49–60.

Matiyasevich, Y. (1970). Diophantine Representation of Recursively Enumerable Predicates. Proceedings of the Second Scandinavian Logic Symposium, North-Holland.

McCarthy, J. (1960). Recursive Functions of Symbolic Expressions and Their Computations by Machine, Part, I. CACM 3, 4, 184–195.

McNulty, G. (1976). The Decision Problem for Equational Bases of Algebras. Annals of Mathematical Logic, 11, 193–259.

Meyer, A. R., Stockmeyer, L. (1973). Word Problems Requiring Exponential Time. Fifth ACM Symposium on Theory of Computing, April, 1–9.

Musser, D. L., A Data Type Verification System Based on Rewrite Rules. 6th Texas Conf. on Computing Systems, Austin, Nov. 1978.

Nash-Williams, C. St. J. A. (1963). On Well-quasi-ordering Finite Trees. Proc. Cambridge Phil. Sot. 59, 833–835.

O'Donnell, M. (1977). Computing in Systems Described by Equations. Lecture Notes in Computer Science 58, Springer Verlag.

Plaisted, D. (1978). Well-Founded Orderings for Proving Termination of Systems of Rewrite Rules. Dept. of Computer Science Report 78–932, University of Illinois at Urbana- Champaign, July.

Post, E. (1947). Recursive Unsolvability of a Problem of Thue. J. Symbolic Logic 12, l–11.

Presburger, M., Uber die Vollstandigkeit eines gewissen Systems der Arithmetik ganzer Zahlen, in welchem die Addition als einzjge Operation hervortritt. Comptes-Rendus du ler CongrCs des MathCmaticiens des Pays Slaves, 1929.

Rosen, B. K. (1973). Tree-Manipulation Systems and Church-Rosser Theorems. JACM 20, 160–187.

Scott, D. (1970). Outline of a Mathematical Theory of Computation. Monograph PRG-2, Oxford University Press.

Seidenberg, A. (1954). A New Decision Method For Elementary Algebra and Geometry. Ann. of Math., Ser. 2, 60, 365–374.

Sethi, R., Testing for the Church-Rosser Property. JACM 21 (1974), 671–679. Erratum JACM 22 (1975), 424.

Staples, J. (1975). Church-Rosser Theorems for Replacement Systems. Algebra and Logic, ed. Crossley, J., Lecture Notes in Math., Springer Verlag, 291–307.

Stickel, M. E. (1975). A Complete Unification Algorithm for Associative-Commutative Functions. 4th International Joint Conference on Artificial Intelligence, Tbilisi.

Tarski, A. (1951). A Decision Method for Elementary Algebra and Geometry. University of California Press, Berkeley.

Tarski, A. (1968). Equational Logic. Contributions to Mathematical Logic, ed. Shiite et al., North-Holland.

Thatcher, J., Wagner, E., Wright, J. (1978). Data Type Specifications: Parameterization and the Power of Specification Techniques. Tenth ACM Symposium on Theory of Computing.

Tserunyan, A. Introduction to Descriptive Set Theory. www.math.uiuc.edu/~anush/Notes/dst_lectures.pdf

Yiannis, N. Moschovakis (2009). Descriptive Set Theory, Professor of Mathematics. University of California, Los Angeles and Emeritus.

CHAPTER 5

DESCRIPTIVE SETS AND INFINITARY LANGUAGES

CONTENTS

5.1   INTRODUCTION

Let us begin with a brief on the analytical hierarchy of formulas. The notation $\Sigma_0^1 = \Pi_0^1 = \Delta_0^1$ indicates the class of formulas in the language of second-order arithmetic with no set quantifiers. This language does not contain set parameters. Each corresponding boldface symbol denotes the corresponding class of formulas in the extended language with a parameter for each real. The language hierarchy is sometimes called projective hierarchy. A formula in the language of second-order arithmetic is defined to be $_{n+1}^1$ if it is logically equivalent to a formula of the form $\exists x_1 \ldots \exists x_k$ where is Π_n^1. A formula is defined to be Π_{n+1}^1 if it is logically equivalent to a formula of the $\forall x_1 \ldots \forall x_k$ where ψ is Σ_n^1. This inductive definition defines the classes Σ_n^1 and Π_n^1 for every natural number n.

Because every formula has a prenex normal form prenex normal for form, every formula in the language of second-order arithmetic is Σ_n^1 or Π_n^1 for some n. Because meaningless quantifiers can be added to any formula, once a formula is given the classification Σ_n^1 or Π_n^1 for some n it will be given the classifications Σ_m^1 and Π_m^1 for all m greater than n. The analytical hierarchy of sets of natural numbers is as follows. A set of natural numbers is assigned the classification Σ_n^1 if it is definable by a Σ_n^1 formula. The set is assigned the classification Π_n^1 if it is definable by a Π_n^1 formula. If the set is both Σ_n^1 and Π_n^1 then it is given the additional classification Δ_n^1. The Δ_1^1 sets are called *hyperarithmetical.* An alternate classification of these sets by way of iterated computable functionals is provided by hyperarithmetical theory. The analytical hierarchy on subsets of Cantor and Baire space are as follows. The analytical hierarchy can be defined on any effective Polish space; the definition is particularly simple for Cantor and Baire space because they fit with the language of ordinary second-order arithmetic. Cantor space is the set of all infinite sequences of 0s and 1s; Baire space is the set of all infinite sequences of natural numbers. These are both Polish spaces.

The ordinary axiomatization of second-order arithmetic uses a set-based language in which the set quantifiers can naturally be viewed as quantifying over Cantor space. A subset of Cantor space is assigned the classification Σ_n^1 if it is definable by a Σ_n^1 formula. The set is assigned the classification Π_n^1 if it is definable by a Π_n^1 formula. If the set is both Σ_n^1 and Π_n^1 then it is given the additional classification Δ_n^1.

A subset of Baire space has a corresponding subset of Cantor space under the map that takes each function from ω to ω to the characteristic function of its graph. A subset of Baire space is given the classification Σ^1_n, Π^1_n, or Δ^1_n if and only if the corresponding subset of Cantor space has the same classification. An equivalent definition of the analytical hierarchy on Baire space is given by defining the analytical hierarchy of formulas using a functional version of second-order arithmetic; then the analytical hierarchy on subsets of Cantor space can be defined from the hierarchy on Baire space. This alternate definition gives exactly the same classifications as the first definition.

From here on we are following the standard developments on the syntax and semantics of infinitary languages with admissible sets since (Barwise, 1967). Because Cantor space is homomorphic to any finite Cartesian power of itself, and Baire space is homeomorphism to any finite Cartesian power of itself, the analytical hierarchy applies to finite Cartesian power of one of these spaces as well. A similar extension is possible for countable powers and to products of powers of Cantor space and powers of Baire space. Let start from where we stated Hanf coding for infinitary language fragments on the preceding chapter. The map $\varphi \mapsto \ulcorner \varphi \urcorner$ from $Frm(\kappa, \lambda)$ into $H(\kappa)$ is easily seen to be one-one and is the required coding map.

Accordingly, we identify $Val(L(\kappa, \lambda))$ with its image in $H(\kappa)$ under this coding map. An L-formula is called a Δ_0-formula if it is equivalent to a formula in which all quantifiers are of the form $\forall x \in y$ or $\exists x \in y$ (i.e., $\forall x (x \in y \to \dots)$ or $\exists x (x \in y \circ \dots))$. An L-formula is a Σ_1-formula if it is equivalent to one that can be built up from atomic formulas and their negations using only the logical operators $\circ$, $\vee$, $\forall x \in y$, $\exists x$. A subset X of a set A is said to be Δ_0 (resp. Σ_1) on A if it is definable in the $\langle A, \in \restriction A \rangle$ by a Δ_{0-} (resp. Σ_{1-}) formula of L.

For example, if we identify the set of natural numbers with the set $H(\omega)$ of hereditarily finite sets in the usual way, then for each $X \subseteq H(\omega)$:

X is Δ_0 on $H(\omega) \Leftrightarrow X$ is recursive; X is Σ_1 on $H(\omega) \Leftrightarrow X$ is recursively enumerable. Thus, the notions of Δ_{0-} and Σ_{1-} set may be regarded as generalizations of the notions of *recursive* and *recursively enumerable* set, respectively. The completeness theorem for L implies that $Val(L)$— regarded as a subset of $H(\omega)$ is recursively enumerable, and hence Σ_1 on $H(\omega)$. Similarly, the completeness theorem for $L(\omega 1, \omega)$ (see §2) implies that $Val(L(\omega 1, \omega))$ — regarded as a subset of $H(\omega_1)$ — is Σ_1 on $H(\omega_1)$. However, this pleasant state of affairs collapses completely as soon as $L(\omega 1, \omega_1)$

is reached. For one can prove Δ_1 and Π_1 and Σ_1 sets; admissible structures; ω_{ck}; recursive well-founded relations; ordinal analysis of trees; Barwise compactness; so on and so forth.

In certain situations the best way to understand the structure of point classes is to understand them in the context of an appropriate inner model that reflects their behavior with a degree of absoluteness.

For instance one could view Σ_1 sets using Godel's constructible universe L, and look for similarities on the structural properties with L. Borel sets can be comprehended by going to any model of set theory with ω is isomorphic to the real ω. Perhaps at the level of $\Pi13$ we should consider sufficiently iterable inner models of a Woodin cardinal. At the level of $\Pi1$, the right inner models are well-founded models of KP – that is to say, admissible structures. To illustrate this, the first theorem we will head towards is Spector-Gandy, which, in its simplest form, states that the $\Pi1$ subsets of ω are precisely those $\Sigma1$ definable over the least admissible structure. The specific theorems such as Spector-Gandy are less important than the method, which is to reduce questions about Borel and projective complexity of sets of real to set theoretical calculations over inner models.

5.2 ADMISSIBLE SETS AND STRUCTURES

A nonempty transitive set A is said to be *admissible* when the following conditions are satisfied: (i) if $a,\ b \in A$, then $\{a,\ b\} \in A$ and $\cup A \in A$; (ii) if $a \in A$ and $X \subseteq A$ is Δ_0 on A, then $X \cap a \in A$; (iii) if $a \in A$, $X \subseteq A$ is Δ_0 on A, and $\forall x{\in}a\exists y({<}x,\ y{>} \in X)$, then, for some $b \in A$, $\forall x{\in}a\exists y{\in}b({<}x,\ y{>} \in X)$.

Condition (ii) the Δ_0-*separation scheme*—is a restricted version of Zermelo's axiom of separation. Condition (iii) a similarly weakened version of the axiom of replacement—may be called the Δ_0-*replacement scheme*.

It is quite easy to see that if A is a transitive set such that $<A,\ \in\ |\ A>$ is a model of ZFC, then A is admissible. More generally, the result continues to hold when the power set axiom is omitted from ZFC, so that both $H(\omega)$ and $H(\omega_1)$ are admissible. However, since the latter is uncountable, the Barwise compactness theorem fails to apply to it.

Given what we now know about infinitary languages, it would seem that $L(\omega1,\omega)$ is the only one to be reasonably well behaved. However, the failure of the compactness theorem to generalize to $L(\omega1,\omega)$ in any useful fashion is a severe drawback as far as applications are concerned. Let us attempt to analyze this failure in more detail. Recall from Chapter 3 that we may code the formulas of a first-order language L as hereditarily finite sets, for example, as members of $H(\omega)$. In that case each finite set of (codes of) L-sentences is also a member of $H(\omega)$, and it follows that the compactness theorem for L can be stated in the form:

Proposition 5.1: If $\Delta \subseteq \mathbf{Snt}(L)$ is such that each subset $\Delta_0 \subseteq \Delta$, $\Delta_0 \in H(\omega)$ has a model, so does Δ.

Proposition 5.2: If $\Delta \subseteq \mathbf{Snt}(L)$ and $\psi \in \mathbf{Snt}(L)$ satisfy $\Delta \vDash \psi$, then there is a deduction D of Σ from Δ such that $D \in H(\omega)$.

Now it is well-known that the above proposition is an immediate consequence of the *generalized completeness theorem* for L. Chapter 3 outlined that the compactness theorem fails on full $(L\omega1,\omega)$. A set $\Gamma \subseteq Snt(\omega1,\omega)$ was constructed such that each countable subset of Γ has a model but Γ does not. In Chapter 3, since the notion of deduction in $L(\omega1,\omega)$ implies deductions are of countable length it follows from the preceding that:

Theorem 5.1: There is a sentence $\psi \in \mathbf{Sent}(\omega1,\omega)$ such that $\Gamma \vDash \psi$, but there is no deduction of ψ in $L(\omega1,\omega)$ from Γ.

Now the formulas of $L(\omega1,\omega)$ can be coded as members of $H(\omega_1)$, and it is clear that $H(\omega_1)$ is closed under the formation of countable subsets and sequences. Accordingly, we might write the above theorem and preceding note on compactness as each $\Gamma_0 \subseteq \Gamma$ such that $\Gamma_0 \in H(\omega_1)$ has a model, but Γ does not; there is a sentence $\Sigma \in \mathbf{Snt}(\omega1,\omega)$ such that $\Gamma \vDash \sigma$, but there is no deduction $D \in H(\omega_1)$ of Σ from Γ. It follows that the propositions 5.1 and 5.2 fail when "L" is replaced by "$L(\omega1,\omega)$" and "$H(\omega)$" by "$H(\omega_1)$." Moreover, it can be shown that the set $\Gamma \subseteq \mathbf{Snt}(\omega1,\omega)$ in and)may be taken to be Σ_1 on $H(\omega_1)$. Thus, the compactness and generalized completeness theorems fail for Σ_1-sets of full $L(\omega1,\omega)$-sentences. The reason why the generalized completeness theorem fails for Σ_1-sets in $L(\omega1,\omega)$ is that essentially $H(\omega_1)$ is not "closed" under the formation of deductions from Σ_1-sets of sentences in $H(\omega_1)$.

However, there fragments that can remedy all that: for example can we replace $H(\omega_1)$ by sets A which are, in some sense, closed under the formation of such deductions, and then to consider just those formulas whose codes are in A. *The well-behaved admissible fragments, so on and so forth were developed for that.* Here is an example of an admissible fragment that is well-behaved: For each countable transitive set A, let L_A

$= \mathbf{Frm}(\mathrm{L}(\omega 1,\omega)) \cap A$. We say that L_A is a *sublanguage* of $\mathrm{L}(\omega 1,\omega)$ if the following conditions are satisfied:

(i) $\mathrm{L} \subseteq \mathrm{L}_A$
(ii) if $\varphi, \psi \in \mathrm{L}_A$, then $\varphi \circ \psi \in \mathrm{L}_A$ and $\neg \varphi \in \mathrm{L}_A$
(iii) if $\varphi \in \mathrm{L}_A$ and $x \in A$, then $\exists x \varphi \in \mathrm{L}_A$
(iv) if $\varphi(x) \in \mathrm{L}_A$ and $y \in A$, then $\varphi(y) \in \mathrm{L}_A$
(v)   if $\varphi \in \mathrm{L}_A$, every subformula of φ is in L_A
(vi) if $\Phi \subseteq \mathrm{L}_A$ and $\Phi \in A$, then $\circ \Phi \in \mathrm{L}_A$.

We say that φ is *deducible* from Δ in L_A if there is a deduction $\boldsymbol{D}$ of φ from Δ in L_A; under these conditions we write $\Delta \vdash_A \varphi$. In general, $\boldsymbol{D}$ will not be a member of A; in order to ensure that such a deduction can be found in A it will be necessary to impose further conditions on A. Let A be a countable transitive set such that L_A is a sublanguage of $\mathrm{L}(\omega 1,\omega)$ and let Δ be a set of sentences of L_A. We say that A (or, by abuse of terminology, L_A) is Δ-*closed* if, for any formula φ of L_A such that $\Delta \vdash_A \varphi$, there is a deduction $\boldsymbol{D}$ of φ from Δ such that $\boldsymbol{D} \in A$. It can be shown that the only countable language which is Δ-closed for *arbitrary* Δ is the first-order language L, for example, when $A = H(\omega)$.

Barwise discovered that there are countable sets $A \subseteq H(\omega_1)$ whose corresponding languages L_A differ from L and yet are Δ-closed *for all Σ_1 sets of sentences* Δ. Such sets A are called *admissible sets*; roughly speaking, they are extensions of the hereditarily finite sets in which recursion theory, and hence proof theory, are still possible. From Barwise's result one obtains the following:

Theorem 5.2: *Let A be a countable admissible set and let Δ be a set of sentences of* L_A *which is Σ_1 on A. If each $\Delta' \subseteq \Delta$ such that $\Delta' \in A$ has a model, then so does Δ.*

The presence of "Σ_1" here indicates that this theorem is a generalization of the compactness theorem for *recursively enumerable* sets of sentences. Another version of the Barwise compactness theorem is the following. Let $\mathbf{ZFC}$ be the usual set of axioms for Zermelo-Fraenkel set theory, including the axiom of choice. Then we have:

Theorem 5.3: *Let A be a countable transitive set such that $A = \langle A, \in \restriction A \rangle$ is a model of $\mathbf{ZFC}$. If Δ is a set of sentences of* L_A *which is definable in A by a formula of the language of set theory and if each $\Delta' \subseteq \Delta$ such that $\Delta' \in A$ has a model, so does Δ.*

A nonempty transitive set A is said to be *admissible* when the following conditions are satisfied:

(i) if $a, b \in A$, then $\{a, b\} \in A$ and $\cup A \in A$;
(ii) if $a \in A$ and $X \subseteq A$ is Δ_0 on A, then $X \cap a \in A$;
(iii) if $a \in A$, $X \subseteq A$ is Δ_0 on A, and $\forall x \in a \exists y (\langle x, y \rangle \in X)$, then, for some $b \in A$, $\forall x \in a \exists y \in b (\langle x, y \rangle \in X)$.

Condition (ii)—the Δ_0-*separation scheme*—is a restricted version of Zermelo's axiom of separation. Condition (iii)—a similarly weakened version of the axiom of replacement—may be called the Δ_0-*replacement scheme*.

Note that if A is a transitive set such that $\langle A, \in | A \rangle$ is a model of **ZFC**, then A is admissible. More generally, the result continues to hold when the power set axiom is omitted from **ZFC**, so that both $H(\omega)$ and $H(\omega_1)$ are admissible.

5.3 BASIC DESCRIPTIVE CHARACTERIZATIONS

Descriptive set theory is mainly concerned with studying subsets of the space of all countable binary sequences. The generalization with uncountable sequences explores properties of Baire and Cantor spaces, equivalence relations and their Boral reducibility. Descriptive set theory does not look the same look that similar the classical, countable case. There one can examine the connections between the stability theoretic complexity of first-order theories and the descriptive set theoretic complexity of their isomorphism relations.

Characterization of well-orderings in $L(\omega 1, \omega_1)$ is interesting to brief on: The base language L here includes a binary predicate symbol $\leq$. Let σ_1 be the usual L-sentence characterizing linear orderings. Then the class of L-structures in which the interpretation of $\leq$ is a well-ordering coincides with the class of models of the $L(\omega 1, \omega_1)$ sentence $\Sigma = \sigma_1 \times \sigma_2$, where $\sigma_2 =_{df} (\forall v_n)_{n \in \omega} \exists x [\vee_{n \in \omega} (x = v_n) \wedge \wedge_{n \in \omega} (x \leq v_n)]$.

Notice that the sentence σ_2 contains an *infinite quantifier*: it expresses the essentially *second-order* assertion that every countable subset has a least member. It can in fact be shown that the presence of this infinite quantifier is essential: the class of well-ordered structures cannot be characterized in any finite-quantifier language. This example indicates that infinite-quantifier languages such as $L(\omega 1, \omega_1)$ behave rather like second-order languages. Many extensions of first-order languages can be *trans-lated* into infinitary languages.

The language $L(\omega 1,\omega)$ plays an important role among infinitary languages, since like first-order languages, it admits an effective *deductive system*. Let us add to the usual first-order axioms and rules of inference the new axiom scheme $\circ\Phi \rightarrow \varphi$ for any countable set $\Phi \subseteq \mathbf{Frm}(\omega 1,\omega)$ and any $\varphi \in \Phi$, together with the new rule of inference.

$$\frac{\varphi_0, \varphi_1, \ldots, \varphi_n, \ldots}{\circ_{n\in\omega} \varphi_n}$$

Now allow deductions to be of countable length. Writing $\vdash^*$ for deducibility in this sense, we then have the

Theorem 5.4: $L(\omega 1,\omega)$-Completeness Theorem. For any $\psi \in \mathbf{Snt}(\omega 1,\omega)$, $\models \psi \Leftrightarrow \vdash^*\psi$

As an immediate corollary this deductive system is *adequate for deductions from countable sets of premises in* $L(\omega 1,\omega)$. That is, we have, for any *countable* set $\Delta \subseteq \mathbf{Snt}(\omega 1,\omega)$

Proposition 5.3: $\Delta \models \psi \Leftrightarrow \Delta\vdash^*\psi$

This completeness theorem can be proved by modifying the usual Henkin completeness proof for first-order logic, or by employing Boolean-algebraic methods. Similar arguments, applied to suitable further augmentations of the axioms and rules of inference, yield analogous completeness theorems for many other finite-quantifier languages.

Turning to the *Löwenheim-Skolem theorem*, we find that the *downward* version has adequate generalizations to $L(\omega 1,\omega)$ (and, indeed, to all infinitary languages). In fact, one can show in much the same way as for sets of first-order sentences that if $\Delta \subseteq \mathbf{Snt}(\omega 1,\omega)$ has an infinite model of cardinality $\geq |\Delta|$, it has a model of cardinality the larger of $\aleph_0$, $|\Delta|$. In particular, any $L(\omega 1,\omega)$-sentence with an infinite model has a countable model. On the other hand, the *upward* Löwenheim-Skolem theorem in its usual form does not hold for all infinitary languages.

For example, the $L(\omega 1,\omega)$-sentence characterizing the standard model of arithmetic has a model of cardinality $\aleph_0$ but no models of any other cardinality. However, all is not lost here, with Hanf numbers there is a remedy for that: define the *Hanf number* **H(L)** of a language L to be the least cardinal κ such that, if an **L**-sentence has a model of cardinality κ, it has models of arbitrarily large cardinality. The existence of **H(L)** is readily established. For each **L**-sentence Σ not possessing models of arbitrarily large cardinality let $\kappa(\sigma)$ be the least cardinal κ such that Σ does not have a model of cardinality κ. If λ is the supremum of all the $\kappa(\sigma)$, then, if a sentence of **L** has a model of cardinality λ, it has models of arbitrarily large cardinality.

Define the cardinals $\mu(\alpha)$ recursively by

$$\mu(0) \quad = \quad \aleph_0$$

$$\mu(\alpha+1) \quad = \quad 2^{\mu(\alpha)}$$

$$\mu(\lambda) \quad = \quad \sum_{\alpha<\lambda} \mu(\alpha), \text{ for limit } \lambda.$$

The following proposition can be proved.

Proposition 5.4: $\mathbf{H}(\mathrm{L}(\omega 1,\omega)) = \mu(\omega_1)$.

5.4   BOOLEAN VALUED MODELS

Beginning from Scott's Isomorphism theorem from Chapter 4, there is a $\mathrm{L}(\infty, \omega)$ version as follows:

Given any structure A, there is an $\mathrm{L}(\infty, \omega)$-sentence Σ such that, for all structures B, $A \cong_p B \Leftrightarrow B \models \sigma$.

Partial isomorphism and (∞, ω)-equivalence are related to the notion of *Boolean isomorphism*. To define this we need to introduce the idea of a Boolean-valued model of set theory. Given a complete Boolean algebra B, the *universe $V^{(B)}$ of B-valued sets*, also known as the *B-extension of the universe V of sets*, is obtained by first defining, recursively on α,

$V_\alpha^{(B)} = \{x\colon x \text{ is a function} \circ \mathrm{range}(x) \subseteq B \circ \exists \eta<\alpha[\mathrm{domain}(x) \subseteq V\eta^{(B)}]\}$
and then setting

$$V^{(B)} = \{x\colon \exists \alpha(x \in V_\alpha^{(B)})\}.$$

Members of $V^{(B)}$ are called *B-valued sets*. It is now easily seen that a B-valued set is precisely a B-valued function with domain a set of B-valued sets. Now let **L** be the first-order language of set theory and let $\mathbf{L}^{(B)}$ be the language obtained by adding to **L** a name for each element of $V^{(B)}$ (we shall use the same symbol for the element and its name). One can now construct a mapping $[\cdot]^{(B)}$ of the (sentences of the) language $\mathbf{L}^{(B)}$ into B: for each sentence Σ of $\mathbf{L}^{(B)}$, the element $[\sigma]^{(B)}$ of B is the "Boolean truth value" of Σ in $V^{(B)}$. This mapping $[\cdot]^{(B)}$ is defined so as to send all the theorems of Zermelo-Fraenkel set theory to the top element 1 of B, for example, to "truth"; accordingly, $V^{(B)}$ may be thought of as a *Boolean-valued model of set theory*. In general, if $[\sigma]^{(B)} = 1$, we say that Σ is *valid* in $V^{(B)}$, and write $V^{(B)} \models \sigma$.

Now each $x \in V$ has a canonical representative x in $V^{(B)}$, satisfying

$x = y$ iff $V^{(B)} \models x = y$

$x \in y$ iff $V^{(B)} \models x \in y$

Two similar structures A, B are Boolean isomorphic, written $A \cong_b B$, if, for some complete Boolean algebra B, we have $V^{(B)} \models A \cong B$, that is, if there is a Boolean extension of the universe of sets in which the canonical representatives of A and B are isomorphic with Boolean value 1. It can then be shown that:

$$A \equiv_{\infty\omega} B \Leftrightarrow A \cong_b B.$$

This result can be strengthened through category-theoretic formulation. For this we require the concept of a(n) (elementary) *topos*. Example areas that present themselves considering the intricacies of infinitary completeness are the Scott's undefinability theorem for $L(\omega 1, \omega_1)$. We will examine that Borel reducibility on uncountable structures is a model theoretically natural way to compare the complexity of isomorphism relations, as was the Boolean reducibility on Chapter 4 concrete descriptive computing.

5.6   ADMISSIBLE SETS AND ORDINALS

Definition 5.1: (Trees) $T \subset \omega^{<\omega} \times \omega^{<\omega}$ is a tree iff
1. $(u, v) \in T$ implies $lh(u) = lh(v)$ (they have the same length as finite sequences); and
2. T is closed under subsequences, in the sense that if $(u, v) \in T$ and $1 < lh(u)$ then $(u|l, v|l) \in T$.

We then let $[T]$, the set of branches through T, be the set of all $(x, y) \in \omega\omega$ such that $\forall n (x|n, y|n) \in T$, where here $x |n = (x(0), x(1), \ldots, x(n-1))$. Let $p[T]$ be the projection of $[T]$–that is to say the set of x for which there exists a y with $(x, y) \in [T]$. We then say that a set is $\Sigma 1$, or analytic, if it equals $p[T]$ for $\sim$.

Notation for $(u, v), (u', v') \in T$, I will write $(u, v) < (u', v')$ if u is strictly extended by u' and v is strictly some tree T extended by v'. I write $(u, v) \perp (u', v')$ if neither extends the other.

Theorem 5.5: $p[T]$ is empty if and only if T has a ranking function. That is to say, some $\rho: T \to \delta$, some ordinal δ, with $\rho(u, v) > \rho(u|l, v|l)$ all $(u, v) \in T, l < lh(u)$.

Theorem 5.6: (Perfect set theorem for $\Sigma \sim 1$.) Let T be a tree. Then either $p[T]$ is countable, or it contains a homeomorphic copy of 2ω, and hence has cardinality $2\aleph 0$.

Proof: This is actually similar to the last proof.

An ordinal α is said to be recursive if there is a recursive well order $<$ of ω with $(\alpha; \in) \sim= (\omega; <)$.

Let $\omega_1 ck$ be the supremum of the recursive well orders.

Definition 5.2: For M a model with of a binary relation, $\in_M$. Let $Sup_0(M)$ be the sup of the ordinals in the well-founded part.

Lemma 5.1: If M is a model of ZFC, then $Sup_0(M)$ is admissible.

Proof: First of all we may assume the well-founded part of M is actually a transitive set. If M is well-founded, then there is nothing to prove, so assume M is ill founded. If the lemma fails, then there would be some set X in the well-founded part of M and some $\Sigma 1$ formula ψ such that at each $x \in X$ there is a least ordinal $\beta x < Sup_0(M)$ such $\psi(x, \beta x)$ and the βx's are unbounded in $Sup_0(M)$. But then M would have the have the ability to define the cut corresponding $Sup_0(M)$ and would be confronted by its own ill-foundedness. Of course assuming $M \models ZFC$ is something of an over kill in the assumptions of the lemma.

Lemma 5.3: A well-founded recursive relation on ω has rank less than $\omega_1 ck$.

Note that a recursive well order of an infinite recursive set will be isomorphic to a recursive well order of ω and hence have order type less than $\omega_1 ck$.

Definition 5.4: A tree is said to be recursive if the set of Gödel codes for its elements is recursive as a subset of ω. A subset of $(\omega^{(\omega)})^n$ is said to be Σ^1 if it is the projection of a recursive tree; it is Π^1 if its complement is Σ^1; and it is Δ^1 if it is both Σ^1 and Π^1.

There is another nice definition of recursive when considering subset of the hereditarily finite sets: A $\subset$ HF is recursive if it is $\Delta 1$ definable over (HF, $\in$).

Definition 5.5: A set of the form Lα is admissible if α is a limit and Lα satisfies $\Sigma 1$ collection. We then also say that α is an admissible ordinal.

Theorem 5.6: $L_{\omega 1ck}$ is admissible.

Proof: (Exercises)

5.7   SET REDUCIBILITY

The standard set notation is applied:

A $\subseteq$ B means that A is a subset of B or is equal to B.

A $\subset$ B means proper subset. Union, intersection and set theoretical difference are denoted, respectively, by A $\cup$ B, A $\cap$ B and A \ B. For larger unions and intersections $\cup i \in I$ Ai, etc. $\wp(A)$ is the power set of A and [A]$<\kappa$ is the set of subsets of A of size $<\kappa$. Usually the Greek letters κ, λ and μ will stand for cardinals and α, β and γ for ordinals, but this is not strict. Also η, ξ, ν are usually elements of $\kappa\kappa$ or 2κ and p, q, r are elements of $\kappa<\kappa$ or $2<\kappa$. cf(α) is the cofinality of α (the least ordinal β for which there exists an increasing unbounded function f: $\beta \rightarrow \alpha$).

Basic notions by S$\lambda\kappa$ is meant $\{\alpha<\kappa|cf(\alpha)=\lambda\}$. A λ-cub set is as subset of a limit ordinal (usually of cofinality > λ) which is unbounded and contains suprema of all bounded increasing sequences of length λ. A set is cub if it is λ-cub for all λ. A set is stationary if it intersects all cub sets and λ-stationary if it intersects all λ-cubsets. Note that C$\subset\kappa$ is λ-cub if and only if C$\cap$S$\lambda\kappa$ is λ-cub and S$\subset\kappa$ is λ-stationary if and only if S $\cap$ S$\lambda\kappa$ is (just) stationary. We denote by f(x) the value of x under the mapping f and by f[A] or just fA the image of the set A under f. Similarly f−1[A] or just f−1A indicates the inverse image of A.

Domain and range are denoted respectively by domf and ranf. If it is clear from the context that f has an inverse, then f−1 denotes that inverse. For a map f: X $\rightarrow$ Y injective means the same as one-to-one and surjective the same as onto. Suppose f: X $\rightarrow$ Y α is a function with range consisting of sequences of elements of Y of length α. The projection prβ is a function

Y$\alpha \to$ Y defined by prβ((yi)i<α) = yβ. For the coordinate functions of f we use the notation fβ =pr$\beta \circ$f for all $\beta<\alpha$. By support of a function f we mean the subset of dom f in which f takes non-zero values, whatever "zero" means depending on the context.

Reductions: Let E1 $\subset$ X2 and E2 $\subset$ Y 2 be equivalence relations on X and Y respectively. A function f: X $\to$ Y is a reduction of E1 to E2 if for all x, y $\in$ X we have that xE1y $\Leftrightarrow$ f(x)E2f(y). Suppose in addition that X and Y are topological spaces. Then we say that E1 is continuously reducible to E2, if there exists a continuous reduction from E1 to E2 and we say that E1 is Borel reducible to E2 if there is a Borel reduction.

Definition 5.6: (Borel Reducibility) Let E and F be equivalence relations on standard Borel spaces X and Y, respectively. E is Borel reducible to F, written E $\leq$B F, if there is a Borel f:X $\to$Y such that xEy $\Leftrightarrow$ f(x)Ff(y).

This means that the points of X can be classified up to E-equivalence by a Borel assignment of invariants that are F-equivalence classes.

f is required to be Borel so that the invariant f(x) has a reasonable computation from x. We shall examine the applications in the following sections and chapters.

5.7.1   ADMISSIBLE TREE RECURSION

Definition 5.7: A tree is said to be recursive if the set of Go̎del codes for its elements is recursive as a subset of ω. A subset of $(\omega^{\omega})^n$ is said to be Σ^1 if it is the projection of a recursive tree; it is Π^1 if its complement is Σ^1; and it is Δ^1 if it is both Σ^1 and Π^1.

There is another nice definition of recursive when talking about subset of the hereditarily finite sets: A $\subset$ HF is recursive if it is Δ1 definable over (HF, $\in$). This agrees with the definition we have used above, though I will not take the time out to prove that.

5.7.2   ADMISSIBLE SET REDUCIBILITY

Here is a more inclusive characterization on set reducibility following a newer brief since this author's ASL publications, from Stukachev (2011) that can complement this chapter's model-theoretic characterization Let M be a structure of a computable signature and let A be an admissible set.

Let M be a structure of a computable signature and let A be an admissible set. For a set *M,* consider the set HF(M) of hereditarily finite sets over *M* defined as follows:

HF(M) = $\cup_n$ HF$_n$(M), $n \in \omega$

HF0(M) = $\{\varnothing\} \cup M$, HF$_{n+1}$(M) = HFn(M) $\cup$ {a | a is a finite subset of HFn(M)}.

For a structure M = $\langle M, \Sigma_M \rangle$ of (finite or computable) signature Σ_M, hereditarily finite superstructure HF(M) = $\langle$HF(M); Σ_M, U, $\in$, $\varnothing \rangle$ is a structure of signature Sig'$_M$ (with HF(M) |= $U(a) \Leftrightarrow a \in M$).

Note: For infinite signature, we assume that σ' contains an additional relation *Sat(x, y)* for atomic formulas under some fixed Gödel numbering.

To have an algebraic reach to diagram definable models let us consider presentability notions for arbitrary structures.

Definition 5.8: *A* presentation *of* M *in* A *is any structure* $\Re$ *such that* $\Re \sim=$ M *and the domain of* $\Re$ *is a subset of A.*

We can treat (the atomic diagram of) a presentation $\Re$ as a subset of *A,* using some Goʺdel numbering of the atomic formulas of the signature of M.

The notion of presentability *of* M *in* A *is the family* PrA(M) *consisting of the atomic diagrams of all possible presentations of* M *in* A:

Comparing to the generic diagram notions from this author over a decade or more, let us examine the following:

PrA(M)={ $\Re$ |$\Re$ *is a presentation of M in A}.*

Denote by M the set Pr HF($\varnothing$)(M) of all presentations of M in the least admissible set.

Definition 5.9: *A* presentation *of* M *in* A *is any structure* $\Re$ *such that* $\Re \sim=$ M *and the domain of* $\Re$ *is a subset of A.*

We can treat (the atomic diagram of) a presentation $\Re$ as a subset of *A,* using some Goʺdel numbering of the atomic formulas of the signature of M.

Proposition 5.5 (author 2015) $\Re$ is a presentation of A in M iff there is an embedding from the generic diagram for $\Re$ to in M and |$\Re$| $\subseteq$ A.

Let us denote the "presentability" *of* M *in* A *with the family* Pr<A(M)> *consisting of the atomic diagrams of all possible presentations of* M *in* A:

Pr<A(M)>= {$\Re$ |$\Re$ *is a presentation of M in A}*

Definition 5.10: Denote by M the set $PrHF(\varnothing)(M)$ of all presentations of M in the least admissible set. X is *Muchnik reducible* to Y $(X \leq_{Aw} Y)$ if, for any $Y \in Y$, there exist binary Σ-operators $F0$ and $F1$ such that $\langle Y, A\backslash Y\rangle \in \delta c(F0) \cap \delta c(F1)$ and, for some $X \in X$, $X = F0\ (Y, A\backslash Y)$ and $A\backslash X = F1$ $(Y, A\backslash Y)$.

For arbitrary cardinal α, let $K\alpha$ be the class of all structures (of computable signatures) of cardinality α. We define on $K\alpha$ an equivalence relation $\equiv_\Sigma$ as follows: for M, N $\in K\alpha$, $M\leq_{\equiv_\Sigma}N$ if $M \leq_{\equiv_\Sigma}N$ and $N \leq_{\equiv_\Sigma}M$.

The structure is $\Sigma(\alpha) = \langle K\alpha/\equiv_\Sigma, \leq_{\equiv_\Sigma}\rangle$—an upper semilattice with the least element, and, for any M, N $\in K\alpha$, where (M, N) denotes the model-theoretic pair of M and N.

Proposition 5.6: *For any structure A, there exists a graph (in fact, a lattice) G_A such that $A \equiv_\Sigma G_A$.*

We can define Σ-reducibility for presentable structures in more specific terms than ordinary algebraic reduces. For example consider the following definition from Stukachev (2011) from the new proviso since this author's 2014 proposition 5.5 to reconcile with these notions.

Definition 5.11: Let A, B be admissible sets. A is said to be Σ-reducible to B $(A \sqsubseteq_\Sigma B)$ if there is a mapping $v: B \to A$ s.t. $v-1$ transforms every A-constructive process to B-constructive. A is said to be weakly Σ-reducible to B $(A \sqsubseteq w_\Sigma B)$ if there is a mapping $v: B \to A$ s.t. $v-1$ transforms every A-constructively generated object to B-constructively generated.

Let A be an admissible set, for example, *Muchnic Reducibility (in fact that are additional reducibility notions that Stukachev presents from Russian mathematematians that are grouped with the same definition).* For structures M, N, denote by $M^r{}_A <= w_\Sigma N$ the fact that $Pr_A(M) \leq_A^r Pr_A(N)$.

Let $K_\Sigma(M) = \{N|N\leq_A M\}$, for example, $K_\Sigma{}^A(M) \subseteq K_w{}^A(M)$, $K \leq_A^w$, Kw being Muchnic reducibility, these inclusions are proper in the case A $= HF(\varnothing)$ (Kalimullin, 2009).

EXERCISES

1. The intersection of two $\Sigma 1$ sets is $\Sigma 1$.
2. The union of two $\Sigma 1$ sets is $\Sigma 1$.
3. Let $T \subset (\omega<\omega)n$ be a recursively enumerable tree.
   a. T cab be written to the form $\{(u, v): \exists k R(u, v, k)\}$, where R is suitably recursive. Now make S consist of (u, w) where w encodes a sequence

(v(0), v(1), ..., v(l − 1), k0, k1, ..., kl−1) such that each (u|i, (v(0), ..., v(i − 1)), ki) ∈ R.)

   b. Show there is a recursive tree S with p[S] = p[T].

4. Let $A = \langle A, \in \restriction A \rangle$ be a model of ZFC. A model $B = \langle B, E \rangle$ of ZFC is said to be a *proper end-extension* of A if (i) $A \subseteq B$, (ii) $A \neq B$, (iii) $a \in A$, $b \in B$, $bEa \Rightarrow b \in A$. Thus, a proper end-extension of a model of ZFC is a proper extension in which no "new" element comes "before" any "old" element. Consider the generalized compactness Theorem.: *Let A be a countable transitive set such that $A = \langle A, \in \restriction A \rangle$ is a model of ZFC. If Δ is a set of sentences of* L_A *which is definable in A by a formula of the language of set theory and if each* $\Delta' \subseteq \Delta$ *such that* $\Delta' \in A$ *has a model, so does* Δ. Prove that Each countable transitive model of ZFC has a proper end extension.

5. Each n we have the following strict containments:

$$\Pi^1_n \subset \Sigma^1_{n+1}$$

$$\Pi^1_n \subset \Pi^1_{n+1}$$

$$\Sigma^1_n \subset \Pi^1_{n+1}$$

$$\Sigma^1_n \subset \Sigma^1_{n+1} \; .$$

6. Prove theorem 5.4.

7. On section 5.7.2 prove that HF(M) *is the least admissible set over* M.

8. Let L be the first-order language without extralogical symbols and let Σ be the L(ω1,ω)-sentence which characterizes the class of finite sets. Suppose that Σ were equivalent to a conjunction $\circ_{i \in I} \sigma_i$ of prenex L(ω1,ω)-sentences σ_i. Then each σ_i is of the form $Q_1 x_1 \ldots Q_n x_n \varphi_i(x_1, \ldots, x_n)$, where each Q_k is ∀ or ∃ and φ_i is a (possibly infinitary) conjunction or disjunction of formulas of the form $x_k = x_1$ or $x_k \neq x_1$.

Hint: Since each σ_i is a sentence, there are only finitely many variables in each φ_i, each φ_i is then equivalent to a first-order formula. The set Δ = {$\sigma_i : i \in I$} have the same models. Compactness considerations implies the proof.

9. Prove proposition 5.3: $\Delta \models \psi \Leftrightarrow \Delta \vdash^* \psi$.

10. Prove theorem 5.6

KEYWORDS

- admissible sets and ordinals
- admissible sets and structures
- admissible tree recursion
- Boolean valued models
- ordinal computability
- set reducibility

REFERENCES

Barwise, J. (1967). *Infinitary Logic and Admissible Sets*. Ph.D. Thesis, Stanford University.

Barwise, Jon. (1975). *Admissible Sets and Structures*, Berlin: Springer-Verlag. Barwise, J. and S. Feferman (eds.). (1985). *Handbook of Model-Theoretic Logics*, New York: Springer-Verlag.

Baumgartner, J. (1974). "The Hanf number for complete $L_{\omega 1,\omega}$ sentences (without GCH)," *Journal of Symbolic Logic*, 39: 575–578.

Bell, J. L. (1970). "Weak Compactness in Restricted Second-Order Languages," *Bulletin of the Polish Academy of Sciences*, 18: 111–114.

Burgess, J. P. *Infinitary languages and descriptive set theory*, PhD Thesis, Univ. of California, Berkeley, 1974.

Chang, C.C. (1968). "Some Remarks on the Model Theory of Infinitary Languages." in *The Syntax and Semantics of Infinitary Languages* (Lecture Notes in Mathematics: Volume 72), J. Barwise (ed.), Springer-Verlag, Berlin, 36–63.

Ellentuck, E. (1976). "Categoricity Regained," *Journal of Symbolic Logic*, 41(3), 639–643.

Gregory, J. Higher Souslin trees and the generalized continuum hypothesis, Journal of Symbolic Logic 41 (1976), no. 3, 663–671.

Halko, A. Negligible subsets of the generalized Baire space $\omega\,\omega 1$, Ann. Acad. Sci. Ser. Diss. 1 Math. 108 (1996).

Hanf, W. P. (1964). *Incompactness in Languages with Infinitely Long Expressions*, Amsterdam: North-Holland.

Harrington, L., Kechris, A. S., Louveau, A. (1990). A Glimm-Eros dichotomy theorem for Borel equivalence relations – Journal of the American Mathematical Society, vol. 3, pp. 903–928. MR 91h:28023.

Huuskonen, T., Hyttinen, T., Rautila, M. (2004). On potential isomorphism and non-structure – Arch. Math. Logic 43, pp. 85–120.

Hyttinen, T., Rautila, M. (2001). The canary tree revisited – J. Symbolic Logic 66, no.4, pp. 1677–1694.

Jech, T. (2003). Set Theory – ISBN-10-3–540–44085–7 Springer-Verlag Berlin Heidelberg New York.

Karp, C. (1965). Finite-quantifier equivalence – Theory of Models (Proc. 1963 Internat. Sympos. Berkeley), pages 407–412. North Holland, Amsterdam.

Karp, C. (1964). *Languages with Expressions of Infinite Length*, Amsterdam: North-Holland.

Karttunen, M. Model theory for infinitely deep languages – Ann. Acad. Sci. Fenn. Ser. A I Math. Dissertations, vol. 64, 1987.

Keisler, H. J. (1974). *Model Theory for Infinitary Logic*, Amsterdam: North-Holland.

Keisler, H. J., and Julia F. Knight. (2004). "Barwise: Infinitary Logic And Admissible Sets," *Journal of Symbolic Logic*, 10(1), 4–36.

Kreisel, G. (1965). "Model-Theoretic Invariants, Applications to Recursive and Hyperarithmetic Operations," in *The Theory of Models*, J. Addison, L. Henkin, and A. Tarski (eds.), Amsterdam: North-Holland, 190–205.

Kueker, D. (1975). "Back-and-forth arguments in infinitary languages," in *Infinitary Logic: In Memoriam Carol Karp* (Lecture Notes in Mathematics: Volume 492), D. Kueker (ed.), Berlin: Springer-Verlag.

Kunen, K. Set Theory/An Introduction to Independence Proofs – North Holland Publishing Company. (1980). Studies in Logic and Foundations of Mathematics; v. 102.

Lopez-Escobar, E. G. K. (1965). "An Interpolation Theorem for Infinitely Long Sentences," *Fundamenta Mathematicae*, 57: 253–272.

Makkai, M. (1977). "Admissible Sets and Infinitary Logic," *Handbook of Mathematical Logic*, J. Barwise (ed.), Amsterdam: North-Holland, 233–282.

Nadel, M. (1985). "$L_{\omega_1,\omega}$ and Admissible Fragments," in J. Barwise and S. Feferman (eds.) 1985, 271–287.

Nagel, E., Suppes, P., Tarski, A. (eds.),"Some problems and results relevant to the foundations of set theory," in E, *Logic, Methodology and Philosophy of Science*, Stanford: Stanford University Press, 123–135.

Platek, R. (1966). *Foundations of Recursion Theory*, Ph.D. Thesis, Stanford University.

Ressayre, J. P. (1977). *Models with compactness properties relative to an admissible language*, Ann. Math. Logic **11** (1977), no. 1, 31–55.

Scott, D. (1965). "Logic with Denumerably Long Formulas and Finite Strings of Quantifiers," *The Theory of Models*, J. Addison, L. Henkin, and A. Tarski (eds.), Amsterdam: North-Holland, 329–341.

Scott, D. (1961). "Measurable Cardinals and Constructible Sets," *Bulletin of the Academy of Polish Sciences*, 9: 521–524.

Scott, D., Tarski, A. (1958). "The sentential calculus with infinitely long expressions," *Colloquium Mathematicum*, 16, 166–170.

Shelah, S., Väänänen, J. Stationary Sets and Infinitary Logic – J Symbolic Logic 65 (2000) 1311–1320.

Silver, J. H. (1980). Counting the number of equivalence classes of Borel and coanalytic equivalence relations – Ann. Math. Logic 18, 1–28.

Silver, J. H. (1980). Counting the number of equivalence classes of Borel and coanalytic equivalence relations – Ann. Math. Logic 18 (1980), 1–28. Models and Games – Cambridge Studies in Advanced Mathematics (No. 132), ISBN: 9780521518123. Not yet published – available from April 2011.

Tarski, A. (1939). "Ideale in völlständigen Mengenkörpern I," *Fundamenta Mathematicae*, 32: 140–150.

Tuuri, H. (1992). Relative separation theorems for Lκ+κ. Notre Dame J. Formal Logic Volume 33, Number 3, 383–401.

Väänänen, J. Games and trees in infinitary logic: A Survey. In M. Krynicki, M. Mostowski and L. S. Tarski's theory of definability: common themes in descriptive by JW Addison.

Väänänen, J. (2008). How complicated can structures be? Nieuw Archief voor Wiskunde. June 117–121.

Vaught, R. (1973). *Descriptive set theory in L_{ω}_{1ω}*, Cambridge Summer School in Mathematical Logic (Cambridge, England, 1971), Springer, Berlin. pp. 574–598. Lecture Notes in Math., Vol. 337.

Vaught, R. (1974/75). Invariant sets in topology and logic. Fund. Math. 82, 269–294.

CHAPTER 6

COMPLEXITY AND COMPUTING

CONTENTS

6.1   INTRODUCTION

Complexity classes have been studied for three decades on areas ranging from computability to more abstract notions. Some complexity classes are:

$$L \subseteq NL \subseteq P \subseteq NP \subseteq PH \subseteq PSPACE$$

where, L: log-space, N: non-deterministic, P: polynomial.

Are the inclusions proper? What is known can be summarized by NL$\neq$ PSPACE?

L: log-space, N: non-deterministic, P: polynomial.

Are the inclusions proper?

What is known can be summarized by NL$\neq$ PSPACE.

This chapter presents the positive forcing and the preceding chapter's applications to concrete and implicit complexity theory based on what the author published around 1985. The areas were further examined to further support what is being called descriptive complexity since 1997 (Nourani 1998, 2003). It defines positive forcing conditions on r.e. sets to the polynomial time computability problems for recursive sets. It is a generalization of the earlier papers to recursive sets. The forcing notion apply positive conditions only and is called Positive Forcing or +forcing. It is shown that r.e. sets can be applied as positive forcing conditions. In the 80's we showed that the P-NP problem is not independent of a sufficiently rich fragment of Universal Peano Arithmetic. Further, it is shown that the polynomial time reducibility of P-NP is a question which positive forcing can determine. The proof relies on our past papers to conclude the existence of a universal forcing companion for the fragment. The P-NP problem is considered as a problem of mutually reducing the two sets. We apply it to get a generalization for our theorems on the time complexity of polynomial time functions to the recursive sets.

The proofs makes no assumptions on the nature of computation, except of course, that the functions are necessarily Diophantine representable. It allows us to conclude that the proof can be extended to the practically computable recursive sets. This chapter briefs how positive forcing can be applied to solve time-complexity question for recursive sets. It is first shown that the problem is not independent of a sufficiently rich fragment of a universal theory of arithmetic. It is then proved that the mutual polynomial time reducibility of P-NP sets are +forced definable sets.

The proof techniques combine major advances in recursion theory, number theory, model theory, and forcing defined on structures. Since the present paper was presented in 1985 there have been a series of structure in complexity papers which might shed further light on where structure can be applied, but with no obvious relation to the present paper. The P-NP problem is defined as a number-theoretic problem. A generic model of arithmetic is defined and a forcing companion for the arithmetic fragment sufficient to represent problems in P and NP as r.e. sets is presented. The techniques and the nature of their composition is novel in itself. It is only once we show that positive forcing conditions can be defined that one can think of Diophantine equations as forcing conditions.

Sections 6.2 and 6.3 introduce a Hilbert like program for viewing computational complexity from the models view point, where the complexity questions are addressed as forcing and number theoretic problems that might have models without basing the techniques on a specific computation measure or system, thereby avoiding a direct computability comparison that has to have a recursive effective procedure or computability model to address. The $I\Sigma_n$ inductive theories that are receiving attention at proof theory are example application areas.

The basic definitions are at times duplicates from a preceding chapter to be self-contained here. The development on addressing the basic concrete complexity questions are contained in the arithmetic hierarchy so far as the descriptive parts are but the direct encoding is not apparent or necessary to address the questions from the modes and forcing viewpoint. This is part because based on what the author has shown around 1985 the basic P-NP problem is not independent of a universal fragment of Peano arithmetic. Section 6.4 presents admissible sets and admissible recursion so far as basic computational complexity is concerned. There with the Kripke Platek recursion on admissible sets complexity questions are addressed presenting important foundational areas the author addressed over the more recent years to revisit p-generic sets on admissible compact formulas. Further topological structures and more on complexity notions on the arithmetic hierarchy and Turing degrees are treated in a following chapter. Section 6.5 presents basic on the arithmetic hierarchy where applicable to concrete descriptive complexity. The theorems in this chapter, unless explicitly stated, are the author's accomplishments on the topic areas.

6.2 FORCING, COMPLEXITY, AND DIOPHANTINE DEFINABILITY

In 1981–1982, the author shows that we can develop a forcing notion which embeds negation is the concept of whether or not a set is a condition for a forcing property. That mathematical basis was applied to computational complexity during 1984–1985 as follows. Whether or not the classes P and NP are the same may be determined from the representation of a solution to the question in terms of their Diophantine sets. Thus, the question P=NP itself is a class of Diophantine sets to which we are seeking a solution. For the question to have an affirmative answer, it is sufficient to show that to a problem in NP there must correspond a problem in P such that their Diophantine representations are solved within deterministic polynomial time of one another. It amounts to proving that there is a polynomial-time computable function, which given the Diophantine representation of the problem in P and its solution, it can transform the encoding to obtain a solution to the Diophantine representation of the corresponding problem in NP. We call such P-NP pair similar. It is only necessary to have one such pair to solve complexity problems.

Any arbitrary problem in NP or P has a Diophantine representation since it is a computable problem. The question of speed of computation reduces to the question whether the two corresponding problems in P and NP, say pi and npi, have similar sets of Diophantine representations. The solution speed would clearly be of the same complexity. This in itself is an unusual result. What is unusual is that the sets P and NP are fixed sets of Diophantine equations, definable in arithmetic. Thus, when building the generic model, even thought new elements are added to the standard model, no new r.e. sets area added. However, the required r.e. sets are enumerated. We prove there are models where P-NP problems are seen within polynomial time.

In view of the above, the question $P \neq NP$ is expressed as a countable collection of $\Sigma 1$ sentences, one for each pair (p, np). The $\Sigma 1$ sentence sates there exist an x such that $p(x) \neq np(x)$. Likewise, the statement that P and NP are polynomially equivalent is a countable collection of Π_1 sentences, stating that for all x, $p(x) = np(x)$. The "="is interpreted as similar in the above sense with the obvious interpretation assigned to its negation '$\neq$'.

The reason for viewing the problem as a number-theoretic problem is precisely because the author's a priori realization that the problem could be expressed as one not independent of a universal fragment of Peano arithmetic. Alternatively, we can pick any arbitrary pair of problems from P and NP time. Take their Diophantine representation to see if they are solved within polynomial time of one another in the above sense. However, it might be impossible to do in practice.

6.3 TECHNICAL PRELIMINARIES

6.3.1 INITIAL MODELS

To prove Gödel's completeness theorem, Henkin style, one proceeds by constructing a model directly from the syntax of the given theory. Our techniques in Nourani (1987, 1992) and the preceding chapters for model building allows us to build and extend models by generalized diagrams. Generalized diagrams were developed by this author to build models with a minimal set of generalized Skolem functions (Nourani, 1987). The minimal set of function symbols are those with which a model can be built inductively. We focus our attention on such models, since they are Initial and computable. Initial models are models that are unique up to isomorphism (ADJ, 1976) on initial algebras and Nourani (1979) on initial models. From these models there is unique homomorphism to all other models in the categories of such models. The formulation in the present paper is with the infinitary logic $L\omega_1\omega$.

6.3.2 ALGEBRAIC AND CANONICAL MODELS

A technical example of algebraic models defined from trees appeared in defining initial algebra's for computational equational theories. Indirection for computing models of equational theories of computing are presented by a pair (Σ, E), where Σ is a signature (of many sorts, for a sort) and E a set of Σ-equations. Let $T<\Sigma>$ be the free tree word algebra of signature Σ. The quotient of $T<\Sigma>$, the word algebra of signature Σ, with respect to the Σ-congruence relation generated by E, will be denoted by $T<\Sigma, E>$, or $T<P>$ for presentation P. $T<P>$ is the "initial" model of the presentation P. The Σ-congruence relation will be denoted by $==P$. One representation such n of $T(P)$ which is nice in practice consists of an algebra of the canonical representatives of the congruence classes. This is a special case of generalized standard models defined further in the present paper. In what follows g t1...tn denotes the formal string obtained by applying the operation symbol g in Σ to an n-tuple t of arity corresponding to the signature of g. Furthermore, gC denotes the function corresponding to the symbol g in the algebra C.

Definition 6.1: Let M be a set and F a finite family of functions on M. We say that (F, M) is a monomorphic pair, provided for every f in F, f is injective, and the sets {Image(f):f in F} and M are pair-wise disjoint.

This definition is basic in defining induction for abstract recursion-theoretic hierarchies and inductive definitions. The definition for generalized standard models with monomorphic pairs. This definition was put forth by the present author around 1982 for the computability problems of initial models.

Definition 6.2: A standard model M, with base M and functionality F, is a transitive model inductively defined by <F, M> provided that <F, M> forms a monomorphic pair.

We will review these definitions in the sections to follow.

Definition 6.3: A model <M, E> of ZF set theory is said to be a standard model iff M is a transitive set, for example, x ε y ε M → x in M, and E the epsilon relation on M. We say that a set M is a standard model iff <M, E> is a standard model.

The following definition is also useful.

Definition 6.4: Let M and N be standard models of ZF. M is said to be an inner submodel of N and N an outer extension of M iff M is a subset of N, and M and N have the exact same ordinals.

Note that our standard models as defined by definition 1.2 by construction are initial, thus have no proper submodels. The standard models by definition is minimal.

6.3.3 GENERIC DIAGRAMS FOR INITIAL MODELS

The generalized diagram were briefed on the preceding chapters. On such diagram the elements of the structure are all represented by a minimal set of function symbols and constants, such that it is sufficient to define the truth of formulas only for the terms generated by the minimal set of functions and constant symbols. Such assignment implicitly defines the diagram. This allows us to define a canonical model of a theory in terms of a minimal family of function symbols.

The minimal set of functions that define a generic diagram are those with which a standard model could be defined by a monomorphic pair. Formal definition of diagrams are stated here, generalized to G-diagrams, and applied in the sections to follow.

Definition 6.5: Let M be a structure for a language L, call a subset X of M a generating set for M if no proper substructure of M contains X, for example, if M is the closure of X U {c(M): c is a constant symbol of L}. An assignment of constants to M is a pair <A, G>, where A is an infinite set of constant symbols in L and G: A → M, such that {G(a): a in A} is a set of generators for M. Interpreting a by g(a), every element of M is denoted by at least one closed term of L(A). For a fixed assignment <A, G> of constants to M, the diagram of M, D<A, G>(M) is the set of basic (atomic and negated atomic) sentences of L(A) true in M. (Note that L(A) is L enriched with set A of constant symbols.)

Thus, initial models could be characterized by their generic diagrams, c.f. preceding chapters. Further practical and the theoretical characterization of models by their G-diagrams are presented in the preceding chapter. That builds the basis for what follows in several sections below, where we show how initial models could appear out of thin air within our formulation. The G-diagram for the model is also defined from the same free trees. The conditions of the theorems are what you expect them to be: that canonical subset be closed under constructor operations, and that operations outside the constructor signature on canonical terms yield canonical terms. Further computational techniques based on the method of G-diagrams was reported by this author at very recent times.

6.3.4 MODELS AND FRAGMENT INDUCTIVE CLOSURE

Given an axiomatization presenting a computing theory for a world of interest, the axiomatization presents a theory T(P). Let F be a family of functions which minimally defines the universe M of the computing model intended for the theory, by inductive definitions on a monomorphic pair with F as the set of functions.. The inductive closure T*(P) is the set of formulae of the language provable from T(P), in conjunction with a family of induction schemas <I(f)> for each f in F, through induction and modus ponens, in a language which includes implication → as a logical symbol and in which negation appears as a Boolean-valued operation only. The definition of monomorphic pairs requires that the functions f in F be injective and the applied sets be pairwise disjoint.

These conditions are exactly what is needed to satisfy the properties of induction schemas as presented by R.M. Robinson's axiomatization of arithmetic and extended to many sorted structure by this author in 1978 such that the induction schemas are well-formed and the sets modeling them are well-founded.

Thus, a theory T#(P) = T(P) + <I[f]: f in F> is defined. Such characterizations are during the most recent times since this authors over two decades ago, are for example are called $I\Sigma_1$. For algebraic applications of computing only the universal reduct of T#(P) is sufficient. We refer to the universal reduct by T*(P), the Universal Inductive Closure (UIC). It is shown in the following section that T*(P) is a generic set for a forcing property to be defined, thus it is a forcing companion.

6.4   POSITIVE FORCING AND INFINITARY MODELS

The basic results proved in this section is that the partially ordered set defined by the power set of T*(P), under the natural subset ordering, defines a forcing property for which T*(P) generates a generic set. Thus, the closure construction contains all one needs to conclude the truth of formulas in the particular computing world defined. The formulation of forcing that we had presented in Davis et al. (1976) and put forth in the present paper is based on model-theory and set-theory within a well-behaved fragment of infinitary logic $L\omega_1 \omega$. The language and some of the formulation, for example the notion of a forcing property is the same as that of Keisler (1973), except for the definition for algebraic theories we have presented here. The Keisler formulation is quite elegant and encodes all the usual notions of forcing.

But from that point on things are not the same. We formulate theories, models, and forcing properties and conditions on algebraic theories and initial models. First, we take a countable language L and then enrich it with a countable supply of constants to name individuals. A well-behaved fragment of $L\omega_1 \omega$ is applied (see appendix for further definitions). Call the enriched language L[C]. We shall use the symbol M for a model and M for the set of individuals in its universe. Throughout this paper our theories do not admit trivial one-point models. Now, we put forth some definitions.

Definition 6.6: A forcing property for a language L is a triple $\Im = (S, <, f)$ such that

 (i) (S, <) is a partially ordered structure with a least element, for example, 0, thus < is not always proper and could be read as < or =;

 (ii) f is a function which associates with each p in S a set f(p) of atomic sentences of L[C];

 (iii) whenever p < q, f(p) is subset of f(q);

 (iv) let l and t be terms of L[C] without free variables and p in S, φ a formula of L[C] with one free variable. Then if (l-=t) is in f(p) then (t=l) is in f(q) for some q > p. φ(t) in f(p) implies φ(l) is in f(q) for some q > p. For some c in C and q > p, (c=l) is in f(q). The elements of S are called conditions for $\Im$.

Now, consider a presentation P, its initial standard model M, obtained by taking the quotient of its free tree algebra (word algebra) with respect to the congruence relation generated by the equations of the presentations on the trees. Enrich the model with the names of the function symbols that can define the G-diagram for the model. In other words, add on Skolem function names that can complete the set and define a G-diagram corresponding to what amounts to the set of functions that can define the model by a monomorphic pair.

Lemma 6.1: ($\wp(T^*(P), \subseteq, f)$, where P is the powered operation, and f an assignment of free variables in T*(P) into M, is a forcing property.

Proof: First we note that such f indeed does associate a set f(p) of atomic sentences of Lc with each p in S; for f defines a function g, g: $\wp(T^*(P))$ →Γ, whereΓ is the set of atomic sentences of Lc. Given any r in $\wp$ (T*(P)), r is a subset of T*(P), therefore, g assigns to each free variable in r, from M, which is assumed to be built out of Lc. Thus, f in turn maps to a set of atomic sentences of Lc by virtue of T*(P) being strictly a set of equational formulas. Thus, f implicitly defines the mapping. Furthermore, ($\wp(T^*(P))$, $\subseteq$) is clearly a partially ordered structure, with the least element being the empty set. To show that (iii) of Definition 6.1 is satisfied: just note that since for any p, q in p, q $\subset$ T*(P), therefore f(p) $\subseteq$ f(q). Checking (iv) takes a bit more work: if (l=t) is in f(p) the (t=l) is in f(q) for some q > p, by the symmetric property of equality in T(P). Next, if (l=t) and φ(l) in f(p) then by substitution of equals in T(P), $\varphi(t)$ is in f(q) for some q > p, q a subset of T*(P). Last, to show that for some c in C and q>p, (c=l) in f(q). This follows because Lc has an infinite supply of names for constants, and by assigning to each representative term in M, equations of the form c=l, for l a term, would be consistent with M and T*(P). The assignment of names to terms can be done in a predetermined manner within the countable fragment such that c=l is in some q > p. This completes the proof of the lemma.

Next, we define positive forcing.

Definition 6.7: The relation p ‖—+ φ read "p positively forces φ" is defined for conditions p an q in S as follows: for an atomic sentence φ, p ‖—+ φ iff φ is in p; for an open formula φ of the form f(X) = g(X), p ‖—+ φ, iff for all c in L[C], where c is an n-tuple of constants form L[C] and all q, p < q implies there is an r such that r > q and r ‖—+ f(c) = g(c). For arbitrary sentence φ, p ‖—+ φ iff p +forces a universal formula logically equivalent to φ.

Definition 6.8: Let $\Im$ = (S, <, f) be a positive forcing property, we say that a subset G of S is positively generic iff

(i)   p in G and q <p implies q in G;
(ii) p, q in G, implies there is an r in G such that p < r and q <r;
(iii) for each sentence φ, there exists p in G, such that either p ‖ + φ or there is no q in S, q >p such that q ‖ + φ.

A special case of the above definition is when S consists of sets of formulas. For such S, we can make a substitution: subset relation for <. Clause (iii) can then be stated:

(iii)' for each sentence φ, there exists p in G such that either p ‖—+ φ or p U {φ} is not a condition for F.

Next, we state the positive forcing theorem that forms the basis for model building with infinitary logic on algebraic trees. We should note that the notion of condition in the present formulation is always relative to a standard model M, in the sense defined by the present paper. That is the model with respect to which we take a UIC. Thus, conditions are always formulae consistent with M and UIC.

Conditions are infinitary in the present formulation, therefore, we are out in a universe where the conditions are infinitary.

Theorem 6.1: Let T(P) be an equational theory, M its standard model obtained by a monomorphic pair. Then $\wp$(T*(P)) is positively generic for T(P).

Proof: We show that S = $\wp(T^*(P))$ satisfies the conditions put forth in definition 6.4, with the subset relation as the ordering on S. To check (i) note that $T^*(P)$ is the maximal element of $\wp(T^*(P))$ and every condition p is a subset of $T^*(P)$. If p is in S, and q > p, then clearly there is a condition r such that r > q and r > p, namely p U q. this follows form the fact that both p and q must be consistent with $T^*(P)$ and further they must be modeled by M. Finally, to show that the third requirement of 6.4 is satisfied: suppose not and consider the condition $T^*(P)$, where $T^*(P)$ does not force φ, but for some q > $T^*(P)$, q ||– φ. One such q is $T^*(P)$ U {φ}, but then by definition of ||–+, $T^*(P)$ ||–+ φ.

Thus, we have exhibited an explicit construction for a generic set. We have further shown that UIC has the genericity property. It is important to note that within the present formulation of forcing, the set to which we force is known. Thus, there is an infinitary set, in principle, that could contain all that various conditions could imply for the forcing property. That is what makes this method and results applicable for computing. That does not mean that one could get his hands on it, however, to trivialize the computational implications. Thus, to state an mathematical result and construction of an infinite set, the forcing companion, does not imply that it could be invoked in computing practice to force a solution to everyday problems, then and there. But it helps to generate models and to figure out what sets are not conditions in proofs. Hence, it has computational significance. Most other notions of forcing do not have forcing companions there by free construction. Thus, in those formulations it is not known to what set things are forced to. We want to show the reader that by the present approach we can generate infinitary models of computational theories.

6.4.1 GENERATING MODELS BY POSITIVE FORCING

A consequence of the genericity property is that it solves problems more general setting of varying infinitary forcing, similar to set-theoretic forcing. We shall make use of the same notation as Barwise and Robinson (1970) to be consistent with the logic literature. Let K be a fixed set of axioms over a signature Σ, for example, let (Σ, K) be a presentation.

Definition 6.9: Let F =(S, <, f) be a forcing property. Let P = (Σ, K) be a presentation. Given q in S, we let KF [q] be the set of sentences positively forced by q. In particular KF [∅], denoted simply by KF, is the universal forcing companion of P.

By the theorem above T*(P) is the universal forcing companion for P. An important special case for T*(P) is when P is Peano arithmetic. Of course, PA is not a universal theory, but by taking iterated Skolem expansions we can in principle get an extension of PA which is universal. Now we see that T*(P) for PA is obtained with a single induction schema. Furthermore, by Theorem 4.5 and definition 6.1, T*(P) is a universal forcing companion for PA. The problems of this nature are generally known to be difficult. Perhaps the techniques presented in this paper could be helpful for other problems in this area. An application area is presented in a following section.

Definition 6.10: A theory is said to be inductive if T is equivalent to a set of ∀∃ sentences, that is sentences of the form $\forall x1...\forall xn \exists x1 \exists. \exists$ xn Y, where Y is quantifier-free. []

Proposition 6.1: T*(P) is inductive.

Proof: Immediate from definitions.

Therefore, inductive closures are inductive theories in the technical sense.

Definition 6.11: Let (M, a)c in C be defined such that M is a model for L and each constant c in C has the interpretation a in M. the mapping c → ac is an assignment of C in M. We say that (M, a)c in C is canonical model for a presentation P iff the assignment c→ a maps C onto M, for example, M=(a:c in C).

Definition 6.12: A generic set is said to generate M iff M is a canonical model and every sentence φ of L[C] which is forced by some p in G holds in M.

Definition 6.13: M is a generic model for a condition p iff M is generated by some generic set G which contains p.

Theorem 6.2: The canonical term initial algebra M is a generic model of T*(P).

Proof: By definition, M is a generic model for a condition p in S iff M is generated by some generic set G containing p. Now, by Theorem 6.10, $\wp(T^*(P))$ generates M, and $\wp(T^*(P))$ certainly contains T*(P) as a condition. Thus, M is a model for T*(P).

6.4.2 FORCING AND COMPUTABILITY

The question P = NP can be reduced to the question of comparing Diophantine representation of problems in a universal fragment of arithmetic, with respect to which P=NP is +forced. The closure construction generates the generic model. The fragment has to be rich enough to express Diophantine equations on polynomials. Let us denote it by D-universal-D for Diophantine. Now for a pair (pi, npi) (their Diophantine representation), either the sets are similar, or they are not a condition for forcing P=NP. If in fact it is the case that P /= NP, then the Diophantine representations of pi and npi, denoted rep(pi) rep(npi), respectively, are not conditions for no pairs pi and npi in their respective classes. Thus, we have the following

Theorem 6.3: The P-NP question is not independent of D-universal fragment of Peano arithmetic.

Proof: By the +-forcing results there is a closure construction for the standard model of the fragment of arithmetic, which is a forcing companion for the fragment. The generic model which is generated for the fragment includes all the required R.E. sets for the P and NP problems as Diophantine sets. Thus, either for all respective pairs of (pi, npi) Diophantine sets are not conditions, in which case P /= NP is +forced. Otherwise, the forcing conditions include one pair (pi, npi), thus, the simultaneous satisfiability of the pair is forced. Consequently, the P-NP question is ‖–+ determined on structures that mode infinitary languages.

Note that there is no inconsistency between arbitrary Diophantine representation of problems such as Boolean satisfiability and an arbitrary algorithm. Thus, it is impossible to assume that no pairs of P-NP representations can form a condition set. Once we achieve Theorem 6.1 the main theorem is obtained applying a similar proof.

Theorem 6.4: There is a generic model where the P=NP question can be +forcing determined.

Proof: By the above theorem it is impossible to assume no P-NP pairs can be condition sets, for they are arbitrary polynomials with different sets of parameters. The proof is then immediate from Theorem 6.1.

It thus becomes apparent that one way to "solve" that problem with an example, is to find a pair pi and npi which have the same Diophantine representation, within a renaming of variables and parameters. But it is not likely to be a small task at all. One can ask whether the above can be generalized to all the R.E. sets. If we limit the sets to the sets with tractable computability the answer is affirmative. For the recursive sets which are practically computable the complexity theorems above generalize. We certainly cannot go beyond the practically computable recursive sets since there are fragments of arithmetic which have strictly exponential complexity, rendering them practically not computable.

Theorem 6.5: If an R.E. set is computable at all, in the sense that there are nondeterministic polynomial time algorithms for its decidability, then the set is deterministically polynomial-time computable.

Proof: Follows direct from theorem 6.1 and 6.2 by noting that the proofs made no assumption on the nature of Diophantine sets, except of course, their being computable. Thus, is an R.E. set is computable at all, it is polynomial time computable.

Subsequent to writing the original version of the above on briefs, few references were brought to the author's attention which may put the chapter on a nicer perspective. For example, DeMillo and Lipton (1980) has been advocating the use of model theory to the P-NP problem in fragments of arithmetic. The techniques we have applied is making use of a particular version of model-theoretic forcing within infinitary logic.

The notion "sufficiently rich fragment" applied by our paper, though not necessary to be narrowed down any further to understand the theorems, could be formulated in theorems similar to ET studied in DeMillo-Lipton (1980). ET is the theory of true universal formulas, consisting of terms formed over +, −, x, exp., min., max, and polynomial time computable predicates. The P-NP theorems in the present paper can perhaps be better understood by studying ET. Other related papers are Joseph-Young (1980). A comparison of various notions of reducibility appears on that paper. The paper of Baker et al. (1975) was an important step in understanding the nature of the problem relative to oracles. Generic notions in recursion theory were applied to the P-NP problem in Ambos-Spies et al. (1984) almost simultaneously as the abstract for the present paper was written (Nourani, 1984). Blum's JACM paper was an important step for defining a machine-independent complexity theory for recursive functions.

6.5   COMPLEXITY CLASSES, MODELS, AND URELEMENTS

There are several definitions of the class if recursive functions on natural numbers, all providing support for Church's thesis on effective computability. Playing recursion on domains other than integers has presented 'new' challenges and insights since late 60's. Kleene's enumeration and the second recursion theorem lift to arbitrary admissible sets. Admissible and Descriptive sets appear to indicate that when various recursive definitions are lifted to other domains than integers, for example admissible sets, there are equivalence breakdown. The most dramatic there being two competing notions of r.e. and admissible sets. The preceding had lead the author to examine specific admissible computability functions called Admissible Computable that is not full recursive admissible sets. At the Summer 1996 Colloquium, Nourani (1996) had put forth descriptive computing principles, defining descriptive computable functions.

Amongst the theorems at is that for A an admissible computable set, A is descriptive computable. Generic diagrams were applied to define admissible computable sets and models. The author had defined computable Functors in Nourani (1996, 1997). The Functors define generic sets from language strings to form limits and models (Nourani, 1954–1995).

The notion of admissible set was introduced by Platek to study recursively regular (i.e., admissible) ordinals. Infinite language categories IFCS were presented and their categorical properties are defined by the author's 1994–95 papers. Computable Functors are defined in a December 1996 paper. By defining admissible generic diagrams we define functors to structures. Starting with infinite language category $L_{\omega1,K}$ we define generic sets on $L_{\omega1,\omega}$. Let L be the language defined on the Keisler $L\omega1,\omega$ fragment. Generic diagrams, denoted by G-diagrams were defined since 1980's to be diagrams for models defined by a specific function set, for Example $\Sigma1$Skolem functions. Urelements and fragment consistent models are applied to computable KPU recursion and Peano models towards specific computability questions. Specific genericity concepts are presented as example applications.

Playing recursion on domains other than integers has presented 'new' challenges and insights since late 60's. Kleene's enumeration and the second recursion theorem lift to arbitrary admissible sets. Admissible and Descriptive sets appear to indicate that when various recursive definitions are lifted to other domains than integers, for example admissible sets, there are equivalence breakdown.

The most dramatic two competing notions are r.e. and admissible sets. The preceding had lead the author to examine specific admissible computability functions called Admissible Computable that is not full recursive admissible sets. Nourani 1980's ASL and AMS papers had stated and proved theorems that well-known computational complexity problems were not independent of certain universal fragments of Peano arithmetic Here we examine specific computable sets in retrospect with respect to the developments here, to explore additional insights. Let A be a transitive set. We say that a set $X \subseteq A$ is a Σ_1 on A if there is a Σ_1 formula which defines X on (A, ε).

X is Σ_1 ~on A if it is defined by a Σ_1 formula with parameters from A, that is if there is a $\Sigma1$ formula θ (x, y1, ..., yn) and parameters b1... bn in A such that $X = \{a$ in A $|(A, \varepsilon) \models \theta$ [a, b1, ..., bn]}. We read X is Σ_1 parameters. The ~ notation on Δ_0 is defined the same. A set is Δ resp. $\Delta1$ on A if both X and A ~ X are Σ resp. Σ_1. That is why the admissibility definition was stated on Σ~. By a Σ reflection we should mean the restriction to θ which are Σ sentences of set theory. We won't be considering strict Σ reflection. If an admissible set A is countable then LA countable. L_A is more convenient than $L_{\omega 1, \omega}$ for model-theoretic purposes since $L_{\omega 1, \omega}$ has $2^{\aleph 0}$ formulas. Kiesler established the categoricity is for these languages. Most sentences of $L_{\omega 1, \omega}$ which describe interesting algebraic structures are already in L_A where A is the smallest admissible set different form H(ω).

From the basic Arithmetical Hierarchy theorem, an n-ary relation R is in the arithmetical hierarchy iff it is recursive or, for some m, can be expressed as $\{<x, ..., xn> | (Q1y1....(Qmym) S(x1, ..., xn, y1, ..., ym)\}$ where Q is either $\forall$ or $\exists$ for 1 <= I <= m, and S is an (n+m)-ary recursive relation.

To have a basic feel for KPU with some primitive recursive example areas is Rathjen (1992). Let KP be the theory resulting from Kripke-Platek set theory by restricting Foundation to Set Foundation. Let G: $\backslash V \rightarrow \backslash V$ ($\backslash V$:= universe of sets) be a $\Delta 0$-definable set function, for example, there is a $\Delta 0$-formula $\varphi(x, y)$ such that $\varphi(x, G(x))$ is true for all sets x, and $\backslash V$ models $\forall x \exists! y$ (x, y). The collection of set functions primitive recursive in G coincides with the collection of those functions which are Σ_1 definable in KP- + Σ_1-Foundation + $\forall x \exists! y\varphi$ (x, y). Let us consider the infinitary language $L_{\omega 1, \omega}$ and carry on admissible sets on L_A where L is $L_{\omega 1, \omega}$. Let us

take the Keisler countable fragment K_A and carry on admissible computability with respect to +forcing (Forcing and the P-NP, written at the Summer Logic colloquium, 1984).

Recall the follow that all statements on K_A generalizes to $L_{\omega1,\omega}$. However, we are more interested in KA, therefore stating the following based
on K_A, and taking reducts on $L_{\omega1,\omega}$ to K_A. At the Summer '96 Colloquium,
Nourani (1996) had put forth descriptive computing principles, defining
descriptive computable functions. Amongst the theorems at is that for A
an admissible computable set, A is descriptive computable. Generic diagrams were applied to define admissible computable sets and models.
The author had defined computable functors in Nourani (1996, 1997) the
functors define generic sets from language strings to form limits and models (Nourani 19954–1995). The notion of admissible set was introduced
by Platek to study recursively regular (i.e., admissible) ordinals. Starting
with the infinite language category $\mathbf{L_{\omega1,K}}$ we define generic sets on $L_{\omega1,\omega}$.
Let L_A be the admissible language defined on the Keisler $L_{\omega1,\omega}$ fragment.
Generic diagrams, denoted by G-diagrams were defined since 1980's to
be diagrams for models defined by a specific function set, for Example Σ_1
Skolem functions.

Definition 6.14: A model is admissible iff its universe, functions, and relations are definable with admissible sets.

Theorem 6.6: Admissible models are obtained by taking a reduct from the
admissible hull to the Skolem hull definable by a generic diagram.

Proof: (e.g., Nourani 2014 volume).

6.5.1   FUNCTORIAL IMPLICIT COMPLEXITY

Let L be an admissible language defined on the Keisler fragment $L_{\omega1,\omega}$. The
functor on the category $L_{\omega1,k}$ might be defined with admissible hom sets.
Thus, we can apply generic diagrams to define admissible computable sets
and models. Let us define computable functors (Nourani, 1996). The functors define generic sets from language strings to form limits and models.

Definition 6.15: A functor F: A $\rightarrow$ B is computable iff there is an effective
procedure for defining the arrow object mapping from A to B.

Definition 6.16: A model is admissible iff its universe, functions, and relations are definable with admissible sets.

Theorem 6.17: There is a generic functor defining an admissible model.

Proof: (Chapter 4).

Let L be an admissible language defined on the Keisler fragment. The functor on the category $L_{\omega1,K}$ might be defined with admissible hom sets.

Theorem 6.18: The generic functors define admissible models, provided the functions and relations in the language fragment are L -admissible.

Proof: (Chapter 4).

Theorem 6.19: There is a generic functor defining an admissible model. A functor on the category $L_{\omega1,K}$ might with be defined admissible hom sets.

Proof: (Exercise 3).

6.5.2. ABSTRACT DESCRIPTIVE COMPLEXITY

From Barwise (1968) on infinitary logic and admissible sets, introduce the infinitary language L_A where A is a countable admissible sets. We are interested in the sublanguages L_A of the language L $(= L_{inf,\, \omega})$ which allow finite strings of quantifiers and arbitrary conjunction and disjunction. We consider formula φ of L to be sets, and the language L_A is L <A>. To insure that L_A is a sensible language, we must require that A satisfy certain closure conditions. An extended language of set theory is applied where in addition to the only relation symbol ε, the language allows additional relation symbols Si, that n_1-ary is the Admissible sets and are defined based on rudimentary transitive sets. The definitions apply a separation principle.

There is a distinction between the relation symbol ε and the metasymbol $\in$.

Definition 6.17: Δ_0-separation, where $\Delta_0 (S_1, \ldots S_k)$ formulas of set theory are defined as a closure as the smallest Y such that

(a)
(i)   for atomic formals θ, both θ and $\neg\, \theta$ are in Y,
(ii) Y is closed under conjunction and disjunctions on formulas,
(iii) if θ is Y then so are $\forall x\, [x\, \varepsilon\, y \rightarrow]$ and $\exists x\, [x\ y \wedge\theta]$, denoted by $\forall x$
       $\varepsilon\, y\, \theta$, and $\exists x\, \varepsilon\, y\, \theta$.

(b) The $\Sigma(S1, \ldots, Sk)$ formulas if set theory form the smallest collection Y closed under (i), (ii), (iii) and

(iv) if θ is in Y, $\exists\, x\, \theta$ is in Y.

(c) The $\Sigma 1(S1, \ldots, Sk)$ formulas of set theory form the smallest collection Y containing the $\Delta_0 (S_1, \ldots S_k)$ formulas and closed under (iv).

Call a formula θ a Δ_0 formula if is a $\Delta_0 ()$ formula, for example, no a additional relation symbols are there.

These classes are important in defining end extensions (Feferman-Kreisler, 1972; Nourani, 1997).

Definition 6.18: A nonempty transitive set A is rudimentary if A satisfies the following

(a) if a, b ε A, then a $\times$ b and TC($\{a\}$) are ε A;

(b) (Δ_0-separation) if θ is any a Δ_0-formula and y is a variable not free in θ, the following is universally true in A: $\exists y \forall x\ [x\ \varepsilon\ y \leftrightarrow x\ \varepsilon\ w \wedge \theta]$.

(c) For any formula θ and variable y of set theory $\theta^{(y)}$ is the Δ_0-formulas obtained from θ, by relativizing all quantifiers in θ to y.

Definition 6.19: A is admissible if A is rudimentary and satisfies the $\underset{\sim}{\Sigma}$-reflection principle: if θ is a Σ-formula and y is a variable not free in θ, then the following is universally true in A: $\theta\ \Delta\ \exists\ y\ [y$ is transitive $\wedge\ \theta^{(y)}]$.

Let A be an admissible set and let L(A) denote the collection of formulas of L<inf, ω> that are in A. Call L<A> an admissible fragment (of L<inf, ω >).

6.5.3 A DESCRIPTIVE COMPUTING EXAMPLE REVISIT

The author presented in 1997 what we refer to by Descriptive Computation applying generic diagrams. We define descriptive computation to be computing with G-diagrams for the model and techniques for defining models with G-diagrams from the syntax of a logical language. Generic diagrams are diagrams definable with a known function set. Thus, the computing model is definable by a generic diagrams with a function set. An example function set might be $\Sigma 1$ Skolem functions. The corresponding terminology (Martin) in set theory refers to sets or topological structure definable in a simple way. We further define H[κ]> is the language usually denoted by L$_{\kappa,\omega}$.

Definition 6.18: A set is Descriptive Computable iff it is definable by a G-diagram with computable functions.

Proposition 6.1: For descriptive computable sets the set $H[\kappa]$ is definable from a generic diagram by recursion.

Theorem 6.20: (Barwise, 1968) Let A be admissible, $A \subseteq H(\omega_1)$.

Theorem 6.21: For A an admissible computable set, A is descriptive computable.

Proof: Since A is admissible, from Barwise (1968) applying Kunens implicit definability, Gentzen system completeness for $L\omega 1,\omega$ and Beth definability, $A \subseteq H[\omega 1]$ can be proved. Thus, by the proposition A is descriptive.

If A is rudimentary and a, b ε A the {a, b} and a $\wedge$ b and a ~ b are in A.

Thus, every finite subset of A is an element of A. In particular $H[\omega] \subseteq$ A. If A is rudimentary, then L_A has the following closure properties.

   (i)   if $\varphi \varepsilon L_A$ then $\neg\varphi$ L_A;
   (ii) if $\varphi \varepsilon L_A$ and a εA then $(\forall v_a \varphi) \varepsilon L_A$;
   (iii) If Γ is a finite subset of L_A then $_\wedge \Gamma \varepsilon L_A$;
   (iv) if $\subseteq L_A$, $\Gamma \varepsilon A$ then $_\wedge \Gamma \varepsilon A$.

Let A be a transitive set. We say that a set $X \subseteq A$ is Σ_1 on A if there is a Σ_1 formula which defines X on $<A, \varepsilon>$.

Theorem 6.22: (Barwise, 1968) If A is $\sim_1$ compact then A is Admissible,

There are completeness, interpolations, definability, and countable compactness from Barwise (1968).

Completeness—The set of valid sentences of $L_{A \text{ is}} \Sigma_1$ on A.

Compactness—Let A be countable. If A is admissible then A is $\sim_1$ compact.

6.6   RUDIMENTS, KPU, AND RECURSION

6.6.1   ADMISSIBLE HULLS

Let M be a structure of finite similarity type, M = (|M |, R1, ..., RI). R within $V_{|M|}$. Inside $\mathbb{V}_| M_|$ we form the next admissible set HYP (M) and call it HYP_M the admissible hull of $\mathcal{M}$. Technically, HYP M) and HYP ((|M

|, R1, ..., Rl)) may differ, for example, the former may not contain urelements at all. $\mathbb{V}$ is the urelement world.

Let U be any set of the elements called urelements. For ordinals α, we define $\Delta_{U\alpha}$ regarding the elements of $|$ M $|$ as urelements, let us place ourselves $\Delta_{U,0} = U$ and $\Delta_{U,\alpha+1} = \wp (\Delta_{U,\alpha})$. $\Delta_{U,\lambda} = U_{\alpha<\lambda} V_{U,\alpha}$.

Definition 6.19: A model is admissible iff its universe, functions, and relations are definable with admissible sets.

Theorem 6.23: (Nourani, 1977) Admissible models are obtained by taking a reduct from the admissible hull to the Skolem hull definable by a generic diagram.

Platek (1966) developed a recursion theory on admissible sets by calling a function F with (with domain and range subset s of A) A-recursive iff its graph is $\Sigma_{\sim 1}$ on A A set $X \subseteq A$ is A-recursive if its characteristic function in A-recursive, and X is A-recursively enumerable. Consistent with Barwise (1968), however we use the term A-recursive, etc. only when A is countable.

Definition 6.20: Let A be admissible and let R be a relation on A. R is A-*r.e. if R is* $_{1\ on}$ A.

(i)   R is A-*recursive if R is* Δ_1 on A.
(ii) R is A-*finite if R ε A.*
(iii) A function f with domain and range subsets of A *is A-recursive if its graph is A-r.e.*

If $A = L(\alpha)$ then we refer to these notions as α-r.e., α-recursive, and α-finite.

Platek's Admissible Recursion.

Let A be an admissible set. Call a function F (with domain and range subsets of A0 A –recursive iff its graph is $\Sigma_{\sim 1}$ on A. A set $X \subseteq A$ is A-recursive if is characteristic function is A-recursive and X is A-recursively enumerable A-r.e.) if it is the range of a recursive function, Kleene's enumeration and the second recursion theorem lift to arbitrary admissible sets. Admissible and Descriptive sets appear to indicate that when various recursive definitions are lifted to other domains than integers, for example admissible sets, there are equivalence breakdown. The most dramatic there being two competing notions of r.e. and admissible sets. The preceding had lead the author to examine specific admissible computability functions called Admissible Computable, that is not full recursive admissible sets.

Nourani 1980's ASL and AMS papers had stated and proved theorems that well-known computational complexity problems were not independent of certain universal fragments of Peano arithmetic Here we examine specific computable sets in retrospect with respect to the developments here, to explore additional insights. Let A be a transitive set. We say that a set $X \subseteq A$ is $\Sigma 1$ on A if there is a $\Sigma 1$ formula which defines X on (A, ε). X is $\Sigma \sim 1$ on A if it is defined by a $\Sigma 1$ formula with parameters from A, that is if there is a $\Sigma 1$ formula θ (x, y1, ..., yn) and parameters b1...bn in A such that $X = \{a \text{ in } A \mid (A, \varepsilon) \models \theta [a, b1, ..., bn]\}$. We read X is $\Sigma 1$ parameters. The $\sim$ notation on $\Delta 0$ is defined the same. A set is Δ resp $\Delta 1$ on A if both X and $A \sim X$ are Σ resp. $\Sigma 1$. That is why the admissibility definition was stated on $\Sigma \sim$. By a Σ reflection we should mean the restriction to θ which are Σ sentences of set theory. We won't be considering strict Σ reflection.

If an admissible set A is countable then LA countable. LA is more convenient than $L\omega 1, \omega$ for model theoretic purposes since $L\omega 1, \omega$ has $2N0$ formulas. Kiesler established the categoricity is for these languages. Most sentences of $L\omega 1, \omega$ which describe interesting algebraic structures are already in LA where A is the smallest admissible set different form $H(\omega)$.

From the basic Arithmetical Hierarchy theorem, an n-ary relation R is in the arithmetical hierarchy iff it is recursive or, for some m, can be expressed as $\{<x, ..., xn> \mid (Q1y10....(Qmym) S(x1, ..., xn, y1, ..., ym)\}$ where Q is either $\forall$ or $\exists$ for $1 <= I <= m$, and S is an (n+m)-ary recursive relation. From KPU Section 3.2, Kleene's enumeration and the second recursion theorem lift to arbitrary admissible sets. Admissible and Descriptive sets appear to indicate that when various recursive definitions are lifted to other domains than integers, for example admissible sets, there are equivalence breakdown. The most dramatic there being two competing notions of r.e. and admissible sets. The preceding had lead the author to examine specific admissible computability functions called Admissible Computable, that is not full recursive admissible, sets.

Let us consider the infinitary language $L\omega 1, \omega$ and carry on admissible sets on LA where L is $L\omega 1, \omega$. Let us take the Keisler countable fragment KA and carry on admissible computability with respect to +forcing (Forcing and the P-NP, written SLK 1984). Recall the following In the following, all statements on KA generalizes to $L\omega 1, \omega$. However, we are more interested in KA, therefore stating the following based on KA, and taking reducts on $L\omega 1, \omega$ to KA.

Theorem 6.24: (Nourani, 1998) There are admissible and L-admissible models for Peano Arithmetic generated by functors on infinitary language fragments, at which well-known computational complexity questions are not independent.

Proof: (Chapter 9).

Definition 6.21: A set is Descriptive Computable iff it is definable by generic diagram with on computable functions.

Theorem 6.25: (Nourani, 1997) For A an admissible computable set, A is descriptive computable.

Theorem 6.26: (Nourani, 1997) Let L be an admissible language defined on the Keisler fragment $L\omega1,\omega$. The functor on the category $L_{\omega1,K}$ might be defined with admissible hom sets.

Proof: (Exercises)

Theorem 6.27: (Nourani, 1997) The generic functors define admissible models, provided the functions and relations in the language fragment are defined L-admissible.

Proof: (Exercises)

Let us consider the infinitary language $L\omega1,\omega$ and carry on admissible sets on LA where L is $L\omega1,\omega$. Let us take the Keisler countable fragment KA and carry on admissible computability with respect to +forcing (Forcing and the P-NP, written SLK 1984). In the following, all statements on KA generalizes to $L\omega1,\omega$.

Definition 6.22: We say that a set is KA admissible iff it is definable with an admissible fragment $S \subseteq KA$.

Lemma 6.2: (Nourani, 1997) p-generic sets are definable with θ in KA such that θ is $\Sigma1$ closed.

Proof: Follows from genercity above, definitions for urelements, and admissible sets from Chapter 4.

Lemma 6.3: (Nourani, 1997) p-generic sets are rudimentary.

Proof: Follows from genercity above, definition here and admissible sets from Chapter 4.

Theorem 6.28: (Nourani, 1997) p-generic sets are KA admissible.

Proof: Follows from genercity above, definition here and admissible sets from Chapter 4.

From (Nourani, 2002) we had the generalization to $L\omega 1, \omega$.

Corollary 6.1: P-generic sets are L-admissible.

Proof: Follows from genercity above, definition here and admissible sets from Chapter 4.

Corollary 6.2: P-generic sets are admissible computable.

Proof: Exercise 4.

Theorem 6.29: (Nourani, ASL) There are models, in particular PA models, which can be obtained definable by P-generic sets, where certain r.e. set independence questions can be addressed.

Theorem 6.30: (Nourani, ASL) There are models, in particular PA models, which can be obtained KA Σ_1 definable, where the consistency for Diophantine equations corresponding to p-generic set descriptions might be determined.

6.6.2 GENERIC CONSISTENCY MODELS

Specific fragment consistency models were developed in Nourani (2005) (since Nourani, 1996). Fragment Consistency models are presented with new techniques for creating generic models. Infinitary positive language categories are defined and infinitary complements to Robinson consistency from the author's preceding paper are further developed to present new positive omitting types techniques to infinitary positive fragment higher stratified consistency. Further classic model-theoretic consequences are presented. The author (Nourani, 1997) had applied generic diagrams and sets from +forcing (Nourani 83) to the We can apply our generic diagrams and sets from to the LA formulas (Barwise 72, Kripke-Platek) and proved that preorders on initial structures are definable by formulas which are preserved by end extensions on fragment consistent models. The author posed a Model-based Hierarchy complexity question since the ASL-UCSD conference 1999 based on the above: Are there models, or in particular PA models, which can be obtained definable by P-generic sets, $\Pi 2$, or polynomial time recognizable predicates, where the consistency for arbitrary Diophantine equations presenting r.e. sets corresponding to P or NP problem descriptions might be determined. The preliminary glimpses to answers to the preceding questions are only in part stated here.

6.7   CONCRETE DESCRIPTIVE COMPLEXITY

In descriptive complexity, instances of algorithmic problems are modeled by finite structures. To relate this to the standard complexity theoretic view of problem instances as binary strings, one has to represent structures by binary strings. The standard representation scheme used in descriptive complexity theory generalizes the adjacency matrix representation of graphs. Note that if we represent a graph by its adjacency, matrix, we implicitly fix an ordering of the vertices of the graph. For example, Grohe (2013) considers a representation scheme canonical if two structures are represented by the same string if and only if they are isomorphic. Is interesting to explore the newer concrete areas to compare to our canonical model notions over two decades.

The "P=NP question" is an example of an inherent inability to determine what can or cannot be computed in a certain amount of computational resource: time, space, parallel time, etc., with only computability bases and tools to express such questions. The only truly effective concrete tool that can be a measure is Cantor's diagonalization argument, that is embedded in part on the preceding sections when considering p-generic sets (Ambos-Spies et al., 1984). Cantor diagonalization is useful for proving hierarchy theorems, for example, that more of a given computational resource enables us to compute more.

The Cantor diagonalization is a clever technique used by Georg Cantor to show that the integers and reals cannot be put into a one-to-one correspondence (i.e., the uncountably infinite set of real numbers is "larger" than the countably infinite set of integers). However, Cantor's diagonal method is more general and applies to any set as described below.

Given any set S, consider the power set $T=\wp(S)$ consisting of all subsets of S. Cantor's diagonal method can be used to show that T is larger than S, for example, there exists an injection but no bijection from S to T. Finding an injection is trivial, as can be seen by considering the function from S to T which maps an element s of S to the singleton set $\{s\}$. Suppose there exists a bijection f from S to T and consider the subset D of S consisting of the elements d of S such that f(d) does not contain d. Since f is a bijection, there must exist an element x of S such that f(x)=D. But by the definition of D, the set D contains x if and only if f(x)=D does not contain x. This yields a contradiction, so there cannot exist a bijection from S to T. Cantor's diagonal method applies to any set S, finite or infinite. If S is a finite set of cardinality n, then $T = \wp(S)$ has cardinality 2^n, which is larger than n. If S is an infinite set, then $T = \wp(S)$ is a bigger infinite set. In particular, the cardinality c of the real numbers $\Re$, which can be shown to be isomorphic to $\wp(N)$, where N is the set of natural numbers, is larger than the cardinality α_0 of N. By applying this argument infinitely many times to the same infinite set, it is possible to obtain an infinite hierarchy of infinite cardinal numbers.

Let us write $A \leq_p B$ to denote a reduction from a problem description A to a problem description B. In complexity a many-to-one polynomial reduction between the languages A and B is a polynomial time computable function f mapping instances of the first problem into instances of the second problem, that $x \in A$ (the answer to x is yes) iff $f(x) \in B$ (the answer to f(x) is yes). However, there are no known techniques for comparing different types of resources, for example, time versus nondeterministic time, time versus space, etc.

Complexity theory typically considers yes/no problems: this is the examination of the difficulty of computing a particular bit of the desired output. Yes/no problems are properties of the input: the set of all inputs to which the answer is yes have the property in question. Rather than asking the complexity of checking if a certain input has a property T, in descriptive complexity we ask how hard it is to express the property T in a formal language logic, descriptive complexity exactly captures the important complexity classes.

Concrete descriptive complexity ranks based on based on expressiveness in logic. For example, encoding to a graph: a logical structure AG = $h\{1, 2,..., v\}$, EGi whose universe is the set of vertices and where EG is the binary edge relation. A graph problem is a set of finite structures whose vocabulary consists of a single binary relation. Similarly, we may think of any problem T in some complexity class C as a set of structures of some fixed vocabulary. Recall that in first-order logic we can quantify over the universe. We can say, for example, that a string ends in a one. The following sentence does this by asserting the existence of a string position x that is the last position and by asserting S(x), for example, the bit at that position is a one: $(\exists x)(\forall y)(y \leq x \land S(x))$. In second-order logic we also have variables Xi that range over relations over the universe. These variables may be quantified. A second-order existential formula (SO∃) begins with second-order existential quantifiers and is followed by a first-order formula.

An important example area is finite model theory where one examines the relationship between logical definability and computational complexity. We want to understand how the expressive power of a logical system such as first-order or second-order logic, or least fixed-point logic. Conversely, one wants to examine computational complexity levels to towards expressive power of logical languages. While computational complexity theory is concerned with the computational resources such as time, space, or the amount of hardware that are necessary to decide a property, descriptive complexity theory asks for the logical resources that are necessary to define it. Many problems of complexity theory can be formulated as problems of mathematical logic, when we can limit ourselves to finite structures.

6.7.1 CONCRETE IMPLICIT COMPLEXITY

Implicit Computational Complexity is a trend to apply logic and formal methods like types and rewriting systems to provide languages for complexity-bounded computation, in particular for polynomial time computing. The goal is to address the computational complexity of programs without referring to a particular machine model and explicit bounds on time or memory, only relying on logical or computational principles that entail complexity properties. Several approaches have been explored for that purpose, like linear logic, restrictions on primitive recursion, rewriting systems, types and lambda-calculus. They often rely on the functional programming paradigm.

Two objectives of ICC are: (i) finding natural implicit logical characterizations of functions of various complexity classes, (ii) to design systems suitable for static verification of programs complexity. In particular, the latter goal explores characterizations that are general enough to include basic known algorithms.

Computational complexity theory aims at classifying computational problems according to their inherent difficulty. The standard way to achieve this classification consists in formalizing a precise execution model (e.g., a Turing machine) and posing explicit bounds on time and memory resources. On the other hand, Implicit Computational Complexity (ICC) aims at studying computational complexity without referring to external measuring conditions or a particular machine model, but only by considering language restrictions or logical/computational principles implying complexity properties. The area of ICC has grown out from several proposals to use logic and formal methods to provide languages for complexity-bounded computation (e.g., polynomial time, logarithmic space computation). ICC methods include, among others, linear logic, typed programming language, second order logic, term ordering. The last decades have seen the development of logical formalisms that characterize functions computable in various complexity classes (polynomial or elementary in time, logarithmic in space).

6.7.2   OVERVIEW TO ARITHMETIC HIERARCHY

Various sets and Turing machines naturally occurring in the theory of computational complexity are shown to be complete on the respective levels of the arithmetical hierarchy. Let us recall some facts and notions concerning the arithmetical hierarchy. Results saying that various assertions concerning computational complexity (e.g., some relativization of the $P = NP$ problem) are independent of formal systems like set theory are obtained as corollaries (Hajek, 1977). Investigation of computational complexity has led to various important properties of Turing machines, not previously studied in Recursion Theory. The position of some natural properties (sets) of Turing schemes in the Kleene's arithmetical hierarchy and show that they are complete on the respective levels. What is the meaning of such results? First, they elucidate the nature of the investigated properties; in particular, we know the minimal number of quantifiers necessary for the definition of our property Compare this with the fact that, in the language of arithmetic, limit of a sequence cannot be defined by less than three quantifiers Second, we exhibit new complete sets naturally occurring in the theory of computational complexity. This gives new evidence to the following observation of Rogers "Almost all arithmetical sets with intuitively simple definitions that have been studied by the above methods have proved to be $\Pi 0n$-complete or Σ_0^n-complete (for some n).

The basic definitions on arithmetical hierarchy for self-containment is stated here. The arithmetical hierarchy assigns classifications to the formulas in the language of first-order arithmetic. The most important classification assigned to a formula is thus the one with the least n, because this is enough to determine all the other classifications The Turing computable sets of natural numbers are exactly the sets at level Δ^0_1 of the arithmetical hierarchy. The recursively enumerable sets are exactly the sets at level Σ^0_1. No oracle machine is capable of solving its own halting problem (a variation of Turing's proof applies). The halting problem for a $\Delta\{0, Y\}_n$ oracle in fact sits in $\Sigma\{0, Y\}_{n+1}$.

The arithmetical hierarchy assigns classifications to the formulas in the language of first-order arithmetic. The classifications are denoted and for natural numbers n (including 0). The Greek letters here are lightface symbols, which indicates that the formulas do not contain set parameters.

If a formula is logically equivalent to a formula with only bounded quantifiers then is assigned the classifications; and the classifications are defined inductively for every natural number n using the following rules:

- If is logically equivalent to a formula of the form, where is, then is assigned the classification.
- If is logically equivalent to a formula of the form, where is, then is assigned the classification.

Also, a formula is equivalent to a formula that begins with some existential quantifiers and alternates times between series of existential and universal quantifiers; while a formula is equivalent to a formula that begins with some universal quantifiers and alternates similarly.

Because every formula is equivalent to a formula in prenex normal form, every formula with no set quantifiers is assigned at least one classification. Because redundant quantifiers can be added to any formula, once a formula is assigned the classification or it will be assigned the classifications for every m greater than n. The most important classification assigned to a formula is thus the one with the least n, because this is enough to determine all the other classifications.

A set X of natural numbers is defined by formula φ in the language of Peano arithmetic if the elements of X are exactly the numbers that satisfy φ. That is, for all natural numbers n,

$$n \in X \Leftrightarrow N \models \varphi(n),$$

where n is the numeral in the language of arithmetic corresponding to n. A set is definable in first order arithmetic if it is defined by some formula in the language of Peano arithmetic.

Each set X of natural numbers that is definable in first order arithmetic is assigned classifications of the form Σn^0, Π_n^0 and Δ_n^0, where $\mathbf{N}$ is a natural number, as follows. If X is definable by a Σn^0 formula then X is assigned the classification Σn^0. If X is definable by a Π_n^0 formula then X is assigned the classification Π_n^0. If X is both Σn^0 and Π_n^0 then X is assigned the additional classification Δ_n^0. Note that it rarely makes sense to speak of Δ_n^0 formulas; the first quantifier of a formula is either existential or universal. So a Δ_n^0 set is not defined by a Δ_n^0 formula; rather, there are both Σn^0 and Π_n^0 formulas that define the set. A parallel definition is used to define the arithmetical hierarchy on finite Cartesian powers of the natural numbers. Instead of formulas with one free variable, formulas with k free number variables are used to define the arithmetical hierarchy on sets of k-tuples of natural numbers.

The following meanings can be attached to the notation for the arithmetical hierarchy on formulas.

The subscript $\mathbb{N}$ in the symbols Σn^0 and $_n^0$ indicates the number of alternations of blocks of universal and existential number quantifiers that are used in a formula. Moreover, the outermost block is existential in Σn^0 formulas and universal in Π_n^0 formulas. The superscript 0 in the symbols indicates the type of the objects being quantified over. Type 0 objects are natural numbers, and objects of type i+1 are functions that map the set of objects of type i to the natural numbers. Quantification over higher type objects, such as functions from natural numbers to natural numbers, is described by a superscript greater than 0, as in the analytical hierarchy, Chapter 5. The superscript 0 indicates quantifiers over numbers, the superscript 1 would indicate quantification over functions from numbers to numbers (type 1 objects), the superscript 2 would correspond to quantification over functions that take a type 1 object and return a number, and so on.

The $_1^0$ sets of numbers are those definable by a formula of the form $\exists n1$, ..., $nk \exists \psi(n1, ..., nk, m)$ where ψ has only bounded quantifiers. These are exactly the recursively enumerable sets.

The set of natural numbers that are indices for Turing machines that compute total functions is Π_2^0. Intuitively, an index e falls into this set if and only if for every M is an S such that the Turing machine with index e halts on input M after S steps." A complete proof would show that the property displayed in quotes in the previous sentence is definable in the language of Peano arithmetic by a Σ_1^0 formula.

Every $_1^0$ subset of Baire space or Cantor space is an open set in the usual topology on the space. Moreover, for any such set there is a computable enumeration of Gödel numbers of basic open sets whose union is the original set. For this reason, Σ_1^0 sets are sometimes called *effectively open*. Similarly, every Π_1^0 set is closed and the sets are sometimes called *effectively closed*.

Let A be a transitive set. We say that a set $X \subseteq A$ is a Σ_1 on a A if there is a Σ_1 formula which defines X on (A, ε). $_X$ is $\Sigma_{\sim 1}$ on A if it is defined by a $\Sigma 1$ formula with parameters from A, that is if there is a Σ_1 formula θ (x, y1, ..., yn) and parameters b1...bn in A such that X ={a in A |(A, ε) $\models \theta$ [a, b1, ..., bn]}. We read X is Σ_1 parameters. The ~ notation on Δ_0 is defined the same. A set is resp. Δ_1 on A if both X and A ~ X are Σ resp. Σ_1.

That is why the admissibility definition was stated on Σ By a Σ reflection we should mean the restriction to θ which are Σ sentences of set theory. We won't be considering strict Σ reflection.

Theorem 6.31: (Arithmetical Hierarchy) An n-ary relation R is in the arithmetical hierarchy iff it is recursive or, for some m, can be expressed as {<x, ..., xn> | (Q1y1....(Q$_m$y$_m$) S(x1, ..., xn, y1, ..., ym)} where Q is either $\forall$ or $\exists$ for 1 <= I <= m, and S is an (n+m)-ary recursive relation.

Theorem 6.32: (Post) The set $\emptyset^{\{(n)\}}$ (the nth Turing jump of the empty set) is many-one complete in Σ_n^0.

The polynomial hierarchy is a "feasible resource-bounded" version of the arithmetical hierarchy in which polynomial length bounds are placed on the numbers involved (or, equivalently, polynomial time bounds are placed on the Turing machines involved). It gives a finer classification of some sets of natural numbers that are at level Δ_1^0 of the arithmetical hierarchy.

The more proof theoretic beginnings to explore bounded complexity were for example (Buss, 1985).

Definition 6.23: A bounded quantifier is a quantifier of the form $(\forall x <= t)$ or quantifiers $(\exists x <= t)$ where t is any term. A sharply bounded quantifier is a bounded quantifier of the form $(x <= |t|)$ or $(\exists x <= |t|)$ $(\forall x)$ and (x) are unbounded.

A hierarchy of bounded formulas are defined by counting alternations of quantifiers, including the sharply bounded quantifiers.

The bounded quantification from Stockmeyer (1977), Meyer and Stockmeyer (1975) to Chandra and Stockmeyer (1976) explores alternating quantifiers. Alternating Turing Machines are like nondeterministic Turing Machines, except that existential and universal quantifiers alternate. Alternation links up time and space complexities rather well, in that alternating polynomial time equals deterministic polynomial space, and alternating linear space equals deterministic exponential time. Such considerations lead to a two-person game complete in polynomial time, and other games complete in exponential time. The next theorem is considered an alternative definition for the polynomial time hierarchy. It states that a predicate belongs to certain level of the polynomial hierarchy iff it is expressible by a formula of Bounded Arithmetic of a certain complexity. We define a hierarchy of bounded formulae by counting alternations of quantifiers, ignoring the sharply bounded quantifiers.

The next theorem can be thought of as an alternative definition for the polynomial time hierarchy. It states that a predicate belongs to certain level of the polynomial hierarchy iff it is expressible by a formula of Bounded Arithmetic of a certain complexity. Let $<x>$ be a vector corresponding to $x1, \ldots, xn$.

Theorem 6.33: (Stockmeyer, Wrathall and Hudgson) Let $i >= 1$. A predicate q is in Σ_i^p; iff there is a Σ_i^p; formula φ such that for all $<n> >= 0$, $q(<n>)$ iff $N \models \varphi(<n>)$.

EXERCISES

1. Let L be a language containing a non-unary relation symbol.
1a. The set of valid first-order sentences over L is r.e. but not co-r.e.
1b. The set of first-order sentences over L that are valid over finite structures is co-r.e. but not r.e.
2. Prove Theorem 6.19: there is a generic functor defining an admissible.
3. Prove Theorem 6.23.
4. P-generic sets are admissible computable.
5. Provide a detailed proof for Lemma 6.2
6. Provide a detailed proof for Lemma 6.27.
7. Prove (a) Corollary 6.1. (b) Corollary 6.2.
8. Prove Theorem 6.26.
9. Prove Theorem 6.27.

KEYWORDS

- **complexity classes**
- **computational complexity**
- **concrete descriptive complexity**
- **diophantine definability**
- **forcing**
- **KPU**
- **models**
- **recursion**
- **rudiments**
- **urelements**

REFERENCES

Abstracts http://www.esf.org/generic/1850/Truss03101.pdf

Addisson, J. (1960). "The Theory of Hierarchies," in Proceedings of the International Congress of Logic, Methodology, and Philosophy of Science, Stanford, (1960). pp. 26–37.

ADJ-Goguen, J. A., Thatcher, J. W., Wagner, E. G., Wright, J. B. "A Junction Between Computer Science and Category Theory," (parts I and II), IBM T. J. Watson Research Center, Yorktown Heights, N. Y. Research Report, RC4526, 1973.

Admissible Models and Peano Arithmetic, ASL, March (1998). Los Angeles, CA. BSL, vol.4, no.2, June 1998.

Aho, A. V., Hopcroft, J. E., Ullman, J. D. (1974). The Design and Analysis of Computer Algorithms, Addison-Wesley, Reading, MA.

Ambos-Spies, Flieschhack, K., Huwing, H. (1984). "P-Generic Sets," Proc. 11th Colloquium on Automata Languages and Programming, Belgium, July (1984). pages 58–64.

Ambos-Spies, K., (1996). Resource bounded genericity. In Computability, Enumerability, Unsolvability (S. B. Cooper et al., eds.), London Math. Soc. Lecture Notes Series 224, Cambridge University Press.

Baker, T., Gill, J., Solovay, R. (1975). "Relativization of the P-NP Question, "SIAM Journal on Computing, 4, December 75, pp. 431–432.

Barwise, J, "Syntax and Semitics of Infinitary Languages, "Springer-Verlag Lecture Notes in Mathematics, vol. 72, (1968). Berlin-Heidelberg-NY.

Barwise, J., (1978). Handbook Mathematical Logic, (Barwise, editor), Studies in Logic and Foundations, Vol., 90, (1978). North-Holland.

Barwise, J., Robinson, A. (1970). "Completing Theories By Forcing", Annals of Mathematical Logic, vol.2, no.2, (1970). 119–142.

Barwise, K., "Implicit Definability and Compactness in Infinitary Languages," in The Syntax and Semitics of Infinitary Languages, Edited by J. Barwise, Springer-Verlag LNM, vol.72, Berlin-Heidelberg, NY.

Blum, M., "A Machine Independent Theory of the Complexity of Recursive Functions," JACM,14, no.2, pages 322–326.

Buss, S. R. (1985). The Polynomial Hierarchy and Fragments of Bounded Arithmetic (Extended Abstract) Department of Mathematics, Princeton University, STOC, 1985.

Chandra, A., Stockmeyer, L. J. Alternation, Foundations of Computer Science, 1976., 17th Annual Symposium on 25–27 Oct. 197698–108ISSN: 0272–5428Houston, TX, USA10.1109/SFCS.1976.4, IEEE.

Computability: Turing, Gödel, Church, and Beyond, MIT Press. A man who loved only numbers.

Davis, M., Matijasevic, Y., Robins, J. (1976). Hilbert's Tenth Problem Is Unlovable—Positive Aspects of A Negative Solution, "Proc. Symposium in Pure Mathematics, vol. 28, 1976.

Demillo, R. A., Lipton, R. J. (1980). "The Consistency of the P-NP and Related Problems with Fragments of Number Theory", Proc.12th ACM SIGACT STOC, 1980.

Feferman, S. Arithmetization of metamathematics in a general setting Fund. Math., 49 (1960), pp. 35–92.

Feferman, S., Kreisel, G. Persistent and invariant formulas relatives to theories of higher types, Bull Amer. Math. Soc., 72, (1966). 480–485.

Garey, M. R., Johnson, D. S. Computer and Intractability—A guide to the Theory of NP-Completeness, W. H. Freeman and Company.

Georg Cantor (1892). "Ueber eine elementare Frage der Mannigfaltigkeitslehre". Jahresbericht der Deutschen Mathematiker-Vereinigung 1890–1891, 1, 75–78 (84–87 in pdf file). (in German).

Grohe, M (2013). Descriptive Complexity, Canonisation, and Definable Graph Structure Theory December 4, 2013, Darmstsadt Mathematics technical report.

Hájek, P. (1977). Arithmetical complexity of some problems in computer science, in: J. Gruska (Ed.), Mathematical Foundations of Computer Science (1977). Lecture Notes in Computer Science, 53, Springer-Verlag, Berlin (1977), pp. 282–287.

Hartmanis, J. (1977). Relations between diagonalization, proof systems and complexity gaps (preliminary version). Proceedings 9th ACM Symp. on Theory of Computing pp. 223–227

Hartmanis, J., Hopcroft, J. E. (1976). Independence results in computer science SIGACT News, 8 (4), pp. 13–24.

Hartmanis, J., Stearns, R. (1965). On the computational complexity of algorithms. Transactions of the American Mathematical Society 117, 285–306.

Hinman, P. G. Recursion Theoretic Hierarchies, Spring-Verlag, 1980.

Hoffmann, P. (1998). The Man Who Loved Only Numbers: The Story of Paul Erdős and the Search for Mathematical Truth. New York: Hyperion, pp. 220.

Homer, S., Admissible Recursion Theory, Recursion Theory and Computational Complexity, C. I. M. E. Summer Schools Volume 79, (2011). pp. 5–28.

Immerman, N., Descriptive Complexity A Logician's Approach to Computability, Notices of AMS, October 1995.

Joseph, D., Young, P. (1980). "Independence Results in Computer Science?," Proc. 12th ACM SIGACT STOC 1980.

Keisler, H. J., "Forcing and the Omitting Types Theorem", Mathematical Association of America, Studies in Mathematics, vol. 8, M. D. Morely editor, pp. 97–133.

Keisler, H. J., (1971). Model Theory for Infinitary Logic, North Holland, Amsterdam, Lawvere, F. W., "Functorial Semantics of Algebraic Theories," Proc. National Academy of Sciences, USA, 1963.

Keisler, H. J., Forcing and the Omitting Types Theorem, Studies in Model Theory, Mathematical Association of America, 96–133, M. D. Morley editor.

Keisler, H. J., Model Theory for Infinitary Logic, North Holland, Amsterdam, 1971.

Knight, J., "Generic Expansions of Structures," JSL, 38, (1973). 561–570.

Ladner, R. E., Lynch, N. A., Selman, A. L. (1975), "A Comparison of Polynomial Time Reducibilties", TCS 1, pages 103–123, North-Holland.

Lectures. www.icm2002.org.cn/B/Schedule_Section01.htm

Mac Lane, S., (1971). Categories for The Working Mathematician, GTM Vol. 5, Springer-Verlag, NY, Heidelberg Berlin, 1971.

Makaai, M. (1981). "Admissible Sets and Infinitary Logics" Handbook Chapter A7, (Barwise, editor), Studies in Logic and Foundations, Vol. 90, 1981.

Marion, J.-Y., Analyzing the implicit complexity of programs. Information and Computation 183(1), 2–18 (2003).

Marion, J.-Y., Ramified Recurrence and Computational Complexity II: Substitution and Poly-Space. In: Bezem, M., Groote, J. F. (eds.) TLCA (1993). LNCS, vol. 664, pp. 486–500. Springer, Heidelberg (1993).

Martin Grohe1, Kreutzer, S., Siebertz, S., RWTH Aachen Universe. Characterizations of Nowhere Dense Graphs, ACM.

Martin, D. A. "Descriptive Set Theory", Handbook, Mathematical Logic, 783–815. J. Barwise, editor, (1978). (Barwise, editor), Studies in Logic and Foundations, Vol. 90, (1978). North-Holland.

Meyer, A., Stockmeyer, L. Inherent Computation Complexity of Certain Arithmetic Expressions. MIT 1976.

Nourai, C. F., "Functorial Consistency," May (1997). AMS 927, Milwaukee, Wisconsin, Special Session on Applications of Model Theory to Analysis and Topology, (1997). Abstract number 927–03–29.

Nourai, C. F.," Functorial Model Theory and Infinitary Language Categories," Proc. ASL, January (1995). San Francisco. See BSL, Vol.2, Number 4, December 1996.

Nourani, C. F., (1983). "Forcing With Universal Sentences," (1981). Proc. ASL, vol. 49, Boston. MA.

Nourani, C. F. "Dense Sets and P-NP Sets," ASL, January1985, JSL December 1985.

Nourani, C. F. "Functorial Model Computing," FMCS, UBC Mathematics Department, Vancouver, Canada, June 2005.

Nourani, C. F. (1977). "Functorial Models, Admissible Sets, and Generic Rudimentary Fragments," March (1997). Summer Logic Colloquium, Leeds, July (1997). BSL, vol. 4, no.1, March (1998). www.amsta.leeds.ac.uk/events/logic97/con.html

Nourani, C. F. (1984). "Positive Forcing and Complexity," Written at SLK, Manchester, England, July.

Nourani, C. F. (1984). "Forcing, Nonmonotonic Logic and Initial Models," Logic Colloquium, Manchester (1984).

Nourani, C. F. (1985). P=NP is +Forced, 821st Meeting of the AMS, August (1985). Laramie, Wyoming, abstracts, vol.6, no.2.

Nourani, C. F. 1985," P-NP is +Forced, 821st Meeting of the AMS, August (1985). Laramie, Wyoming, abstracts, vol.6, no.2.

Nourani, C. F. (1995). "Functorial Model Theory, Generic Functors and Sets," January 16, (1995). International Congress, Logic, Methodology, and Philosophy of Science, Florence, Italy, August 1995.

Nourani, C. F. (1996). "Functorial admissible models," Spring ASL March 97, MIT. BSL, Vol.3, 1997.

Nourani, C. F. (1996). "Functorial Model Theory and Generic Fragment Consistency Models," October, AMS-ASL, San Diego, January 1997.

Nourani, C. F. (1996). "Computable functors and generic models diagrams," December (1996). ICM, Berlin. Functorial Computability and Generic Definable Models, International Congress, Mathematicians, Berlin, August 18–27, (1998).

Nourani, C. F. (1996). "Functorial Model Theory and Generic Fragment Consistency Models," October (1996). AMS-ASL, San Diego, January 1997.

Nourani, C. F. (1997). "Functorial Consistency," May (1997). AMS 927, Milwaukee, Wisconsin, (1997). Abstract number 927–03–29.

Nourani, C. F. (1998). "Functorial Metamathematics," Maltsev Meeting, Novosibirsk, Russia, November (1998). math.nsc.ru/conference/malmeet/thesis.htm Kleene, S., Mathemtical Logic, Wiley, 1967.

Nourani, C. F. (1998). "Admissible Models and Peano Arithmetic," ASL, March (1998). Los Angeles, CA. BSL, vol.4, no.2, June 1998.

Nourani, C. F. (2002). "Functorial Models and Implicit Complexity International CONGRESS of Mathematicians ICM2002, Beijing, China, August 20–28. http://www.logic.univie.ac.at/cgi-bin/abstract/show.pl?new=706f88f57c1b9897b5204f820cc55793

Nourani, C. F. (2006). "Functorial Models, Generic Sets and Recursion Urelements" Brief Abstract, SLK Netherlands, July 2006.

Nourani, C. F., (1993). "Computation on Diagrams," Logic Colloquium, England.

Nourani, C. F., "Descriptive Computing-The Preliminary Definition," Summer Logic Colloquium, July (1996). San Sebastian Spain. See AMS April (1997). Memphis.

Nourani, C. F., "Functorial Generic Filters," July 2005.

Nourani, C. F., "Functorial Model Theory and Infinite Language Categories," September, (1994). Presented to the ASL, January (1995). San Francisco.

Nourani, C. F., "Functorial Models and Implicit Hierarchy Degrees, October (2002). Algebra and Discrete Mathematics Under the Influence of Models, 26–31 July, Hattingen-www.esf.org/euresco/03/pc03101

Nourani, C. F., "Functorial String Models," ERLOGOL-2005: Intermediate Problems of Model Theory and Universal Algebra, June 26–July 1, State Technical University/Mathematics Institute, Novosibirsk, Russia. www.nstu.ru/science/conf/erlogol-2005.phtml www.ams.org/mathcal/info/ 2005_jun26-jul1_novosibirsk.html

Nourani, C. F., "Higher Stratified Consistency and Completeness Proofs," http://www.logic.univie.ac.at/cgibin/abstract/show.pl?new=e049a2efe0c1a4b7a3ddaa11a75d8152. April 2003; SLK (2003). Helsinki August 14–20; http://www.math.helsinki.fi/logic/LC2003/abstracts/

Nourani, C. F., "Infinite Language Categories Limit Topology and Categorical Computing, Brief Overview," MFPS, Boulder Colorado, June 1996.

Nourani, C. F., (1955). "Functorial Model Theory, Generic Functors and Sets, January 16, (1995). International Congress, Logic, Methodology, and Philosophy of Science, Florence, Italy, August 1995.

Nourani, C. F., (1983). "Forcing with Universal Sentences and Initial Models," Technical Report, University of Pennsylvania, 1981–82, Annual Meeting of the Association for Symbolic Logic, December (1983). Proc. in Journal of Symbolic.

Nourani, C. F., (1984). "Forcing the P-NP Question," Extended Abstract, Written July 22–1984 at the Logic Colloquium, England.

Nourani, C. F., (1988). "R.E. Set Forcing and Polynomial Time Computability," Informal Appearance, Structure in Complexity, Ithaca, NY.

Nourani, C. F., (1994). "Functorial model theory and infinitary language categories, September 1994," Association for Symbolic Logic, San Francisco, January (1995). see Association for Symbolic Quarterly, Summer (1996). for recent abstract.

Nourani, C. F., (1995). "Functorial Model Theory and Infinite Language Categories," September (1994). Presented to the ASL, January, San Francisco.

Nourani, C. F., (2002). On Admissible Computable sets and P-genericity, written at SLK, Helsinki, 2003.

Nourani, C. F., Henkin, L., (1949). "The Completeness of First Order Functional Calculus," Journal of Symbolic Logic," vol. 14, Structure in Complexity, (1986). Berkeley, CA.

Nourani, C. F. (1997). "Functorial Models, Admissible Sets, and Generic Rudimentary Fragments, March (1997). Summer Logic Colloquium, Leeds, July (1997). BSL, vol. 4, no.1, March (1998). www.amsta.leeds.ac.uk/events/logic97/con.html

Platek, R. (1966).Foundations of recursion theory, Doctoral dissertation and Supplement, Stanford.

Rathjen, M., (1992). A Proof-Theoretic Characterization of the Primitive Recursive Set Functions, The Journal of Symbolic Logic, Vol. 57, No. 3 (Sep., 1992), pp. 954–969.

Rogers, H., (1959). The Theory of Recursive Functions and Effective Computability. MIT, Hardback 362 pages, ISBN 0262018993.

Rudnev, V. A., (1986). A universal recursive function on admissible sets, Algebra and Logic, July–August 1986. Volume 25, Issue 4, pp. 267–273.

Shoenfield, J., (1967). Mathematical Logic, Addison-Wesley.

Stockmeyer, Larry (1976). "The Polynomial-Time Hierarchy", Theoretical Computer Science, 3, 1–22.

Thather, J. W., Wagner, E. G., Wright, J. B. (1979). "Notes On Algebraic Fundamentals For Theoretical Computer Science," IBM T. J. Watson Research Center, Yorktown Heights, NY, Reprint from Foundations Computer Science III, part 2, Languages, Logic, Semantics, J. de-Bakker and J. van Leeuwen, editors, Mathematical Center Tract 109, 1979.

Turing, A. (1937). On computable numbers, with an application to the Entscheidungs problem. Proceedings of the London Mathematical Society 2, 230–265.

Wilkie, A., Paris, J. (1984), "On the scheme of induction for bounded arithmetic formulas", Logic Colloquium '84, Proc. of an ASL Conference in Manchester, England, North-Holland.

CHAPTER 7

ARITHMETIC HIERARCHY AND ENUMERATION DEGREES

CONTENTS

7.1   INTRODUCTION

This chapter is an overview to Turing Degrees, arithmetic hierarchy, and enumerability degrees towards a structure for reducibility and abstract ordinal computability notions. Arithmetic hierarchy is presented based on decidable sets, language recognition, definability degrees, Turing Jumps enumeration definability. Newer classical computability theory examines the degree structures that arise from reducibilities on the power set $\wp(\omega)$ of the natural numbers: we say that a set A is "reducible" to a set B if there is a way to "compute." membership in A from membership information about B. There are several natural formalizations of this idea, giving different reducibilities. Such a reducibility is always reflexive and transitive and thus induces a preorder on $\wp(\omega)$. The equivalence classes of sets reducible to each other are usually called "degrees," and the preorder on the sets induces a partial order on the degrees. The most commonly studied reducibility is Turing reducibility: a set A is Turing reducible to a set B if there is an algorithm that, on any input x, determines whether $x \in A$ in finitely many steps and making finitely many membership queries to B. A natural extension of Turing reducibility is enumeration reducibility: a set A is enumeration reducible to a set B if there is an algorithm to enumerate all elements of A from any enumeration of all the elements of B. Specific infinitary languages are studied followed by admissible set recursion and Kripke Platek sets.

The isomorphism types for sets are examined for the Turing degree spectrums. A standard technique is to use countable ordinals for natural numbers, and countable sets play the role of finite sets. For example, assuming that every subset of $\omega 1$ is amenable for $L\omega 1$ or $\backslash V = L$, or that $\wp(\omega) \subseteq L$. An approach for studying degrees is to embed the structure in a richer context that reveals more specifics on the enumeration degrees, introduced by (Friedberg and Rogers, 1959). This structure is developed from a weaker form of relative computability: a set A is enumeration reducible to a set B if every enumeration of the set B can be effectively transformed into an enumeration of the set A.

More structures on Turing degrees are examined based on Kallimulin pairs where we invoke KPU recursion urelements with isomorphism types towards lifts on K-pairs with direct product filters to examine degrees. This author's publications on generic set computability date back to the 1984 Summer logic colloquiums on pairing with generic sets to Peano arithmetic fragment models when on new postdoc appointments at 1984 onto KPU characterizations at the Summer logic colloquium 2003.

Here are the basic practical computability characterizations on the arithmetic hierarchy:

For $1 \leq n < \omega$, a set is Σ_n^0 if it is computably enumerable relative to $\varnothing^{(n-1)}$. A set is Π_n^0 if the complement is Σ_n^0. A set is Δ_n^0 if it is both Σ_n^0 and Π_n^0. In everyday computability sense:

$\Delta_1^0 = \{$decidable sets$\}$

$\Sigma_1^0 = \{$Recognizable Sets$\}$, for example, recognizable languages.

$\Sigma_{n+1}^0 = \{$sets recognizable in some $B \in \Sigma_n^0\}$ based on Oracle Σ_{n+1}^0 reducibility: Say that A is decidable in B if there is an oracle TM M with oracle B that decides A. A "Turing Reduces" to B.

$\Sigma_{n+1}^0 = \{$sets decidable in some $B \in \Sigma_n^0\}$

$\Delta_{n+1}^0 = \{$sets decidable in some $B \in \Sigma_n^0\}$

$\Pi_n^0 = \{$complements of sets in $\Sigma_n^0\}$

7.2   TURING DEGREES AND ISOMORPHISM TYPES

Alan Turing's 1938 thesis was the beginning to a legacy to investigate hierarchies along ordinals that lead to a characterization for computation know now as Turning machine.

Theorem 7.1: (Turing, 1936) There exists a Turing machine M_U the Universal Turing Machine, that given input (i, x) simulates the i-th Turing machine with input x: $M_U (i, x) = \varphi_i(x)$.

There are alternative definitions; {n: the nth Turing machine relative to X halts} {φ: φ is provable from PA and the diagram of X}. The alternative definitions are equicomputable, in fact recursively isomorphic. (Exercises). The simplest definition for the arithmetic hierarchy is: "The arithmetically definable subsets of the natural numbers are those generated by iterating the Turing jump and closing under relative definability". The arithmetically definable sets appear in a hierarchy based on counting the number of applications of the jump.

Notation: Write $X \geq_T Y$ when Y is computable from X. Let $\rightleftharpoons$ denote "if and only if."

Theorem 7.2: (Diagonalization) For all X,' X ' $>_T X$

Corollary 7.1: For all X and Y, if $X \geq_T Y$ then $X' \geq_T Y.'$

Therefore, the function $X \mapsto X'$ is a $>_T$ -increasing and $\geq_T$ is an order-preserving function.

Definition 7.1: Consider the relation c.e. in between Turing degrees defined by: x is c.e. in u if there are sets $X \in x$ and $U \in u$, such that X is c.e. in U.

Definition 7.2: The Turing degree of the set A is the equivalence class containing A: $dT(A) = \{B|B\equiv_T A\}$. $dT(A) \leq dT(B) \rightleftharpoons A \leq_T B$

Theorem 7.3: (Slaman) $A \leq e \; B \rightleftharpoons (\forall X - total)(B \leq e \; X \Rightarrow A \leq e \; X)$.

Proof: ($\Rightarrow$) From the transitivity of $\leq e$.

($\Leftarrow$) Suppose that $A \leq e \; B$. Then we construct a B-regular enumeration f, s.t. $A \leq e \; \langle f \rangle$, but $\langle f \rangle$ is total and $B \leq e \; \langle f \rangle$. Contradiction.

The Turing degree spectrum of a countable structure A is the set of all Turing degrees of isomorphic copies of A. The Turing degree of the isomorphism type of A, if it exists, is the least Turing degree in its degree spectrum. We show that there are elements with isomorphism types of arbitrary Turing degrees in each of the following classes: countable fields, rings, and torsion-free Abelian groups of any finite rank. We also show that there are structures in each of these classes the isomorphism types of which do not have Turing degrees.

Enumeration reducibility is an interesting concept that has an abstract recursion presentation as follows.

Enumeration reducibility is denoted by $A \leq e \; B$ if there exists an effective procedure that, given any enumeration of B, computes an enumeration A. Let $W\alpha$ be the c.e. set with index α

Definition 7.3: The operator $\Gamma: 2^N \rightarrow 2^N$ is an enumeration operator,

if: $x \in \Gamma(A) \rightleftharpoons (\exists D \subseteq A)(x \in \Gamma(D) \; \& \; D\text{-finite})$ (Γ is compact),

There is a total computable function h, s.t. $\Gamma(W\alpha) = Wh_{(\alpha)}$ (Γ is effective).

An important goal of computable algebra is to understand how algebraic properties of structures interact with their computability theoretic properties. While in algebra and model theory isomorphic structures are often identified, in computable model theory, they can have very different algorithmic properties. Here, we study Turing degrees of isomorphism types of algebraic structures from some well-known classes. We consider only countable structures for computable (usually finite) languages. The universe A of an infinite countable structure A can be identified with the set ω of all natural numbers. Furthermore, we often use the same symbol for the structure and its universe (for the definition of a language and a structure).

Let $\{Aj, j \in \omega\}$ be a sequence of structures contained in a structure B. Then, by $\Pi j \in \omega\ Aj$, is considered the smallest substructure of B containing Aj for all j. This is not the usual definition of a product, rather it corresponds to a weak direct product, and, more importantly, it expresses a rather intuitive way of putting structures together. More specifically, we will be looking at products of number fields and function fields and at products of rings contained in a number field. In the case of fields, we fix an algebraic closure of Q, a rational function field over a finite field of characteristic $p > 0$ or over Q, as required, and we can set B to be this algebraic closure. In the case of subrings of a number field, the number field itself is a natural choice for B.

Example Specifics are sets non-recursive sets to consider are: generic, random, diagonally non-recursive, solutions to Post's problem, sets of minimal Turing degree. in the Baire Category, random sets in measure theory. Let us say that a set X is Turing reducible to (computable in) a set Y, in symbols $X \leq_T Y$, if X can be computed by an algorithm with Y in its oracle. Turing reducibility is the more basic notion, in terms of which Turing degree is defined. We say that the sets X and Y are Turing equivalent, or have the same Turing degree, if $X \leq_T Y$ and $Y \leq_T X$. We use $\equiv_T$ for Turing equivalence. We also write $\deg(X) = \deg(Y)$ or $Y \in \deg(X)$ instead of $X \equiv_T Y$. (Rogers, 1967).

The complexity measure of structures here are identified with their atomic diagrams. The atomic diagram of a structure A is the set of all quantifier-free sentences in the language of A expanded by adding a constant symbol for every $a \in A$, which are true in A. The Turing degree of A, $\deg(A)$, is the Turing degree of the atomic diagram of A. Hence, A is computable (recursive) if and only if $\deg(A) = 0$. Alternative definition is to call a structure computable if it is only isomorphic to a computable one. We also say that a set or a procedure is computable (effective), relative to B, sometimes said computable in B, if it is computable relative to the atomic diagram of B. The Turing degree spectrum of a countable structure A is $DgSp(A) = \{\deg(B): B \sim= A\}$.

A countable structure A is automorphically trivial if there is a finite subset X of the domain A such that every permutation of A which restricts to the identity on X is an automorphism of A. If a structure A is automorphically trivial, then all isomorphic copies of A have the same Turing degree. It was shown in Harizonav and Miller (2007) that if the language is finite, then that degree must be 0. On the other hand, Knight (1986) proved that for an automorphically non-trivial structure A, $DgSp(A)$ is closed upwards, that is, if $b \in DgSp(A)$ and $d > b$, then $d \in DgSp(A)$. Hirschfeldt et al. (2002) further state that for every automorphically non-trivial structure A, the degree spectrums of specific additional structures there, coincides with $DgSp(A)$. Since the Turing degree of a structure is not invariant under isomorphism. From 1981, we have the following complexity measures of the isomorphism type of a structure (Richter, 1981).

Definition 7.4: (i) The Turing degree of the isomorphism type of A, if it exists, is the least Turing degree in $DgSp(A)$. (ii) Let α be a computable ordinal. The α-th jump degree of a structure A is, if it exists, the least Turing degree in $\{deg(B)(\alpha): B \sim= A\}$.

The 0-th jump degree of A coincides with the notion of the degree of the isomorphism type of A.

Theorem 7.3: (Richter, 1981) Let T be a theory in a finite language L such that there is a computable sequence A0, A1, A2, ... of finite structures for L, which are pairwise non-embeddable. Assume that for every set $X \subseteq \omega$, there is a countable model AX of T such that $A_X \leq_T X$, Aj is embeddable in $A_X \Leftrightarrow j \in X$. Then, for every Turing degree d, there is a countable model of T whose isomorphism type has degree d.

Theorem 7.4: (Richter, 1981) Let C be a class of countable structures in a finite language L, closed under isomorphisms. Assume that there is a computable sequence $\{Ai, i \in \omega\}$ of computable (possibly infinite) structures in C satisfying the following conditions.

(a) There exists a finitely generated structure $A \in C$ such that for all i $\in \omega$, $A \subset Ai$.

(b) For any $X \subseteq \omega$, there is a structure A_X in C such that $A \subset A_X$ and $A_X \leq_T X$, and for every $i \in \omega$, $\sigma: Ai \hookrightarrow A_X$, $\sigma|A = id$ if and only if $i \in X$.

(c) Suppose that any A_X is isomorphic to some structure B under isomorphism $\tau: A_X \leftrightarrow B$. Consider a pair of structures Ai, Aj such that exactly one of them embeds in B via Σ with $(\tau -1 \circ \sigma)|A = id$. Then, there is a uniformly effective procedure with oracle B for deciding which of the two structures embeds in B.

Thus, for every Turing degree d, there is a structure in C with isomorphism type of degree d.

The hierarchy of definability applies to calibrate to address whether there is an object, such as a real number, which can be produced using methods, principles, techniques of Type A and which satisfies Property B, or whether principles of Type A and be used to settle questions of Type B. The recursion theoretic view is not always reconciled with the definability on sets taken for granted. Recursion theoretic view is that at least with respect to basic known principles, both areas can be formulated and settled by directly considering the nature of "Type A," as Turing did with the nature of computation, without complete reliance on the specific formalization of theories.

7.3 ARITHMETIC HIERARCHY ON INFINITARY LANGUAGES

A preview to arithmetic hierarchy was briefed in Chapter 6. Here, we begin to state specific characterizations based on specific infinitary languages. Kleene and Mostowski independently defined what is now called the arithmetical hierarchy. Davis and Mostowski independently defined what is now called the hyperarithmetical hierarchy, extending the arithmetical hierarchy through the "computable" ordinals. To address examples with language recognition problems, let ATM = { (M, w) | M is a TM that accepts string w }. A technique for showing that a language L is undecidable by showing that if L is decidable, then so is ATM. One reduces ATM to the language $L_{ATM} \leq L$.

Let K be a subclass of Mod($\mathcal{L}$) which is closed under isomorphism. Vaught showed that K is $\Sigma\alpha$ (respectively, $\Pi\alpha$) in the Borel hierarchy iff K is axiomatized by an infinitary $\Sigma\alpha$ (respectively, $\Pi\alpha$) sentence. We prove a generalization of Vaught's theorem for the effective Borel hierarchy, for example, the Borel sets formed by union and complementation over c.e. sets. This result says that we can axiomatize an effective $\Sigma\alpha$ or effective $\Pi\alpha$ Borel set with a computable infinitary sentence of the same complexity. This result yields an alternative proof of Vaught's theorem via forcing. We also get a version of the pull-back theorem from Knight et al. (2007) which says, if Φ is a Turing computable embedding of K$\subseteq$Mod($\mathcal{L}$) into K' $\subseteq$Mod($\mathcal{L}'$), then for any computable infinitary sentence φ in the language

$\mathcal{L}$, we can find a computable infinitary sentence φ * in $\mathcal{L}'$ such that for all A∈K, A⊨φ * iff Φ(A)⊨φ, where φ * has the same complexity as φ.

Hyperarithmetical hierarchy, extending the arithmetical hierarchy. For a computable ordinal α ≥ ω, a set is Σ^0_α if it is c.e. relative to $\varnothing(\alpha)$. A set is Π0α if the complement is Σ^0_α. A set is Δ^0_α if it is both Σ^0_α and Π^0_α. There are only countably many computable ordinals, and there are only countably many hyperarithmetical sets. With newer insights since the preceding chapters (Johnson Carson et.al., 2011) and consider computable enumerability as a concrete glance.

The following section examines the arithmetical hierarchy in the setting of ω_1 computability. Note that here the computability notions deploy fragments on infinitary logic with only finite variables. An example proposition for the specific infinitary logic from the above authors is the following with the ordinary version being from Ash and Knight (2000).

Proposition 7.1: Let A be an L-structure, let φ(x) be a computable infinitary L-formula. If φ(x) is computable Σα (or computable Πα), for α≥ 1, then the relation defined by φ(x) in A is Σ^0_α (or Π^0_α) relative to A, uniformly in φ.

Proof: (Carson et al., 2012) If φ(x) is a quantifier-free formula of $L_{\omega1,\omega}$, then to decide, for a countable tuple a in A whether A ⊨ φ(a), we find a countable substructure A′ containing a and check whether A′ ⊨ φ(a). The relation defined by φ(x) is computable relative to A. Suppose φ(x) is computable Σ_1, the disjunction of a c.e. set of formulas (∃ui)ψi(x, ui), where ψi is a quantifier-free formula of $L_{\omega1\omega}$. For a in A, we have A ⊨ φ(a) iff there exist i and a tuple bi such that ψi(x, bi). The relation defined by φ(x) is $\Sigma_{1\_}$ definable in (Lω1, A), so it is c.e. relative to A. If φ(x) is computable Π1, the conjunction of a c.e. set of formulas (∀ui)ψi(x, ui), we can see that the relation defined by φ(x) is the complement of a relation c.e. relative to A.

We proceed by induction onα> 1. Let φ(x) be computable Σα, the disjunction of a c.e. set of formulas (∃u)ψi(ui, x), where each ψi is computable Πβ for some 1 ≤ β < α. The formula (∃u) ψi(ui, x) is computable Σβi for some βi < α. By the Induction Hypothesis, the relation defined by ψi(ui, x) is $\Pi^0_{\beta i}$ relative to A, uniformly. Then the relation defined by (∃ui)ψi(x, ui) is the union over the possible i and bi of the relations b ∈ A → A ⊨ ψi(x, bi). This is $\Sigma^0\alpha$.

If $\varphi(x)$ is computable $\Pi\alpha$, we consider $neg(\varphi)$, which has the dual form and is logically equivalent to the negation of φ. Now, $neg(\varphi)$ is computable $\Sigma\alpha$, so the relation that it defines is $\Sigma^0\alpha$ relative to A. The relation defined by φ itself is the complement of the relation defined by $neg(\varphi)$, so it is $\Pi^0\alpha$.

At this point a forward to KPU on the following sections might be helpful for further insights.

7.3.1 COMPUTABILITY, HIERARCHY, AND INFINITARY LANGUAGES

Definition 7.5: The arithmetically definable subsets of the natural numbers are those generated by iterating the Turing jump and closing under relative definability.

Here are two example definitions for computable enumerability from Johnson Carson et al. (2011). Suppose R is a relation of countable arity α. R is computably enumerable if the set of ordinal codes for sequences in R is definable by a Σ_1 formula in $(L_{\omega1}, \in)$. R is computable if it is both c.e. and co-c.e. Here we further examine the specific setting on the arithmetic hierarchy in the setting of $L_{\omega1}$ computability from the above since Greenberg and Knight (2011).

Definition 7.6: Suppose R is a relation of countable arity α. R is computably enumerable if the set of ordinal codes for sequences in R is definable by a $\Sigma1$ formula in it is both c.e. and co-c.e. $(L_{\omega1}, \in)$

Proposition 7.2: R is computable if it is both c.e. and co-c.e. formula in $(L_{\omega1}, \in)$.

Proof: (Exercises).

Assuming that $P(\omega) \subseteq L_{\omega1}$, Gödel's bases give us a computable 1–1 function g from the countable ordinals onto $L\omega1$, such that the relation $g(\alpha) \in g(\beta)$ is computable.

So, computing in $\omega1$ is essentially the same as computing in $L_{\omega1}$. As in the standard setting, we have a c.e. set of codes for $\Sigma1$ definitions. We write $W\alpha$ for the c.e. set with index α. Here is a definition of the arithmetical hierarchy that resembles the definition of the effective Borel hierarchy (Vanden Boom, 2007) Let R be a relation. R is Σ_0 and Π_0 if it is computable. R is Σ^0_1 if it is c.e.; R is Π^0_1 if the complementary relation, $\neg$ R, is c.e.

For countable $\alpha > 1$, R is Σ^0_α if it is a c.e. union of relations, each of which is Π^0_β for some $\beta<\alpha$; R is Π^0_α if $\neg$R is Σ^0_α.

Let us examine the Jump operation at the $L_{\omega1}$ setting to have a basic feel for the following sections.

Definition 7.7: Define the halting set as $K = \{\alpha: \alpha \in W\alpha\}$. For an arbitrary set X, $X' = \{\alpha: \alpha \in W\alpha X\}$. $X(0) = X$. $X(\alpha+1) = (X(\alpha))'$.

Write Δ^0_n for $\varnothing n-1$ for $1 \leq n < \omega$. Write Δ^0_α for $\varnothing \alpha$ for $\alpha \geq \omega$.

Here is the definition for arithmetic hierarchy for $L_{\omega1}$ from Johnson et al. (2011) that resembles the Borel hierarchy.

Definition 7.8: Let R be a relation. R is Σ_0 and Π_0 if it is computable. R is Σ^0_1 if it is c.e.; R is Π^0_1 if the complementary relation, $\neg$ R, is c.e.

For countable $\alpha > 1$, R is Σ^0_α if it is a c.e. union of relations, each of which is $\Pi0\beta$ for some $\beta<\alpha$; R is Π^0_α if $\neg$R is Σ^0_α. This definition does not rely on oracles: membership of an element into a Σ^0_α set occurs if and only if that element is a member of one of the lower $\Pi0\beta$ sets. So membership into a Σ^0_α set uses in formation from a single lower level. There are alternate definitions resembling the hyperarithmetic hierarchy that relies on oracles Δ^0_α to get information from all lower levels simultaneous.

There are alternate definitions but the membership questions are not the same. For example, on the above definition, membership of an element into a Σ^0_α set occurs if and only if that element is a member of one of the lower Π^0_β sets. So membership into a Σ^0_α set uses information from a single lower level set. With alternate definitions membership of an element into a Σ^0_α may use a Δ^0_α oracle to get information from all lower levels.

Definition 7.9: A standard Borel space is a Polish space equipped just with its σ-algebra of Borel sets.

Any Borel subset of a Polish space X. The space F(X) of closed subsets of a Polish space X.

Following the above authors Carson et.al. let us consider computability in the setting where countable ordinals play the role of natural numbers, and countable sets play the role of finite sets. It will be beneficial for us to assume that every subset of $\omega1$ is amenable for $L_{\omega1}$. So we assume V = L, or at least $P(\omega) \subseteq L$. The reason for this assumption is that, since $L\omega1$ is the domain for our model of computation, it will be important for our purposes that for all $A \subseteq L_{\omega1}$, $L_{\omega1}$ will be closed under the function $x \rightarrow A \cap x$.

In what follows we review some basic definitions and known results towards the arithmetical hierarchy in two different ways, one resembling the standard definition of the hyperarithmetical hierarchy, and the other resembling the definition of the effective Borel hierarchy. From Johnson et al. (2012) we glimpse on an important result that a relation R on a computable structure A is relatively intrinsically $\Sigma\alpha$ if and only if it is defined in A by a computable $\Sigma\alpha$ formula. There are two versions of the result, one for each set of definitions. Below, we first say what it means for a set or relation to be computably enumerable. We then define the computable sets and relations, and the computable functions.

Definition 7.10: A relation $R \subseteq (L\omega1)n$ is computably enumerable, or c.e., if it is defined in $L\omega1$ by a $\Sigma1$-formula $\varphi(c, x)$, with finitely many parameters—the formula is finitary, with only existential and bounded quantifiers.

A relation $R \subseteq (L\omega1)n$ is computable if R and the complementary relation $\neg R$ are both computably enumerable. A (partial) function f: $(L\omega1)n \to L\omega1$ is computable if its graph is c.e. When we work with these definitions, we soon see that computations involve countable ordinal steps. Computable functions are defined by recursion on ordinals—the $\Sigma1$ definition for a function f says that there is a sequence of steps leading to the value of f at a given ordinal α. Thus, it might be appropriate to use the term "recursive" instead of "computable" in this setting. There is a conjecture, due to D. A. Martin, mathematically codifying the view that all notions of relative definability extending relative computability appear in the logical hierarchy based on first order quantification over the finite sets.

7.3.2 COMPUTABILITY ON INFINITARY LANGUAGES

$\varphi(x)$ is computable Σ_0 and computable Π_0 if it is a quantifier-free formula of $L_{\omega1,\omega}$. For $\alpha > 0$, $\varphi(x)$ is computable $\Sigma\alpha$ if $\varphi \equiv (\exists u)\psi i(u, x)$, c.e. where each ψi is a countable conjunction of formulas, each computable $\Pi\beta$ for some $\beta < \alpha$.

$\varphi(x)$ is computable $\Pi\alpha$ if $\varphi \equiv (\forall u)\psi i(u, x)$, where each c.e. ψi is a countable disjunction of formulas, each computable Σ_β for some $\beta < \alpha$.)

Lemma 7.1: (L-arithmetic computability)

Let A be an L-structure, and let $\varphi(x) \Sigma\alpha$ (be a computable $\Pi\alpha$ relative to A L-formula. Then the relation defined by $\varphi(x)$ in A is $\Sigma0\alpha$ ($\Pi0\alpha$).

Here is an example definition for computability for infinitary $L_{\omega 1}$ formulas from (Calvert, Carson, Knight et.al. 2011):

$\varphi(x)$ is computable $\Sigma 0$ and computable $\Pi 0$ if it is a quantifier-free formula of $L_{\omega 1, \omega}$.

For $\alpha > 0$, $\varphi(x)$ is computable $\Sigma\alpha$ if $\varphi \equiv (\exists u)\psi i(u, x)$, c.e.

where each ψi is a countable conjunction of formulas, each computable Π_β for some $\beta < \alpha$. $\varphi(x)$ is computable $\Pi\alpha$ if $\varphi \equiv (\forall u)\psi i(u, x)$, where each c.e. ψi is a countable disjunction of formulas, each computable Σ_β for some $\beta < \alpha$.

Proposition 7.3: Let A be an L-structure, and let $\varphi(x)$ be a computable $\Sigma\alpha$ (computable $\Pi\alpha$) L-formula. Then the relation defined by $\varphi(x)$ in A is $\Sigma^0\alpha$ ($\Pi^0\alpha$) relative to A.

Proposition 7.4: (Kleene-Post) $A \leq T\ B \Leftrightarrow A \leq$ c.e. B & $A^\wedge \leq$ c.e. B: $\wedge$ is complementation.

Proposition 7.5: Let D_T be the set of all Turing degrees. $(D_T, \leq)$ is a partial order.

Proposition 7.6: $A \leq T\ B \Rightarrow A \leq$ c.e. B.

Proof: (Exercises)

7.3.3 ENUMERATION DEGREES

The main focus in degree theory, established as one of the core areas in Computability Theory, is to understand a mathematical structure, which arises as a formal way of classifying the computational strength of an object. The most studied examples of such structures are that of the Turing degrees, DT, based on the notion of Turing reducibility, as well as its local substructures, of the Turing degrees reducible to the first jump of the least degree, DT ($\leq 0T'$), and of the computably enumerable degrees, R. In investigating such a mathematical structure among the main question that we ask is: how complex is this structure. The complexity of a structure can be inspected from many different aspects: how rich is it algebraically; how complicated is its theory; what sets are definable in it; does it have nontrivial automorphisms.

The question about definability, in particular, is interrelated with all of the other questions, and can be seen as a key to understanding the natural concepts that are approximated by the corresponding mathematical formalism. Research of the Turing degrees has been successful in providing a variety of results on definability. For the global theory of the Turing degrees, among the most notable results is the definability of the jump operator by Salman and Shore (1999). The method used in the proof of this result, as well as many other definability results in DT, leads Salman and Woodin to conjecture that every definable set in second order arithmetic is definable in DT. Another approach for understanding a structure, often used in mathematics, is to place this structure in a richer context, a context which would reveal new hidden relationships.

The most promising candidate for such a larger context is the structure of the enumeration degrees, introduced by Friedberg and Rogers (1959). This structure is induced by a weaker form of relative computability: a set A is enumeration reducible to a set B if every enumeration of the set B can be effectively transformed into an enumeration of the set A. The degree structure is usually an upper semilattice. The induced structure of the enumeration degrees, De, is an upper-semilattice with jump operation and least element. The Turing degrees can be embedded in the enumeration degrees via the standard embedding ι which maps the Turing degree of a set A to the enumeration degree of $A \oplus A$.

This embedding preserves the order, the least upper bound and the jump operation. The range of ι is therefore a substructure of De, which is isomorphic to D_T. Definability in the enumeration degrees has had its successes as well. Kalimullin has shown that the enumeration jump is definable in De. The method used in his proof is significantly less complex than that used to prove the corresponding result in the Turing degrees. The definition of the enumeration jump is based on the first order definability of a relativized version of the notion of a Kalimullin pair, K-pair.

7.3.4 ENUMERATION DEFINABILITY AND TURING JUMPS

The arithmetically definable subsets of the natural numbers are those generated by iterating the Turing jump and closing under relative definability.

The arithmetically definable sets appear in a natural hierarchy based on counting the number of applications of the jump.

Theorem 7.5: (Salman and Steel) Suppose that F is a Borel function, which is increasing and order preserving. Then, there is a B such that for all $X \geq_T B$, one of the following conditions holds.

F(X) is equicomputable with X(k).

F(X) can compute Xω the first order theory of arithmetic relative to X.

The study of degree structures has been one of the central themes in computability theory. Although the main focus has been on the structure of the Turing degrees and its local substructure, of the degrees below the first jump of the least degree, significant work has been done to examine the properties of an extension of the Turing degrees, the structure of the enumeration degrees. The following in view of nonrecursive nature of compactness arguments and in infinitary combinatorics is an interesting development.

Enumeration reducibility introduced by Friedberg and Rogers arises as a way to compare the computational strength of the positive information contained in sets of natural numbers. A set A is enumeration reducible to a set B if given any enumeration of the set B, one can effectively compute an enumeration of the set A. The induced structure of the enumeration degrees De is an upper semilattice with least element and jump operation. As we mention above, this structure can be viewed as an extension of the structure of the Turing degrees, due to an embedding $\iota: D_T \rightarrow D_e$ which preserves the order, the least upper bound and the jump operation.

Let {Aj, j ∈ ω} be a sequence of structures contained in a structure B. Then, by Πj∈ω Aj, we mean the smallest substructure of B containing Aj for all j.

This is not the usual definition of a product, but in the examples in this paper, it corresponds to the usual weak direct product, and, more importantly, it expresses a rather intuitive way of putting structures together. More specifically, we will be looking at products of number fields and function fields and at products of rings contained in a number field. In the case of fields, we fix an algebraic closure of Q, a rational function field over a finite field of characteristic p > 0 or over Q, as required, and we can set B to be this algebraic closure. In the case of subrings of a number field, the number field itself is a natural choice for B.

7.3.5 *AUTOMORPHISMS AND ENUMERATION DEGREES*

We say that a set X is Turing reducible to (computable in) a set Y in symbols $X \leq_T Y$, if X can be computed by an algorithm with Y in its oracle. Turing reducibility is the more basic notion, in terms of which Turing degree is defined. We say that the sets X and Y are Turing equivalent, or have the same Turing degree, if $X \leq_T Y$ and $Y \leq_T X$. We use $\equiv_T$ for Turing equivalence. We also write $\deg(X) = \deg(Y)$ or $Y \in \deg(X)$ instead of $X \equiv T\ Y$. The local structure of the enumeration degrees, consisting of all degrees below the first enumeration jump of the least enumeration degree, Ge, can therefore in turn be seen as an extension of the local structure of the Turing degrees.

The two structures, D_T and De, as well as their local substructures, are closely related in algebraic properties, definability, and techniques to study them. Therefore, insight is gained from proofs in one of the structures, towards the similar properties of the other.

Definition 7.11: A set A is enumeration reducible ($\leq$e) to a set B if there is a c.e. set Φ such that: $A = \Phi(B) = \{n \mid \exists u(\langle n, u \rangle \in \Phi \ \& \ Du \subseteq B)\}$, where Du denotes the finite set with code u under the standard coding of finite sets. We will refer to the c.e. set Φ as an enumeration operator.

A set A is enumeration equivalent ($\equiv$e) to a set B if $A \leq$e B and B $\leq$e A. The equivalence class of A under the relation $\equiv$e is the enumeration degree de(A) of A. The structure of the enumeration degrees $\langle De, \leq \rangle$ is the class of all enumeration degrees with relation $\leq$ defined by de(A) $\leq$ de(B) if and only if A $\leq$e B. It has a least element 0e = de($\varnothing$), the set of all c.e. sets. We can define a least upper bound operation, by setting de(A) $\vee$ de(B) = de(A $\oplus$ B). The enumeration jump of a set A is defined by (Cooper 1989).

Definition 7.12: The enumeration jump of a set A is denoted by Je(A) and is defined as KA $\oplus$ $\overline{KA}$, where KA = $\{\langle e, x \rangle \mid x \in \Phi e\ (A)\}$. The enumeration jump of the enumeration degree of a set A is de(A)$'$ = de(Je(A)).

A proof technique which arises from the study of the structure of the enumeration degrees is the use of the following notion. Definability in the enumeration degrees has had its successes as well. Kalimullin (2003) has shown that the enumeration jump is definable in De. The method used in his proof is significantly less complex than that used to prove the corresponding result in the Turing degrees. The definition of the enumeration jump is closer to the much-sought natural definition, see Shore, and is based on the first order definability of a relativized version of the notion of a Kalimullin pair, K-pair. Here we will give an alternative proof of the definability of the enumeration jump, which does not use relativization and we see as more natural in a sense that will be made precise.

Let A^ denote the set A's complement.

Definition 7.12a: (Kalimullin, 2003) A pair of sets of natural numbers A and B is a K-pair over a set U if there is a set W ≤e U such that: A×B⊆W &A^×B^⊆W^.

The notion of a K-pair over U, originally known as a U-e-ideal, was introduced and used by Kalimullin (2003) to prove the definability of the jump operation in the global structure De. Kalimullin (2003) proves that the property of being a K-pair over U is degree theoretic and first order definable in the global structure De. A pair of sets A and B form a K-pair over a set U if and only if their degrees a = de(A) and b = de(B) and u = de(U) satisfy the property:

$$K(a, b, u) \rightleftharpoons \forall x[(a \lor x \lor u) \land (b \lor x \lor u) = x \lor u].$$

A pair of enumeration degrees is called a K-pair over a degree u if they contain representatives which form a K-pair over a representative of u in the sense of definitions above.

Definition 7.13: Say that a set of natural numbers A is semi-recursive if there is a total computable selector function sA: N × N → N, such that for any x, y ∈ N, sA(x, y)∈{x, y} and whenever {x, y}∩A≠∅, sA(x, y)∈A.

In terms of structure, the enumeration degrees of a semi-recursive A, and its complement behave as a minimal pair in a very strong sense. Arslanov et al. (2003) showed that for every set of natural numbers X: (de(A) ∨ de(X)) ∘ (de(A) ∨ de(X)) = de(X). This statement cannot be reversed. In fact the class of enumeration degrees for which the statement can be reversed is on K-pairs.

Kalimullin (2003) has shown that the enumeration degrees of K-pairs are precisely the degrees which satisfy the strong minimal-pair property. He shows that the property of being a K-pair is degree theoretic and first order definable in the global structure. A pair of sets {A, B} is a K-pair if and only if ∀x ∈ De[x = (x ∨ de(A)) ∧ (x ∨ de(B))]. Thus, we can lift the notion of a K-pair to the enumeration degrees.

A pair of enumeration degrees a and b shall be called a K-pair if every member of a forms a K-pair with every member of b. By K(a, b) we will denote the first order formula, which is true of a and b if and only if they form a K-pair. K-pairs have been proven useful for coding structures in the local structure of the enumeration degree. It has been shown for instance, that using countable K-systems, systems of nonzero e-degrees such that every pair of distinct degrees forms a K-pair, that every countable distributive semi-lattice can be embedded below every nonzero Δ^0_2 enumeration degree. Example application is the following based on the Turing degrees.

The first order definability of the total enumeration degrees would then follow, if it were true that maximality is the additional structural property needed to capture K-pairs of the form $\{A, A^\wedge\}$. If this were true than we can further argue that the definition of the enumeration jump as follows:

Definition 7.14: Consider the relation c.e. in between Turing degrees defined by: x is c.e. in u if there are sets $X \in x$ and $U \in u$, such that X is c.e. in U.

Definability, automorphisms and enumeration degrees were more recent times examined on Ganchev and Soskova (2014). The enumeration degrees are an upper semi-lattice with a least element and jump operation. They are based on a positive reducibility between sets of natural numbers, enumeration reducibility, introduced by Friedberg and Rogers in 1959. The Turing degrees have a natural isomorphic copy in the structure of the enumeration degrees, namely the substructure of the total enumeration degrees.

With partial functions on the set of the natural numbers N, let $\{\varphi(n)\}$ i$\in\omega$, } be the listings of the Turing computable functions on n arguments. Here i is the code of the Turing machine Mi which computes $\varphi(n)$. The Turing jump for a set based on definability can be defined as follows:

Definition 7.15: The Turing jump of the set A: $A' = KA = \{x \mid x \in dom(\varphi Ax)\}$.

Proposition 7.7: Let x and u be Turing degrees such that u is nonzero. Then x is c.e. in u if and only if there is a K-pair $\{A, A\}$ such that $de(A) \le_e \iota(u)$ and $\iota(x) = de(A) \vee de(A^\wedge)$.

Proof: (Ganchev and Saskova, 2014) Suppose that x is c.e. in u. Let $X \in x$ and $U \in u$ be sets, such that X is c.e. in U. X is c.e. in U if and only if $X \le_e U \oplus U$. Consider the K-pair $\{LX, RX\}$. By Proposition 7.7, $LX \le_e X \le_e U \oplus U$ and $LX \oplus RX \equiv_e X \oplus X$. Thus, $de(LX) \le de(U \oplus U) = \iota(u)$ and $de(LX) \oplus de(RX) = de(X \oplus X) = \iota(x)$. Suppose $\iota(x) = de(A) \vee de(A)$ for some K-pair $\{A, A\}$ such that $de(A) \le \iota(u)$. Again let $X \in x$ and $U \in u$. Then $A \le_e U \oplus U$ and hence A is c.e. in U. On the other hand $A \oplus A \equiv_e X \oplus X$, hence $A \equiv_T X$. Thus, x is c.e. in u.

A long-standing question of Rogers is whether the substructure of the total enumeration degrees has a natural first order definition. The first advancement towards an answer to this question was made by Kalimullin (2003). A special class of pairs of enumeration degrees, K-pairs, were discovered. He showed that this class has a natural first order definition in De. Building on this result, he proved the first order definability of the enumeration jump operator and consequently obtained a first order definition of the total enumeration degrees above $0'e$. Ganchev and Soskova (2014) showed that when we restrict ourselves to the local structure of the enumeration degrees bounded by $0'e$, the class of K-pairs is still first order definable. In Saskove and Saskove (2009) the same author groups investigated maximal K-pairs and showed that within the local structure the total enumeration degrees are first order definable as the least upper bounds of maximal K-pairs.

The question of the global definability of the total enumeration degrees is explored with certain consequences, regarding the automorphism problem in both degree structures. Soskova (2014) explores definability, automorphisms and enumeration degrees. The sense on what the Dt coding parameters have to do with the Σ-compact Ulremenet definable descriptive sets (author's brief at VSL 2014 to Soskova). There are automorphism bases there that can be explored for lifting on K pairs.

Further. They are based on a positive reducibility between sets of natural numbers. The sense on what the Dt coding parameters have to do with the Σ-compact Ulremenet definable descriptive sets (author's brief at VSL 2014 to Soskova). There are automorphism bases there that can be explored for lifting on K pairs. The Turing degrees have a natural isomorphic copy in the structure of the enumeration degrees, namely the substructure of the total enumeration degrees. A long-standing question of Rogers is whether the substructure of the total enumeration degrees has a natural first order definition. The first advancement towards an answer to this question was made by Kalimullin, that discovered the existence of a special class of pairs of enumeration degrees, K-pairs, and showed that this class has a natural first order definition in De.

Building on this result, he proved the first order definability of the enumeration jump operator and consequently obtained a first order definition of the total enumeration degrees above $0'e$. Ganchev and Soskova (2014) showed that when we restrict ourselves to the local structure of the enumeration degrees bounded by $0'e$, the class of K-pairs is still first order definable. In maximal K-pairs within the local structure the total enumeration degrees are first order definable as the least upper bounds of maximal K-pairs. Let us brief on the notion of a K-pair and discuss basic properties to present an alternative first order definition of the enumeration jump.

Definition 7.16: Let A and B be sets of natural numbers. The pair $\{A, B\}$ is a Kalimullin pair (K-pair) if there is a c.e. set W, such that:

$$A \times B \subseteq W \ \& \ A \times B \subseteq W.$$

As a first example of a K-pair consider a c.e. set U and an arbitrary set of natural numbers A. Then U and A form a K-pair via the c.e. set $U \times N$. K-pairs of this sort are considered trivial and we will not be interested in them. A K-pair $\{A, B\}$ is nontrivial if A and B are not c.e.

Non-trivial K-pairs exist. As anticipated, if A is semi-recursive, then $\{A, A\}$ is a K-pair. Indeed let sA be the selector function for A and let n, if $sA(n, m) = m$ $sA(n, m) = m$, if $sA(n, m) = n$.

Now consider the c.e. set $W = \{(sA(n, m), sA(n, m)) \mid n, m \in N\}$ and notice that $A \times A \subseteq W$ and $A \times A = A \times A \subseteq W$. Kalimullin (2003) has shown that the enumeration degrees of K-pairs are precisely the degrees which satisfy the strong minimal-pair property described in the previous section. He shows that the property of being a K-pair is degree theoretic and first order definable in the global structure. A pair of sets $\{A, B\}$ is a K-pair if and only if $\forall x \in De[\ x = (x \vee de(A)) \wedge (x \vee de(B))]$.

Thus, we can lift the notion of a K-pair to the enumeration degrees. A pair of enumeration degrees a and b shall be called a K-pair if every member of a forms a K-pair with every member of b. By $K(a, b)$ we will denote the first order formula, which is true of a and b if and only if they form a K-pair.

Proposition 7.8: Let A and B be a nontrivial K-pair.
   (1) $A \leq_e B$ and $A \leq_e B \oplus Je(\varnothing)$. Similarly $B \leq_e A$ and $B \leq_e A \oplus Je(\varnothing)$;
   (2) The enumeration degrees de(A) and de(B) are incomparable and quasiminimal, for example, the only total degree bounded by either of them is 0e. (3) The class of enumeration degrees of sets that form a K-pair with A is an ideal.

7.3.6 AUTOMORPHISMS, ENUMERATION JUMPS, AND K-PAIRS

The notion of a K-pair over U, originally known as a U-e-ideal, was introduced and used by Kalimullin (2003) to prove the definability of the jump operation in the global structure De. In Kalimullin (2003) proves that the property of being a K-pair over U is degree theoretic and first order definable in the global structure De. A pair of sets A and B form a K-pair over a set U if and only if their degrees a = de(A) and b = de(B) and u = de(U) satisfy the property:

$$K(a, b, u) \rightleftharpoons \forall x[(a \lor x \lor u) \land (b \lor x \lor u) = x \lor u].$$

Definition 7.17: A pair of enumeration degrees is called a K-pair over a degree u if they contain representatives which form a K-pair over a representative of u in the sense of K-pairs over 0e have been proven useful for coding structures in the local structure of the enumeration degrees. It has been shown for instance, that using countable K-systems, systems of nonzero e-degrees such that every pair of distinct degrees forms a K-pair, that every countable distributive semi-lattice can be embedded below every nonzero Δ^0_2 enumeration degree Ganchev and Soskova show that the theory of G_e is computably isomorphic to first order arithmetic, by using K-systems to code standard models of arithmetic.

In the last few years since 2014 Soskov has initiated the study of a further extension of the enumeration degrees: the ω-enumeration degrees, Dω. This structure is an upper semi-lattice with jump operation, where the building blocks of the degrees are of a higher type – sequences of sets of natural numbers. The structure of the enumeration degrees has a definable copy in the ω-enumeration degrees, hence a similar relationship can be seen between De and Dω, as the one described between D_T and De. Here we can define a local structure, Gω, as well, consisting of the degrees bounded by the first ω-enumeration jump of the least degree.

The structure of the ω-enumeration is based on ω-enumeration reducibility, a reducibility which combines two notions: enumeration reducibility and uniformity. A unique phenomenon to ω-enumeration reducibility is the existence of the so called almost zero sequences, sequences whose every member when considered in isolation is equivalent to the corresponding member of the least ω-sequence, $\varnothing\omega$, but whose ω-enumeration degrees (called the almost zero or a.z. degrees) are nonzero. The class of a.z. degrees can be viewed as representing purely the notion of nonuniformity on the one hand, and as a class representing a new type of "lowness" property, which is not as usually defined by domination or the strength of their image under the jump operator.

In this article we study the notion of K-pairs of ω-enumeration degrees in G_ω, inspired by Kallumilin's K-pairs of enumeration degrees. K-pairs of ω-enumeration degrees are defined by Ganchev and Soskova (2014) in terms of their structural properties rather than their set-theoretic properties. There they show that one can distinguish between two types of K-pairs, ones that can be constructed by extending a K-pair over $\varnothing n$ with respect to enumeration reducibility to a sequence, and a second type which consists of almost zero sequence.

7.3.7 K-PAIRS AND THE DEFINABILITY OF THE ENUMERATION JUMP

In this section we will define the notion of a K-pair, give examples of this notion, discuss basic properties and give an alternative first order definition of the enumeration jump.

Definition 7.18: Let A and B be sets of natural numbers. The pair {A, B} is a Kalimullin pair1 (K-pair) if there is a c.e. set W, such that:

$$A \times B \subseteq W \& A \times B \subseteq W.$$

As a first example of a K-pair consider a c.e. set U and an arbitrary set of natural numbers A. Then U and A form a K-pair via the c.e. set $U \times N$. K-pairs of this sort are considered trivial and we will not be interested in them. A K-pair {A, B} is nontrivial if A and B are not c.e. Non- A} is a trivial K-pairs exist. As anticipated, if A is semi-recursive and let:

$$S_A{}^\wedge(n, m) = m, \text{ if } S(n, m) = n \text{ and } S_A{}^\wedge(n, m) = n \text{ if } S(n, m) = n$$

Now consider the c.e. set W = {(sA(n, m), sA(n, m)) | n, m $\in$ N} and notice that $A \times A \subseteq W$ and $A \times A = A \times A \subseteq W$. Kalimullin has shown that the enumeration degrees of K-pairs are precisely the degrees which satisfy the strong minimal-pair property described in the previous section. He shows that the property of being a K-pair is degree theoretic and first order definable in the global structure. A pair of sets $\{A, B\}$ is a K-pair if and only if $\forall x \in De[\ x = (x \vee de(A)) \wedge (x \vee de(B))]$.

Thus, we can lift the notion of a K-pair to the enumeration degrees. A pair of enumeration degrees a and b shall be called a K-pair if every member of a forms a K-pair with every member of b. By $K(a, b)$ we will denote the first order formula, which is true of a and b if and only if they form a K-pair.

Proposition 7.9: Let A and B be a nontrivial K-pair.

(1) $A \leq_e B$ and $A \leq_e B \oplus Je(\varnothing)$. Similarly $B \leq_e A$ and $B \leq_e A \oplus Je(\varnothing)$;

(2) The enumeration degrees de(A) and de(B) are incomparable and quasiminimal, for example, the only total degree bounded by either of them is 0e.

(3) The class of enumeration degrees of sets that form a K-pair with A is an ideal.

Kalimullin's original term for this notion is e-ideal. To show that the jump is definable, Kalimullin then introduces a relativized version of a K-pair. If U is a set of natural numbers then A and B form a K-pair over U if there is a set $W \leq_e U$ such that $A \times B \subseteq W$ and $\overline{A} \times \overline{B} \subseteq \overline{W}$. His proof of the definability of K-pairs also relativizes: a pair of sets $\{A, B\}$ is K-pair over a set U if and only if $\forall x \in De[\ x \vee de(U) = (x \vee de(U) \vee de(A)) \wedge (x \vee de(U) \vee de(B))]$.

A triple of degrees a, b and c, such that each pair $\{a, b\}$, $\{a, c\}$ and $\{b, c\}$ is a nontrivial K-pair (over u) will be called a K-triple (over u). The first order definition of the enumeration jump given by Kalimullin is equivalent to the following. For every enumeration degree $u \in De$, u' is the greatest enumeration degree, which can be represented as the join of a K-triple over u. Here, we give an alternative definition of the enumeration jump, which does not use the relativized version of a K-pair and is in that sense simpler. The proof of this result is for the most part an application of the properties of K-pairs above.

Theorem 7.6: For every nonzero enumeration degree $u \in De$, u' is the largest among all least upper bounds $a \vee b$ of nontrivial K-pairs $\{a, b\}$, such that $a \leq u$.

7.4 ENUMERATION COMPUTABILITY MODELS

7.4.1 URELEMENT AND DEGREES

From Barwise (1968) on infinitary logic and admissible sets, introduce the infinitary language LA where A is a countable admissible sets. We are interested in the sublanguages L_A of the language L (= Linf, ω) which allow finite strings of quantifiers and arbitrary conjunction and disjunction. We consider formulas η of L to be sets, and the LA is $L \cup A$. To insure that LA is a sensible language, we must require that A satisfy certain closure conditions. An extended language of set theory is applied where in addition to the only relation symbol ε, the language allows additional relation symbols S_i, that are ni-ary. Admissible sets are defined based on rudimentary transitive sets. The definitions apply a separation principle known as there is a distinction between the relation symbol ε and the met symbol $\in$.

The Kripke-Platek Sets are defined as follows. The language of set theory to consist of a single binary relation symbol $\in$, with x=y defined by $\forall z(z \in x \leftrightarrow z \in y)$. A formula is said to be Δ_0 if every quantifier is bounded, for example, of the form $\exists x \in y$ or $\forall x \in y$, where these are interpreted in the usual way. A formula is said to be Σ_1 (resp. Π_1) if it is of the form $\exists y \, \varphi$ (resp. $\forall y \, \varphi$), where φ is Δ_0. The classes Σn and Πn are defined analogously.

The axioms of KP are as follows:

1. Extensionality: $x=y \rightarrow (x \in w \rightarrow y \in w)$
2. Pair: $\exists x(x=\{y, z\})$
3. Union: $\exists x \, (x = \cup y)$
4. Δ_0 separation: $\exists x \forall z(z \in x \leftrightarrow z \in y \circ \varphi(z))$ where φ is Δ_0 and x does not occur in φ
5. Δ_0 collection: $\forall x \in z \exists y \varphi(x, y) \rightarrow \exists w \forall x \in z \exists y \in w \varphi(x, y)$, where φ is Δ_0
6. Foundation: $\forall x \, (\forall y \in x \, \varphi(y) \rightarrow \varphi(x)) \rightarrow \forall x \, \varphi(x)$, for arbitrary φ

In 2 and 3, "x = {y, z}" and "x = $\cup$ y" abbreviate the usual representations in the language of set theory. In 4–6, the formula φ may have free variables other than the ones shown. The foundation axiom as presented here is classically equivalent to the assertion that every nonempty definable class of sets has a $\in$-least element. The theory KP-arises if one replaces the foundation schema with the single instance expressing foundation for sets, where $\varphi(x)$ is just the formula $x \in z$. Below we will consider the restriction of the foundation schema to Πn formulas, and we will use Πn-Foundation to denote this restriction.

Let L denote the constructible hierarchy of sets. Ordinals α such that $L\alpha$ models KP are called admissible, with ω being the least such. We will use $KP\omega$ ($KP\omega$-, etc.) to denote the result of adding an axiom of infinity
$\exists x \ (\varnothing \in x \circ \forall y \in x \ (y \cup \{y\} \in x))$ to the corresponding theories above. The least admissible ordinal above ω is the least non-recursive ordinal, also called the Church-Kleene ordinal, ω_{ck}.

Definition 7.19: Δ_0-separation, where $\Delta 0$ (S1, ...Sk) formulas of set theory are defined as a closure as the smallest Y such that

  (a) (i) for atomic formulas θ, both θ and $\neg$ are in Y, (ii) Y is closed under conjunction and disjunctions on formulas, (iii) if θ is Y then so are $\forall x \ [x \ \varepsilon \ y \rightarrow \theta]$ and $x \ [x \ \varepsilon \ y \wedge \theta]$, denoted by $\forall x \ \varepsilon \ y \ \theta$, and $\exists x \ \varepsilon \ y \ \theta$.
  (b) The Σ(S1, ..., Sk) formulas if set theory form the smallest collection Y closed under (i), (ii), (iii) and (iv) if θ is in Y, $\exists x \ \theta$ is in Y.
  (c) The $\Sigma 1$(S1, ..., Sk) formulas of set theory form the smallest collection Y containing the Δ_0 (S1, ...Sk) formulas and closed under (iv).

These classes are important in defining end extensions (Feferman-Kriesel, 1972; Nourani, 1997).

Definition 7.20: A nonempty transitive set A is rudimentary if A satisfies the following:

if a, b ε A, then a b and TC({a}) are ε A ($\Delta 0$-separation) if θ is any a $\Delta 0$ -formula and y is a variable not free in θ, the following is universally true in A: $\exists y \forall x \ [x \ \varepsilon \ y \leftrightarrow x \ \varepsilon \ w \wedge \theta]$.

For any formula θ and variable y of set theory θ (y) is the $\Delta 0$-formulas obtained from θ, by relativizing all quantifiers in θ to y.

Definition 7.21: A is admissible if A is rudimentary and satisfies the Σ~-reflection principle: if θ is a Σ-formula and y is a variable not free in θ, the following is universally true in A: $\theta \Delta \exists y \ [y \text{ is transitive} \wedge \theta \ (y)]$.

Let A be an admissible set and let L(A) denote the collection of formulas of L<inf, ω> that are in A. Call L<A> an admissible fragment (of L<inf, ω >).

Theorem 7.7: (Barwise, 1968) Let A be admissible, $A \subseteq H \ (\omega 1)$.

Theorem 7.8: For A an admissible computable set, A is descriptive computable.

Proof: Since A is admissible, from Barwise (1968) applying Kunens implicit definability, Gentzen system completeness for $L\omega1,\omega$ and Beth definability, $A \subseteq H[\omega1]$ can be proved. Thus, by the proposition A is descriptive.

If A is rudimentary and a, b ε A the {a, b} and a $\cup$ b and a ~ b are in A.

In particular $H[\omega] \subseteq A$. If A is rudimentary, then L_A has the following closure properties:

(i) *if φε* L_A *then* ¬φ ε L_A

(ii) if φ ε L_A and aε A then ($\forall$ va φ)εL_A

(iii) If Γ is a finite subset of L_A then $\circ$ Γ εL_A

(iv) if Γ$\subseteq L_A$, Γε A then $\circ$Γε A.

Let A be a transitive set. Say that a set X$\subseteq$A is Σ_1 on A if there is a Σ_1 formula which defines X on <A, ε>.

Theorem 7.9: (Barwise, 1968) If A is ~1 compact then A is Admissible.

There are completeness, interpolations, definability, and countable compactness from Barwise (1968).

Completeness—The set of valid sentences of L_A is $\Sigma1$ on A.

Compactness—Let A be countable. If A is admissible then A is Σ~1 compact.

7.4.2 RUDIMENTS, KPU, AND RECURSION

To further examine the model theoretic properties that are KPU reducibility interest areas let us consider the following view to KPU from Stuekachev (2011):

Let Sig′ = Sig $\cup$ {U1, $\in$2, $\varnothing$}, where Sig is a (finite) signature.

Definition 7.22: The class of Δ_0-formulas of signature Sig′ is the least class of formulas containing all atomic formulas of signature σ′ and closed under $\circ$, $\vee$, ¬, $\exists x \in y$ and $\forall x \in y$.

Definition 7.21: A structure A of signature Sig′ is called an admissible set if:

1. A $\models$ KPU

2. Ord A = {a | A $\models$ Ord(a)} is well-founded

HF(M) For a set M, consider the set HF(M) of hereditarily finite sets over M defined as follows:

HF(M) = $\cup_n$HF$_n$(M), where n$\in\omega$

$HF_0(M) = \{\emptyset\} \cup M$, $HF_{n+1}(M) = HF_n(M) \cup \{a \mid a$ is a finite subset of $HF_n(M)\}$.

For a structure $M = \langle M, Sig_M \rangle$ of (finite or computable) signature σ, hereditarily finite superstructure

$HF(M) = \langle HF(M); Sig_M, U, \in, \emptyset \rangle$ is a structure of signature Sig' (with $HF(M) \models U(a) \Longleftrightarrow a \in M$).

Remark: in the case of infinite signature, we assume that σ' contains an additional relation Sat(x, y) for atomic formulas under some fixed Gödel numbering. HF(M) is the least admissible set over M.

A mapping $F: P(A)n \to P(A)$ $(n \in \omega)$ is called a Σ-operator if there is a Σ-formula $\Phi(x0,..., xn-1, y)$ of the signature Sig_A with parameters from A such that, for all $S0,..., Sn-1 \in P(A)$,

$F(S0, ..., Sn-1) = \{ a \mid \exists a0, ..., an-1 \in A (\bigwedge_{i<n} ai \subseteq Si \wedge A \models \Phi(a0, ..., an-1, a))\}$. i<n

An operator $F: P(A) \to P(A)$ is strongly continuous in $S \in P(A)$, if

For any $a \subseteq F(S)$, $a \in A$, there exists $a' \subseteq S$, $a' \in A$, s.t. $a \subseteq F(a')$.

For operator $F: P(A)n \to P(A)$, $\delta c(F)$ is the set of elements of $P(A)$ n in which F is strongly continuous. A set $S \in P(A)n$ is called aΣ*-set if $S \in \delta c(F)$ for any Σ-operator $F: P(A)n \to P(A)$.

Note that in HF(M) any subset is a Σ*-set.

Let A be an admissible set, and let $X, Y \subseteq A$ be some objects. Informally, object X is A-constructively generated relative to object Y if there is an A-constructive process which transforms Y to X, and object is A-constructive if it is a result of some A-constructive process with no input data.

Definition 7.23: X is A-constructively generated from Y if there is a Σ-operator F on A s.t. $Y \in \delta c(F)$ and $X = F(Y)$. X is A-constructively generated if it is A-constructively generated from $\emptyset$, for example, there is a Σ-formula $\Phi(x)$ with parameters from A s.t. $X = \Phi^A$.

A structure M is called A-constructivizable, or Σ-definable in A, if there is an A-constructively generated isomorphic copy $C \simeq M$.

Let M be a structure of relational computable signature $\langle Pn0, ..., Pnk, ... \rangle$ and let A be an admissible set.

Applying a definition from Erchov that is a computable piece-meal definability sequence called Σ-definable. For a countable structure M, the following are equivalent:

(i) M is constructivizable (computable);

(ii) M is Σ-definable in HF($\emptyset$).

For arbitrary structures M and N, we denote by M $\leq_\Sigma$N the fact that M is Σ-definable in HF(N).

Structure A is called 'sΣ-definable' in HF(B) (denoted as A $\subseteq$ sΣB) if A is HF(B)-constructively generated, for example, $A \subseteq$ HF(B) is a Σ-subset of HF(B), and all the signature relations and functions of A are Δ-definable in HF(B).

A structure M is called A-constructivizable, or Σ-definable in A, if there is an A-constructively generated isomorphic copy C $\simeq$ M.

For a countable structure M, the following are equivalent:

(i) M is constructivizable (computable);

(ii) M is Σ-definable in HF($\varnothing$).

For arbitrary structures M and N, we denote by M $\leq$ΣN the fact that M is Σ-definable in HF(N).

7.4.3 EFFECTIVE REDUCIBILITIES ON STRUCTURES

Reducibility for arbitrary cardinalities are defined as follows:

for arbitrary cardinal α, let Kα be the class of all structures (of computable signatures) of cardinality $\leq$ α. We define on Kα an equivalence relation $\equiv$Σ as follows:

for M, N $\in$ Kα,

M$\equiv$ΣN if M$\leq_\Sigma$N and N$\leq_\Sigma$M. The structure S$_\Sigma$(α) = $\langle$Kα/ $\equiv$Σ, $\leq_\Sigma\rangle$ is an upper semilattice with the least element, and, for any M, N $\in$ Kα, where (M, N) denotes the model-theoretic pair of M and N. [M]$_\Sigma$V [N]$_\Sigma$= [(M, N)]$_\Sigma$.

Proposition 7.5: For any structure A, there exists a graph (in fact, a lattice) GA such that A $\equiv_\Sigma$GA.

Let A be an admissible set, let r denote the relation Muchnic reducibility: For structures M, N, denote by:

M $\leq_r^A$ N the fact that Pr^A(M) $\leq_r^A$ Pr^A(N). $\leq_r$ denotes $\leq_r$ HF(Ø)

In Chapter 5, we have briefed how the Pr presentability notion corresponds to generic diagram definability.

Definition 7.24: Let A, B be admissible sets.

A is said to be Σ-reducible to B (A $\sqsubseteq$ΣB) if there is a mapping ν: B $\rightharpoonup$ A s.t. ν−1 transforms every A-constructive process to B-constructive.

A is said to be weakly Σ-reducible to B (A $\sqsubseteq$wΣB) if there is a mapping v: B $\rightharpoonup$ A s.t. v−1 transforms every A-constructively generated object to B-constructively generated.

Theorem 7.10: (Stukachev) If A ⊑ΣB then, for any structures M, N;

M ⊑ΣAr N ⇒ M ⊑ΣBr N where r is Muchnic reducibility.

Definition 7.25: A model is admissible iff its universe, functions, and relations are definable with admissible sets.

Let M be a structure of finite similarity type, M = (|M |, R1, ..., Rl). Regarding the elements of | M | as urelements, let us place ourselves within V|M|. Inside V| M |we form the next admissible set HYP (M) and call it HYPM the admissible hull of M. Technically, HYP M) and HYP ((|M |, R1, ..., Rl)) may differ, for example, the former may not contain urelements at all. V is the element world.

Let U be any set of the elements called urelements. For ordinals α, we define $V_{U, \alpha}$ by $V_{U,0} = U$ and $V_{U, \alpha+1} = \wp (V_U, \alpha)$. $V_U, \lambda = \cup_\alpha <\lambda V_{U, \alpha}$.

Theorem 7.11: (Author, 1977) Admissible models are obtained by taking a reduct from the admissible hull to the Skolem hull definable by a generic diagram.

Platek (1966) developed a recursion theory on admissible sets by calling a function F with (with domain and range subset s of A) A-recursive iff its graph is Σ~1 on A. A set $X \subseteq A$ is A-recursive if its characteristic function in A-recursive, and X is A-recursively enumerable. Consistent with Barwise 1(968), however we use the term A-recursive etc. only when A is countable.

Definition 7.26: Let A be admissible and let R be a relation on A;

R is A-r.e. if R is 1 on A.

R is A-recursive if R is 1 on A.

R is A-finite if R ε A.

A function f with domain and range subsets of A is A-recursive if its graph is A-r.e.

If A = L(α) then we refer to these notions as α-r.e., α-recursive, and α-finite.

Platek's Admissible Recursion as is defined as follows:

Let A be an admissible set. Call a function F (with domain and range subsets of A0 A −recursive iff its graph is Σ~1 on A. A set $X \subseteq A$ is A-recursive if is characteristic function is A-recursive and X is A-recursively enumerable A-r.e.) if it is the range of a recursive function. Kleene's enumeration and the second recursion theorem lift to arbitrary admissible sets. Admissible and Descriptive sets appear to indicate that when various recursive definitions are lifted to other domains than integers, for example admissible sets, there are equivalence breakdown. The most dramatic there being two competing notions of r.e. and admissible sets. The preceding had lead this author to examine specific admissible computability functions called Admissible Computable, that is not full recursive admissible sets since Nourani (2007).

To examine Turing degrees further considering the above sections, note that the Turing degrees can be embedded in the enumeration degrees via the standard embedding ι which maps the Turing degree of a set A to the enumeration degree of $A \oplus A^\wedge$. This embedding preserves the order, the least upper bound and the jump operation. The range of ι is therefore a substructure of De, which is isomorphic to DT. Recall definition 7.4:

A pair of sets A and B form a K-pair over a set U if and only if their degrees $a = de(A)$ and $b = de(B)$ and $u = de(U)$ satisfy the property:

K(a, b, u): $\forall x[(a \vee x \vee u) \wedge (b \vee x \vee u) = x \vee u]$.

7.4.4   ADMISSIBLE FRAGMENTS AND LIFTS AND K-PAIRS

The sense on what the Dt coding parameters have to do with the Σ-compact Ulremenet definable descriptive sets (author's brief at VSL 2014 to Soskova).

There are automorphism bases there that can be explored for lifting on K pairs.

Lemma 7.2: (Author July 2014, VSL Vienna). For L_A an admissible language, every $L_A \Sigma_1$_compact admissible fragment set and its complement (with same Turing degrees) from a Kalimullen pair.

Theorem 7.12: (Author, December 2014) Rudimentary fragment pair sets define a natural K-pair reducibility ordering based on the Turing degree of the pair's isomorphism types.

Proof: The isomorphism types on rudimentary fragments are identified by the direct product on the fragment admissible models.

What is Dt Turing degree on the set $D = \Sigma 1 - $ compact Urelements?

Let L be an admissible language defined on the Keisler fragment Lω1,ω. The functor on the category Lω1, K might be defined with admissible hom sets.

Thus, we can apply generic diagrams to define admissible computable sets and models. Let us define computable functors (Nourani 1996). The functors define generic sets from language strings to form limits and models.

Definition 7.27: A functor F: A $\rightarrow$ B is computable iff there is an effective procedure for defining the arrow object mapping from A to B.

Definition 7.28: A model is admissible iff its universe, functions, and relations are definable with admissible sets.

Theorem 7.13: There is a generic functor defining an admissible model.

Let L be an admissible language defined on the Keisler fragment $L_{u'1,u'}$. The functor on the category $L_{\omega1,K}$ might be defined with admissible hom sets.

Theorem 7.14: The generic functors define admissible models, provided the functions and relations in the language fragment are L-admissible.

Theorem 7.15: There is a generic functor defining an admissible model.

Let L_A be an admissible language defined on the Keisler fragment L$_{\omega1,\omega}$. The term fragment is not inconsistent with the terminology applied in categorical logic. A subclass F of class of all formulas of an Infinitary language is called a fragment, if (a) for each formula ϕ in F all the subformulas of ϕ also belong to F; and (b) F is closed under substitution: if ϕ is in F, t is a term of L, x is a free variable in ϕ, the $\phi(x/t)$ is in F.

Definition 7.29: By a fragment of L$_{\omega1,}$ we mean a set L<A> of formulas such that

1. every formula of L belongs to L<A>;
2. L<A> is closed under $\neg$, $\exists$ x, and finite disjunction;
3. if $\phi(x)$ ϵ L<A> and t is a term then $\phi(t)$ ϵ L<A>;
4. if ϕ ϵ L<A> then every subformula of ϕ ϵ L<A>.

Subformulas are defined by recursion from the infinite disjunction by sub (VΦ) = $\cup$ sub (ϕ) $\cup$ {V taken over the set of formulas in Φ; with the basis defined for atomic formulas} by sub(ϕ) = {ϕ}; and for compound formulas by taking the union of the subformulas quantified or logically connected with the subformula relation applied to the original formula as a singleton. L is a countable language for first order logic.

Let C be a countable set of new constants symbols and form the first order language K by adding to L the constants c in C. Let K<A> be the set of formulas obtained from formulas φ in L<A> by replacing finitely many free variables by constants c in C. Let $L_{\omega 1, K}$ be the least fragment of $K_{\omega 1, \omega}$ which contains L<A>. Each formula φ in K<A> contains only finitely many c in C. The same symbol M is used for the model and its set of elements. (M, a<c>)cεC is called a canonical model for K iff the assignment c $\rightarrow$ a<c> maps C onto M, where a<c> is the model element corresponding to c.

The functor on the category $L_{\omega 1, K}$ might be defined with admissible hom sets. Functorial models and admissible sets with urelements are applied to provide techniques to generate models were computational complexity questions might be addressed and answer implicit at models. Specific functorial admissible modes are presented where important computational complexity questions are not independent. Computable Functors are defined since Nourani (1996). By defining admissible generic diagrams we define functors to admissible structures.

Starting with infinite language category $L_{\omega 1, K}$ (Nourani, 1995) we define generic sets on $L_{\omega 1, \omega}$. Let L be the admissible language (Barwise, 1972) defined on the Keisler $L_{\omega 1, \omega}$ fragment (Nourani, 1995). Generic diagrams, denoted by G-diagrams, were defined in the author's papers referenced to be diagrams for models defined by a specific function set, for example Σ_1 Skolem functions. The models are applied towards a functorial glimpse at complexity.

Definition 7.30: A set is Descriptive Computable iff it is definable by generic diagram with on computable functions.

Theorem 7.15: (Author, 1977) For A an admissible computable set, A is descriptive computable.

Definition 7.31: A functor F: A $\rightarrow$ B is computable iff there is an effective procedure for defining the arrow object mapping from A to B.

Definition 7.32: A model is admissible iff its universe, functions, and relations are definable with admissible sets.

Theorem 7.16: (Author, 1977) There is a generic functor defining an admissible model.

Theorem 7.17: (Author, 1977) Let L be an admissible language (Barwise, 1972) defined on the Keisler fragment $L_{\omega 1,K}$. The functor on the category $L_{\omega 1,K}$ might be defined with admissible hom sets.

Theorem 7.18: (Author, 1977) The generic functors define admissible models, provided the functions and relations in the language fragment are L-admissible.

From Nourani (1999) admissible models are obtained by taking a reduct from the admissible hull to the Skolem hull definable by a generic diagram. The techniques are applied to obtain specific fragment models with morphisms. Nourani's (1980) ASL and AMS papers had stated and proved theorems that well-known computational complexity problems were not independent of certain universal fragments of Peano arithmetic. Here we examine specific computable sets in retrospect with respect to the developments here, to explore additional insights. Let A be a transitive set.

We say that a set $X \subseteq A$ is a $\Sigma 1$ on A if there is a $\Sigma 1$ formula which defines X on (A,). X is $\Sigma{\sim}1$ on A if it is defined by a $\Sigma 1$ formula with parameters from A, that is if there is a $\Sigma 1$ formula θ (x, y1, …, yn) and parameters b1…bn in A such that X ={a in A |(A,) $\models \theta$ [a, b1, …, bn]}. We read X is $\Sigma 1$ parameters. The ~ notation on $\Delta 0$ is defined the same. A set is resp. $\Delta 1$ on A if both X and A ~ X are Σ resp. $\Sigma 1$. That is why the admissibility definition was stated on $\Sigma{\sim}$. By a Σ reflection we should mean the restriction to θ which are Σ sentences of set theory. We won't be considering strict Σ reflection.

If an admissible set A is countable then LA countable. LA is more convenient than $L\omega 1,\omega$ for model-theoretic purposes since $L\omega 1,\omega$ has 2N0 formulas. Kiesler established the categoricities for these languages. Most sentences of $L\omega 1,\omega$ which describe interesting algebraic structures are already in LA where A is the smallest admissible set different form H(ω).

From the basic Arithmetical Hierarchy theorem, an n-ary relation R is in the arithmetical hierarchy iff it is recursive or, for some m, can be expressed as {<x, …, xn> | (Q1y10….(Qmym) S(x1, …, xn, y1, …, ym)} where Q is either $\forall$ or $\exists$ for 1 <= I <= m, and S is an (n+m)-ary recursive relation. From KPU Section 3.2, Kleene's enumeration and the second recursion theorem lift to arbitrary admissible sets. Admissible and Descriptive sets appear to indicate that when various recursive definitions are lifted to other domains than integers, for example admissible sets, there are equivalence breakdown. The most dramatic there being two competing notions of r.e. and admissible sets. The preceding had lead the author to examine specific admissible computability functions called Admissible Computable, that is not full recursive admissible sets.

7.5 COMPUTABLE CATEGORICAL TREES

Families of computable and computably enumerable sets play an important role in computable model theory. For example, one method to show that there is a computable structure which is computably categorical but not relatively so, is to prove that there is a family of computable sets which has a unique Friedberg enumeration and which is discrete but not effectively discrete. Such a family of sets can be coded into a graph which has the desired properties. It is natural to ask if the distinction between computable categoricity and relative computable categoricity extends to higher levels of the hyperarithmetic hierarchy with respect to computable models. That is, for a computable ordinal alpha, is it the case that there is a computable structure which is Δ_0 α-categorical but not relatively: the answer to this question is yes, at least for computable successor ordinals (Heidelberg workshop).

A computable structure A is computably categorical if there is only one computable structure isomorphic to it, up to computable isomorphism (thus, if B is computable and B is isomorphic to A, there must exist a computable isomorphism from B to A). Algebraic characterizations of this property for such structures as Boolean algebras and linear orders have previously been found by Dzgoev, Goncharov, and Remmel. We give an algebraic property equivalent to computable categoricity for trees of finite height (our language includes partial order but not meet, and our trees are allowed to be infinite branching). We explain the three different ways in which such a tree can fail to satisfy this property, and show why such trees cannot be computably categorical. Conversely, we prove computable categoricity for any tree of finite height in which these three conditions do not occur.

The computable dimension of a structure is the number of computable structures isomorphic to it, up to computable isomorphism. Thus, all computably categorical structures have computable dimension 1, and we prove that all other trees of finite height have computable dimension omega. However, in certain cases it remains open whether one can diagonalizable effectively against finitely many computable presentations of the same tree, for example, whether the tree has effectively infinite dimension. We will discuss the trees for which this question is not solved.

The work on abstract at Heidelberg (Lempp, et al.) prove that no tree of infinite height can be computably categorical, either in the language of partial orders or in the language with an infimum function. To do so, we consider several cases for trees of height omega, and then one single case which covers all trees of height greater than omega.

The general conjecture is that in each language, as long as Kruskal's Theorem holds for trees in that language, the criterion for computable categoricity is parallel to the one for partial orders. Lempp, McCoy, Solomon, et al.; have conjectured that in the language with an infimum function, one simply replaces the notion of embedding's of a partial order with the notion of embedding's preserving the infimum function. Similarly, joint work between Kogabaev, Kudinov, suggests that the same principle holds for Itrees, for example, trees with a downward-closed subset distinguished by a unary predicate I.

The Itree can be generalized to the notion of a labeled tree, in which every node is labeled by an element of a quasiorder. Since Kruskal's Theorem holds for labeled trees, it is reasonable to conjecture that the principle described above extends to labeled trees as well. In practice, however, the question is more complicated, since we have sometimes encountered cases in which the quasiorder is not computable, although the labeled tree is

7.5.1 ENUMERATIONS MODEL THEORY

A brief here on the hyperarithmetical sets. Write φe for the partial function computed by program number e on an effective list. Similarly, we write φXe for the partial function computed using program number e with oracle X. We write $\varphi e(n) \downarrow$ if program e eventually halts, given input n. Similarly, we write $\varphi Xe(n) \downarrow$ if interactive program number e eventually halts, given oracle X and input n.

The halting set is $K = \{e: \varphi e(e) \downarrow\}$. This set is computably enumerable (or c.e.) but not computable.

For an arbitrary set X, the jump is $X' = \{e: \varphi Xe(e) \downarrow\}$. This set is c.e. relative to X but not computable relative to X.

We can iterate the jump to get $X(n)$, for all $n \in \omega$.

$$X(0) = X$$

$$X(n+1) = (X(n))'$$

We can continue the iteration process through the computable ordinals.

$$\text{Let } X(\omega) = \{<n, x>:x \in X(n)\}.$$

$$X(\alpha+1) = (X(\alpha))'$$

for limit α, $X(\alpha)$ represents $\{< \beta, x >: x \in X(\beta)\}$.

To make this precise, we must code the ordinals by natural numbers, using Kleene's system of ordinal notation. We shall ignore this point.

Stephen Kleene and Andrzej Mostowski independently defined what is now called the arithmetical hierarchy. Martin Davis and Andrzej Mostowski independently defined what is now called the hyperarithmetical hierarchy, extending the arithmetical hierarchy through the "computable" ordinals. We define the two hierarchies in a way that is uniform.

- Arithmetical Hierarchy. For $1 \leq n < \omega$, a set is $\Sigma 0n$ if it is computably enumerable relative to $\varnothing(n-1)$. A set is $\Pi 0n$ if the complement is $\Sigma 0n$. A set is $\Delta 0n$ if it is both $\Sigma 0n$ and $\Pi 0n$.
- Hyperarithmetical hierarchy, extending the arithmetical hierarchy. For a computable ordinal $\alpha \geq \omega$, a set is $\Sigma 0\alpha$ if it is c.e. relative to $\varnothing(\alpha)$. A set is $\Pi 0\alpha$ if the complement is $\Sigma 0\alpha$. A set is $\Delta 0\alpha$ if it is both $\Sigma 0\alpha$ and $\Pi 0\alpha$. There are only countably many computable ordinals, and there are only countably many hyperarithmetical sets.

The infinitary logic $L\omega 1\omega$ allows countable disjunctions and conjunctions, but only finite tuples of quantifiers (Kieselr, 1973). Keisler allows formulas with infinitely many free variables, however, for this specific example area (the our) the formulas will have only finitely many free variables (as in [1]). There is no prenex normal form for formulas of $L\omega 1\omega$. In general, we cannot bring quantifiers to the front. However, we can bring negations inside, and this results in a kind of normal form. We restrict our attention to formulas in this normal form. We classify these formulas as $\Sigma \alpha$ or $\Pi \alpha$ for countable ordinals α.

Classification of formulas of $L\omega 1\omega$

1. $\varphi(x)$ is $\Sigma 0$ and $\Pi 0$ if it is finitary quantifier-free,

2. for a countable ordinal $\alpha > 0$,
- $\varphi(x)$ is $\Sigma\alpha$ if it is a countable disjunction of formulas $(\exists u)\psi(x, u)$, where ψ is $\Pi\beta$ for some $\beta < \alpha$,
- $\varphi(x)$ is $\Pi\alpha$ if it is a countable conjunction of formulas $(\forall u)\psi(x, u)$, where ψ is $\Sigma\beta$ for some $\beta < \alpha$.

Negations. For each formula φ (in normal form), there is a formula in normal form that is logically equivalent to the negation of φ. This formula, denoted by neg(φ), is obtained by switching disjunction/exists with conjunction/for all, and switching atomic formulas with their negations.

The computable infinitary formulas are formulas of $L\omega_1\omega$ in which the infinite disjunctions and conjunctions are over c.e. sets. To make this precise, we would assign indices to the formulas based on Kleene's system of ordinal notation, as is done in Shoenfield (1993). Since the infinite disjunctions and conjunctions in a computable infinitary formula are c.e., the formula is in some way "comprehensible." We classify computable infinitary formulas as computable $\Sigma\alpha$ or computable $\Pi\alpha$, for computable ordinals α. Recall that for finite $n \geq 1$, a set is $\Sigma 0n$ in the arithmetical hierarchy just in case it is definable in the standard model of arithmetic by a finitary Σn formula. This fact extends as follows.

Proposition 7.6: For computable ordinals $\alpha \geq 1$, a set is $\Sigma 0\alpha$ in the hyperarithmetical hierarchy just in case it is definable in the standard model of arithmetic by a computable $\Sigma\alpha$ formula. The computable infinitary formulas are also connected with the effective Borel hierarchy (Vanden Boom, 2007).

Theorem 7.19: (Vanden Boom). Let K be a class of structures, all with universe ω, and all having a fixed computable language. Suppose K is closed under isomorphism. Then K is $\Sigma\alpha$ in the effective Borel hierarchy iff it is axiomatized by a computable $\Sigma\alpha$ sentence.

EXERCISES

1. Prove that the alternate definitions to 7.1 are recursively isomorphic. (Hint) First show that they are similar computable.
2. Theorem: A language A is recognizable
    if and only if there is a decidable predicate R(x, y) such that:
    A = { x | $\exists y$ R(x, y) }.
3. Prove lemma 7.1 (Hint: induction on α.)

4. Prove that, sets of minimal Turing degree is a non-recursive set.
5. What is Dt Turing degree on the set D=Σ1-compact Urelements?
6. Prove theorem 7.3.
7. Prove theorem 7.4.
8. Let R be a relation:
   (a) R is Σ0 and Π0 if it is computable.
   (b) R is Σ01 if it is c.e.;
   (c) R is Π01 if the complementary relation, ¬ R, is c.e.
9. Prove that A ≤T B ⇒ A ≤c.e. B.
10. Let D_T be the set of all Turing degrees. Prove that $(D_T, \leq)$ is a partial order.

KEYWORDS

- **arithmetic hierarchy**
- **computable categorical trees**
- **enumeration computability models**
- **isomorphism types**
- **Turing degrees**

REFERENCES

Andrea Cantini, (1985). On weak theories of sets and classes which are based on strict Π1-reflection. Zeitschrift fu ̈r mathematische Logik und Grundlagen der Mathematik, 31, 321–332.

Artin, E. (1967). Algebraic numbers and algebraic functions (Gordon Breach, New York, 1967).

Ash, C. J., Jockusch, Jr., C. G., Knight, J. F. (1990). 'Jumps of orderings,' Trans. Amer. Math. Soc. 319, 573–599

Ash, C. J., Knight J. F., Mannasse, M., Slaman, T. (1989). Generic copies of countable structures, Anns. of Pure and Appl. Logic, vol 42, 195–205.

Ash, C. J., Knight, J. F. (2000). Computable structures and the hyperarithmetical hierarchy, vol. 144, Studies in Logic and the Foundation of Mathematics (Elsevier, Amsterdam, 2000).

Ash, C. J., Knight, J. F., Mannasse, M., Slaman, T. (1989). Generic copies of countable structures," Annals of Pure and Appl. Logic, vol. 42, pp. 195–205.

Baer, R. (1937). Abelian groups without elements of finite order,' Duke Math. J. 3, 68–122.

Calvert, W., Harizanov, V., Shlapentokh, A. (2006). Turing Degrees of Isomorphism Types of Algebraic Objects, J. London Math. Soc. Page 1–14. London Mathematical Society.

Chevalley, C. (1951). Introduction to the theory of algebraic functions of one variable, Mathematical Surveys VI. (American Mathematical Society, New York, 1951).

Coles, R. J., Downey, R. G., Slaman, T. A. (2000). 'Every set has a least jump enumeration,' J. London Math. Soc. 62, 641–649.

Cooper, S. B. (1990). Enumeration reducibility, nondeterministic computations and relative computability of partial functions, Recursion theory week, Oberwolfach 1989, Lecture notes in mathematics (Heidelberg) (K. Ambos-Spies, G. Muler, and G. E. Sacks, eds.), vol. (1432). Springer-Verlag, 1990, pp. 57–110.

Csima, B. (2004). 'Degree spectra of prime models,' J. Symbolic Logic 69, 430–442.

Degrees of Presentability of Structures in Admissible Sets, Sobolev Institute of Mathematics Novosibirsk, Russia "Algebra and Mathematical Logic" Kazan,' September 25–30, 2011.

Downey, R. D. (1997). 'On presentations of algebraic structures,' Complexity, logic, and recursion theory, Lecture Notes in Pure and Applied Mathematics 187 (ed. A. Sorbi; Marcel Dekker, New York, 1997) 157–205.

Downey, R., Knight, J. F. (1992). 'Orderings with αth jump degree 0(α),' Proc. Amer. Math. Soc. 114, 545–552.

Ershov, S., Goncharov, S., Nerode, A., Remmel, J. B. (1998). Elsevier, Amsterdam, 13, 3–114.

Fried, M. D., Jarden, M. (1986). Field arithmetic, 1st edn., vol. 11, Ergebnisse der Mathematik und ihrer Grenzgebiete (Springer, Berlin, 1986).

Friedberg, R. M., Rogers, Jr. H. Reducibility and completeness for sets of integers, Z. Math. Logik Grundlag. Math. 5 (1959), 117–125.

Fuchs, L., (1973). Infinite Abelian groups, vol. 36, Pure and Applied Mathematics (Academic Press, New York).

Ganchev, H. A., Saskova, M. I. (2014). Definability via Kalimulin Pairs in the structure id the enumeration degrees. Technical Report, Faculty of Mathematics, Sofia University, Department of Mathematics.

Ganchev, H. A., Soskova, M. I. (2012). Cupping and definability in the local structure of the enumeration degrees, J. Symbolic Logic vol. 77, no. 1, pp. 133–158.

Greenberg, N., Knight J. F., Computable structure theory in the setting of ω1, Proceedings of first EMU workshop, to appear. paper for Proceedings of first EMU workshop.

Harizanov, V. S. (1998). Pure computable model theory, vol. 1, Handbook of Recursive Mathematics (eds Yu. L. Ershov, S. S. Goncharov, A. Nerode and J. B. Remmel; Elsevier, Amsterdam, 1998) 3–114.

Harizanov, V. S. (2002). 'Computability-theoretic complexity of countable structures,' Bull. Symbolic Logic 8, 14, 457–477.

Harizanov, V., Miller, R. (2007). 'Spectra of structures and relations,' J. Symbolic Logic.

Hirschfeldt, D. R., Khoussainov, B., Shore, R. A., Slinko, A. M. (2002). Degree spectra and computable dimension in algebraic structures,' Ann. Pure Appl. Logic 115(16), 71–113.

Janusz, G. J. (1973). Algebraic number fields, Lecture Notes in Pure and Applied Mathematics 55, Academic Press, New York.

Jesse Johnson, (2011). Department of Mathematics University of Notre Dame, 2011 ASL North American Meeting, March 26, 2011.

Jockusch, Jr. C. G., Soare, R. I. (1994). 'Boolean algebras, Stone spaces, and the iterated Turing jump,' J. Symbolic Logic 59, 1121–1138.

Kalimullin, I. Sh., (2003). Definability of the jump operator in the enumeration degrees, Journal of Mathematical Logic, vol. 3, pp. 257–267.

Khisamiev, A. N., (2004). 'On the upper semilattice LE,' Siber. Math. J. 45 (2004) 211–228 (in Russian); 173–187, (English translation).

Kleene, S., Post, E. L. (1954). The upper semi-lattice of degrees of unsolvability. Annals of Mathematics, 59, 379–407.

Knight, J. Miller, M. Vanden Boom, 2007) "Turing computable embedding's," J. Symb. Log. 72, No. 3, 901–918.

Knight, J. F. (1986). 'Degrees coded in jumps of orderings,' J. Symbolic Logic 51, 20, 1034–1042.

Lachlan, A. H., Lower bounds for pairs of recursively enumerable degrees. In Proceedings of the London Mathematical Society, volume 16 of Lecture Notes in Computer Science.

Lang, S. Algebraic number theory, (Addison–Wesley, Reading, 1970). 21. D. Marker, Model theory: an introduction, Graduate Texts in Mathematics vol. 217, (Springer, New York, 2002). 22.

Marker, D. (2002). Model theory: an introduction, Graduate Texts in Mathematics vol. 217, (Springer, New York, 2002).

Millar, T. S., (1999). Pure recursive model theory, Handbook of Computability Theory Studies in Logic and the Foundations of Mathematics vol. 140, (ed. E. R. Griffor; Elsevier, Amsterdam, 1999) 23, 507–532.

Oates, S., (1989). Jump degrees of groups, Ph.D. Dissertation, University of Notre Dame, 24.

Rabin, M. O. 'Computable algebra, general theory and theory of computable fields,' Trans. Amer. Math. New York, 1999.

Richter, L. J. (1981). 'Degrees of structures,' J. Symbolic Logic 46, 27, 723–731. H. Rogers, Jr, Theory of recursive functions and effective computability (McGraw–Hill, New York, 1967). 28.

Rogers, H. Jr., Theory of recursive functions and effective computability, McGraw-Hill Book Company, New York, 1967.

Shlapentokh, A. (2000). 'Hilbert's tenth problem over number fields, a survey,' Hilbert's tenth problem: relations with arithmetic and algebraic geometry, Contemporary Mathematics 270 (eds. J. Denef, L. Lipshitz, T. Pheidas and J. Van Geel) American Mathematical Society, Providence, RI, 107–137.

Shoenfield, J. R. (1993). Recursion Theory. Berlin: Springer-Verlag, 43–48.

Shoenfield, J. R. (1993). The Arithmetical Hierarchy. Recursion Theory, Berlin: Springer-Verlag, 43–48 (http://projecteuclid.org/euclid.lnl/1235423987).

Soare, R. I. Recursively enumerable sets and degrees (Springer, Berlin, 1987).

Theodore A. Slaman, (2012). The Hierarchy of Definability: An Extended Thesis, University of California, Berkeley, November 11, 2012.

Vanden Boom, M., (2007). The Effective Borel Hierarchy, Fund. Math., vol 195, pp.269–289.

FURTHER READING

Alexey Stukachev (2011). Degrees of Presentability of Structures in Admissible Sets, Sobolev Institute of Mathematics Novosibirsk, Russia, "Algebra and Mathematical Logic" Kazan,' September 25–30.

Arslanov, M. M., Cooper, S. B., Kalimullin, I. Sh. (2003). Splitting properties of total enumeration degrees, Algebra and Logic 42, 1–13.

Baleva, V. (2006) The jump operation for structure degrees, Arch. Math. Logic 45, pp. 249–265.

Barwise, J. (1975). Admissible Sets and Structures, Springer, Berlin–Heidelberg–New York.

Cooper, S. B. (1989). Enumeration reducibility, nondeterministic computations and relative computability of partial functions, Recursion theory week, Oberwolfach 1989, Lecture notes in mathematics (Heidelberg) (K. Ambos-Spies, G. Muler, and G. E. Sacks, eds.), vol. (1432). Springer-Verlag, 1990, pp. 57–110.

Cooper, S. B., McEvoy, K. (1985). On minimal pairs of enumeration degrees, J. Symbolic Logic 50, no. 4, 983–1001.

Ershov, Yu. L. (1996). Definability and Computability, Consultants Bureau, New York–London–Moscow.

Ershov, Yu. L., Puzarenko, V. G. and Stukachev, A. I. (2011). HF-computability, In S. B. Cooper and A. Sorbi (eds.): Computability in Context: Computation and Logic in the Real World, Imperial College Press/World Scientific, pp. 173–248.

Friedberg, R. M., Rogers, Jr. H. (1959). Reducibility and completeness for sets of integers, Z. Math. Logik Grundlag. Math. 5, 117–125.

Ganchev, H. A., Soskova, M. I. (2012). Interpreting true arithmetic in the local structure of the enumeration degrees, to appear in J. Symbolic Logic.

Giorgi, M., Sorbi, A., Yang, Y. (2006). Properly $\Sigma 02$ enumeration degrees and the high/low hierarchy, J. Symbolic Logic 71, 1125–1144.

Jockusch, C. G., (1968). Semirecursive sets and positive reducibility, Trans. Amer. Math. Soc. 131, 420–436.

Jockusch, C. G., Owings, J. (1990). Weakly semirecursive sets, J. Symbolic Logic 55, no. 2, 637–644.

Kalimullin, I. Sh. (2003). Definability of the jump operator in the enumeration degrees, Journal of Mathematical Logic 3 (2003), 257–267.

Lachlan, A. H., Shore, R. A. (1992). The n-rea enumeration degrees are dense, Arch. Math. Logic 31), 277–285.

McEvoy, K. (1985). Jumps of quasi-minimal enumeration degrees, J. Symbolic Logic 50, 839–848.

Nies, A., Shore, R. A., Slaman, T. A. (1998). Interpretability and definability in the recursively enumerable degrees, Proc. London Math. Soc. 77, 241–249.

Nourani, C. F. (2014). Lifts on K-pairs with KPU, brief note at VSL, Vienna, 2014 to M. Soskova on plenary presentation.

Rogers Jr., H. (1967). Theory of recursive functions and effective computability, McGraw-Hill Book Company, New York.

Rozinas, M., (1978). The semi-lattice of e-degrees, 1978 Recursive functions (Ivanovo), Ivano. Gos. Univ., 1978, Russian, pp. 71–84.

Saskova, I. N., Soskova, A. A. (2009). A jump inversion theorem for the degree spectra, J. Log. Comput. 19, pp. 199–215.

Slaman, T. A. (2005). Global properties of the Turing degrees and the Turing jump, Computational Prospects of Infinity. Part I: Tutorials (C. Chong, Qi Feng, T. A. Slaman, W. H. Woodin, and Y. Yang, eds.), World Scientific, 2005, pp. 83–101.

Slaman, T. A. (2012). The Hierarchy of Definability: An Extended Thesis, University of California, Berkeley. November 11, 2012.

Sorbi, A. (1975). The enumeration degrees of the $\Sigma 02$ sets, Complexity, Logic and Recursion Theory (New York) (A. Sorbi, ed.), Marcel Dekker, pp. 303–330.

Soskov, I. N. (2000). A jump inversion theorem for the enumeration jump, Arch. Math. Logic 39, 417–437.

Soskova, M. (2014). Definability, automorphisms and enumeration degrees, VSL, Vienna, July 2014.

Stukachev, A. I. (1997). Uniformization property in hereditary finite superstructures, Sib. Adv. Math. 7, 1, pp. 123–132.

Stukachev, A. I. (2002). Σ-admissible families over linear orders, Algebra Logic 41, 2, pp. 127–139.

Stukachev, A. I. (2004). Σ-definability in hereditary finite superstructures and pairs of models, Algebra Logic 43, 4, pp. 258–270.

Stukachev, A. I. (2007). Degrees of presentability of structures, I, Algebra Logic 46, 6, pp. 419–432.

Stukachev, A. I. (2010). A Jump Inversion Theorem for the semilattices of Σ-degrees, Sib. Adv. Math. 20, 1, pp. 68–74.

Stukachev, A. I. (2010). Σ-definability of uncountable structures of c-simple theories, Siberian Mathematical Journal 51, pp.515–524.

Stukachev, A. I. (2011). Effective Model Theory via the Σ-Definability Approach, Lecture Notes in Logic 41 (in print).

Vanden Boom, M., (2007). The effective Borel hierarchy, Fund. Math., vol 195, pp. 269–289.

CHAPTER 8

PEANO ARITHMETIC MODELS AND COMPUTABILITY

CONTENTS

8.1   INTRODUCTION

Gödel proves the consistency of the Axiom of Choice and the Generalized Continuum Hypothesis with the axioms of set theory, solving one half of Hilbert's 1st Problem. At the ICM in Bologna, Hilbert claims that the work of Ackermann and von Neumann constitutes a proof of the consistency arithmetic (Gödel, 1930) in Königsberg in von Gödel Neumann's presence. Von Neumann independently derives the Second Incompleteness Theorem as a corollary. Hilbert (1931) suggested new rules to avoid Gödel's incompleteness obstacles: finitary versions of the ω-rule. Here are some fundamental beginnings.

Theorem 8.1: (Gödel's Second Incompleteness Theorem). If T is a consistent axiomatizable theory containing PA, then T⊬ Cons(T).

The Peano arithmetic consequences were that Either PA is inconsistent or the deductive closure of PA is not a complete theory.

Theorem 8.2: (Presburger, 1929). There is a weak system of arithmetic that proves its own consistency ("Presburger arithmetic").

(a) If T is inconsistent, then T ⊢ φ for all φ.
(b) If N is the standard model of the natural numbers, then Th(N) is a complete extension of PA (but not axiomatizable).

Theorem 8.3: (Gentzen 1936). Let T ⊇ PA such that T proves the existence and well-foundedness of (a code for) the ordinal ε0. Then T ⊢ Cons(PA).

8.2   RECURSION ON ARITHMETIC FRAGMENTS

A degree is called r.e., recursively enumerable, if it contains a recursively enumerable set. Every r.e. degree is less than or equal to $0'$ but not every degree less than $0'$ is an r.e. degree. Here are the sequents for r.e. degrees. Sacks (1964) proves that the r.e. degrees are dense. Lachlan and Yates (1966) prove that there are two r.e. degrees with no greatest lower bound in the r.e. degrees and that there is a pair of nonzero r.e. degrees whose greatest lower bound is 0. Every finite distributive lattice can be embedded into the r.e. degrees. In fact, the countable atomless Boolean algebra can be embedded in a manner that preserves suprema and infima (Thomason, 1971).

We say $A \leq_T B$ *if there is machine M such that* with B as oracle M computes A.

Definition 8.1: A Turing functional Φ is an r.e. set of triples of the form $\langle x, y, \sigma \rangle$ where $x, y \in \mathbb{N}$ and $\Sigma \in \mathbb{N}*$ satisfying monotonicity and consistency.

We can further "rewrite" Σ as a pair of finite sets P, N such that $P \cap N = \varnothing$.

$A \leq T B$ iff for some Turing functional Φ, $\Phi(B) = A$.

Since 1944 Post's work, people focus mainly on degrees. Turing degrees and r.e. degrees. Friedberg and Muchnik presented the priority method to solve Post's problem, which asks if there is an intermediate r.e. degree, for example, $0 < a < 0'$. Later Shoenfield and Sacks devised the "infinite injury" method to show jump inversion theorems. In 1970's, Lachlan developed the priority method more.

The fundamental starts are from Hilbert's First Problem: *The Continuum Hypothesis.* "What is the cardinality of the real numbers?" Hilbert's Second Problem. *Consistency of Arithmetic.* "Is there a finitistic proof of the consistency of the arithmetical axioms?" to the Hilbert's Tenth Problem. *Solvability of Diophantine Equations.* Hilbert's Tenth Problem in its original form was to find an algorithm to decide, given a multivariate polynomial equation with integer coefficients, whether it has a solution over the integers. In 1970, Matiyasevich, building on work by Davis, Putnam and Robinson, proved that no such algorithm exists, for example, Hilbert's Tenth Problem is undecidable. "Is there an algorithm that determines whether a given Diophantine equation has a solution or not?" A Diophantine equation is an equation of the form $anxn + an{-}1xn{-}1 + \ldots + a0 = 0$. Is there an algorithm that determines given $\langle an, \ldots, a0 \rangle$ as an input whether the Diophantine equation $anxn + an{-}1xn{-}1 + \ldots + a0 = 0$ has an integer solution? The negative answer was presented at Davis-Putnam-Robinson-Matiyasevich; 1950–1970). To fast forward 1981 to 2014, this author proves and equaitonal incompleteness on $I\Sigma_1$ based on the above Nourani and EATS (1981) to Eisentra (2014) presents analogs areas for this problem that is not completely forgotten yet.

Asking the same question for polynomial equations with coefficients and solutions in other commutative rings. The biggest open problem in the area is Hilbert's Tenth Problem over the rational numbers. The author presents some subrings R of the rationals that have the property that Hilbert's Tenth Problem for R is Turing equivalent to Hilbert's Tenth Problem over the rationals. There are closed semiring characterizations that this author stipulates can benefit from the above (author's private communications at VSL Vienna, 2014).

8.3   GÖDEL HIERARCHY AND ARITHMETIC FRAGMENTS

8.3.1   ARITHMETIC, MODELS, AND PRIORITIES

Being a model-theorist this author has not taken the proof-theorists Plight for studying complexity. The techniques are often hard to process on a glimpse but important to reach the real grind on computation. However, that is, the areas having to do with Gödel's incompleteness, Hilbert's reaches to a 10th problem we must state the basics here for completeness to expand the horizons to what richer contexts there are, for example Diophantine definability be that a model-theoretic juncture. To that end let us expand on arithmetic.

Considering Gödel and ZF, let T1 and T2 be theories. We say T1 < T2 iff T2 proves the consistency of T1. Then we get the "Gödel hierarchy" as follows: measurable cardinal > ZFC > Second Order Arithmetic Z2>... bounded arithmetic.

Between Z2 and bounded arithmetic are all the subsystems of Z2:

RCA0: $\Sigma 1$-induction and $\Delta 0$ -comprehension for $\varphi \in \Delta 0$, $\exists X \forall n(n \in X \leftrightarrow \varphi(n))$.

WKL0: RCA0 and essentially compactness.

ACA0: RCA0 and for φ arithmetic, $\exists X \forall n(n \in X \leftrightarrow \varphi(n))$.

First order Peano arithmetic became the paly ground for complexity of recursion theoretic theorems since 1980, this author including (EATCS, 1980) 'recursion theory' the constructions are with checked by induction priority arguments, or α-recursion.

Assume the language has exponential function and satisfies PA− + B$\Sigma 1$.

Let IΣ n denote the induction schema for $\Sigma 0n$-formulas; and BΣ n denote the Bounding Principle for $\Sigma 0n$ formulas. (Kirby and Paris, 1977) $\cdots$ $\Rightarrow$ IΣ n+1 $\Rightarrow$ BΣ n+1 $\Rightarrow$ IΣ n $\Rightarrow$ ...

(Slaman 2004) IΔn $\Leftrightarrow$ BΣ n. Turing reducibility $\leq$T and Turing degrees.

We say A$\leq$ T B if there is a Turing machine M such that with B as oracle M computes A.

Definition 8.2: A Turing functional Φ is an r.e. set of triples of the form $\langle x, y, \eta \rangle$ where $x, y \in N$ and $\in N*$ satisfying monotonicity and consistency.

We can further "rewrite" η as a pair of finite sets P, N such that $P \cap N = \varnothing$.

$A \leq_T B$ iff for some Turing functional Φ, $\Phi(B) = A$.

Friedberg and Muchnik devised the priority method to solve Post's problem that asks if there is an intermediate r.e. degree, for example, $0 < a < 0'$. Followed by

Shoenfield and Sacks that presented the "infinite injury" method to treat jump inversion theorems. More specifics on priorities were developed by Lachlan (1970).

8.3.2 *PEANO AND BOUNDED COMPUTATIONS*

In computable structure theory, one asks questions about complexity of structures and classes of structures. For a particular countable structure M, how hard is it to build a copy? Can that be done effectively? How hard is it to describe M, up to isomorphism, distinguishing it from other countable structures? For a class K, how hard is it to characterize the class, distinguishing members from non-members? How hard is it to classify the elements of K, up to isomorphism. Knight (2014) describes some results on these questions, obtained by combining ideas from computability, model theory, and descriptive set theory. Of special importance are formulas of special forms.

Definition 8.3: The language of arithmetic consists of:

- A 0-ary function symbol (i.e., a constant) 0,
- A unary function symbol S,
- Two binary function symbols $+, \cdot$,
- Two binary relation symbols $=, <$,
- For each n, infinitely many n-ary predicate symbols Xni.

We often abbreviate $\neg(x = y)$ by $x \neq y$ and sometimes $\neg(x < y)$ by $x \nless y$. We write $x \leq y$ as an abbreviation for $x < y \lor x = y$ and $s+t$, $s \cdot t$ as "abbreviations" for $+st$ and $\cdot st$.

We intend these symbols to represent their usual meanings regarding arithmetic. S is the successor operation. The predicate symbols Xni intentionally have no fixed meaning; their purpose is so that if we prove a formula φ containing one of them then not only have we proven $\varphi[\psi/Xni]$

(the formula where we replace Xni with the formula ψ) for any ψ in our language, we have proven $\varphi[\psi/Xni]$ for any formula in any extension of the language of arithmetic.

Definition 8.4: P consists of formulas:

$\forall x(x = x)$,

$\forall x \forall y(x = y \rightarrow \varphi[x/z] \rightarrow \varphi[y/z])$ where φ is atomic and x and y are substitutable for z in φ,

$\forall x(Sx \neq 0)$,

$\forall x \forall y(Sx=Sy \rightarrow x=y)$,

$\forall x \forall y(x<Sy \leftrightarrow x \leq y)$,

$\forall x(x \not< 0)$,

$\forall x \forall y(x<y \lor x=y \lor y<x)$,

$\forall x(x+0=x)$,

$\forall x \forall y(x+Sy=S(x+y))$,

$\forall x(x \cdot 0=0)$,

$\forall x \forall y(x \cdot Sy=x \cdot y+x)$.

The second equality axiom is a bit subtle. In particular, note that φ is allowed to contain x or y, so we can easily derive $x=y \rightarrow x=x \rightarrow y=x$ (taking φ to be $z = x$) and $y=x \rightarrow y=w \rightarrow x=w$ (taking φ to be $z = w$). Finally, in order to prove anything interesting, we need to add an induction scheme.

Definition 8.5: The axioms of arithmetic, Γ_{PA}, consist of P− plus, for every formula φ and each variable x, the formula $\varphi[0/x] \rightarrow \forall x(\varphi \rightarrow \varphi[Sx/x]) \rightarrow \forall x\varphi$. We write PA⊢$\Gamma \Rightarrow \Sigma$ if Fc ⊢$\Gamma_{PA}\Gamma \Rightarrow \Sigma$ and HA⊢$\Gamma \Rightarrow \Sigma$ if Fi ⊢$\Gamma_{PA}\Gamma \Rightarrow \Sigma$. PA stands for Peano Arithmetic while HA stands for Heyting arithmetic.

Definition 8.6: The numerals are the terms built only from 0 and S. If n is a natural number, we write n for the numeral given recursively by: • 0 is the term 0, • n+1 is the term Sn.

Example theorems here are that:

Theorem 8.4: HA proves that addition is commutative: $\forall x \forall y \ x + y = y + x$.

Proof: By induction on x.(Exercises)

By similar arguments, HA (and so also PA) proves all the standard facts about the arithmetic operations. These systems are strong enough to engage in sensible coding of more complicated, but finite, objects. The details of how to accomplish the coding are tedious. A brief example for what it means to code something in the language of arithmetic is as follows. One of the first things to code is the notion of a finite sequence of natural numbers. Let us name a function π which is an injective map from finite sequences to natural numbers. The range of π should be definable; that is, there should be a formula φπ such that HA ⊢ φπ(n) when n = π(σ) for some Σ and HA ⊢ ¬φπ(n) when n is not in the range of π. Then we need the natural operations on sequences to be definable.

Coding sequences is crucial to the power of HA because we can carry out induction along sequences. In particular, this lets us define exponentiation: we say x=yz if there is a sequence η of length z such that η (0)=y, η (i+1) = η (i)·y for each i, and the last element of η is equal to x. And once we have done this, we could define iterated exponentiation, and so on.

Once we can code sequences, it also becomes much easier to define other notions, since we can use sequences to combine multiple pieces of information in a single number. For instance, we could define a finite group to consist of a quadruple ⟨G, e, +G, −1⟩ where G is a number coding a finite set, e is an element of G, +G and −1 are numbers coding finite sets of pairs, and then write down a long formula describing what has to happen for this quadruple to properly define a group. EFA is the weak fragment of Peano Arithmetic based on the usual quantifier-free axioms for 0, 1, +, exp, together with the scheme of induction for all formulas in the language all of whose quantifiers are bounded. In other words, almost all of conventional combinatorics, number theory, finite group theory, and so on can be coded up and then proven, not only inside PA, but in a comparatively small fragment of PA.

Might be worthwhile here to have a glimpse at more concrete descriptive complexity that examines coding sequences on computation strings. Towards descriptive reducibility let us brief on what is concrete descriptive complexity, for example, on the (Grohe 2014) account that instances of algorithmic problems are modeled by finite structures. To relate this to the standard complexity one can view the problem instances as binary strings. The standard techniques apply adjacency matrix representation of graphs. An adjacency matrix representation implicitly fixes an ordering of the vertices of the graph. The same is true for other standard representations. Grohe proceeds to call a representation scheme "canonical" if two structures are represented by the same string if and only if they are isomorphic. This present author's abstract complexity had applied canonical models on term algebras from concrete to topos since 1985. But that is not concerned with the specific computation grains. The analogy can be only terminological for the time being.

However, if a linear order of the vertices of a structure is explicitly given, the problem disappears. For example, we can represent an ordered graph by the adjacency matrix of the graph where the rows and columns are ordered according to the linear order of the vertices of the graph.

We say that an algorithm decides the property P of τ-structures or of ordered $\tau \cup \{||\}-$ structures if it decides the language L(P)

Definition 8.7: We write $\forall x < y\ \varphi$ as an abbreviation for $\forall x(x < y \rightarrow \varphi)$ and $\exists x < y\ \varphi$ as an abbreviation for $\exists x(x < y \circ \varphi)$.

Note that $PA \vdash \neg \forall x < y\ \neg\varphi \leftrightarrow \exists x < y\ \varphi$ and $PA \vdash \neg\exists x < y\ \neg\varphi \leftrightarrow \forall x < y\ \varphi$. These are what are know as *bounded quantifiers*. As we will see, formulas in which all quantifiers are bounded behave like quantifier-free formulas. We call other quantifiers unbounded.

Because HA can describe sequences in a single number, there is no real difference between a single quantifier $\exists x$ and a block of quantifiers of the same type, $\exists x1\ \exists x2 \cdots \exists xn$—anything stated with the latter could be coded and expressed with a single quantifier. Furthermore, all the coding necessary can be done using only bounded quantifiers. Therefore we will generally simply write a single quantifier, knowing that it could stand for multiple quantifiers of the same type.

Definition 8.8: The Δ_0 formulas are those in which all quantifiers are bounded. Σ_0 and Π_0 are alternates names for Δ_0.

The $\Sigma n+1$ formulas are formulas of the form $\exists x\varphi$ (possibly with a block of several existential quantifiers) where φ is Πn. The $\Pi n+1$ formulas are formulas of the form $\forall x\varphi$ (possibly with a block of several universal quantifiers) where φ is Σn. In particular, the truth of $\Delta 0$ formulas is computable, in the sense that given numeric values for the free variables in $\Delta 0$, we can easily run a computer program which checks in finite time whether the formula is true (under the intended interpretation in the natural numbers). By the same argument that shows every formula is equivalent in Fc to a prenex formula, PA shows that every formula is equivalent to a formula with its unbounded quantifiers in front, which must be Σn or Πn for some n.

Lemma 8.1: f t is a closed term then there is a natural number k such that HA⊢t=k.

Now let us consider r.e. sets with models and Σ_0^1 on Peano models.

Let M be a model of PA +BΣ1. In M we can carry on arithmetic. For example, every a∈M has a unique binary expansion $a(0), a(1), ..., a(l-1)$.

Proposition 8.1: Let M be a model of PA +BΣ1. A set A⊂M is r.e iff A is Σ_0^1-definable in M with parameters.

Proof: Follows from the above definitions.

Let us conclude this section with Ramsey's theorem, Ramsey the British mathematician that only lived to 26, has become a basis to a well-studied area called the Ramsey theory. His accomplishment for computability was a lemma along the true goal in the paper, solving a special case of the decision problem for first-order logic. Within some sufficiently large systems, however disordered, there must be some order. Intuition behind Ramsey's: If we color pairs of natural numbers in two colors (Red and Blue), then there is an infinite subset H ⊂ N, such that any pair formed by elements in H is colored by the same color. These problems are studied to extensive terms for graph complexity at combinatorics.

For A ⊆ N, let [A]n denote the set of all n-element subsets of A.

Theorem 8.5: (Ramsey, 1930) Suppose f:[N]n →{0,1, ..., k−1}.Then there is an infinite set H ⊆ N which is f-homogeneous, for example, f is constant on [H]n. If we think of f as a k-coloring of the n-element subsets of N, then all n-element subsets of H have the same color.

8.4   MORE ON ADMISSIBLE AND FINITE MODELS

In this section, me examine constructive and finite models for descriptive computability to further address reducibility areas. This area was addressed in more recent times on abstract and unpublished briefs (Nourani, 2013–2014) on term algebra models on Topological structures, admissible Hull, and Hausdorf spaces written to P. Ecklund at Umea, Sweden, independent with the developments at Russian school based on Ershov.

Let M be a structure of finite similarity type, M = (|M |, R1, …, Rl). Regarding the elements of | M | as urelements, let us place ourselves within V|M|. Inside V| M |we form the next admissible set HYP (M) and call it HYPM the admissible hull of M. Technically, HYP M) and HYP ((|M |, R1, …, Rl)) may differ, for example, the former may not contain urelements at all. V is the element world.

To study algebraic reducibility based on signatures to canonical models let us examine the following from (Stukachev 2010) towards the Erchov hierarchy.

A mapping F: P(A)n → P(A) (n ∈ ω) is called a Σ-operator if there is a Σ-formula Φ(x0,…, xn−1, y) of the signature σA with parameters from A such that, for all S0,…, Sn−1 ∈ P(A), F(S0, …, Sn−1) = { a | ∃a0, …, an−1 ∈ A

(ai ⊆ Si ∘ A |= Φ(a0, …, an−1, a))}. i<n.

An operator *F: P(A) → P(A)* is *strongly continuous* in *S ∈ P(A)*, if for any a⊆F(S), *a∈A*, there exists a' ⊆S, *a' ∈A*, s.t. a⊆F(a'). For operator *F: P(A)n → P(A)*, δ$_c$(F) is the set of elements of *P(A)n* in which *F* is strongly continuous. A set S∈P(A)n is called a Σ∗-set if S∈δ$_c$(F) for any Σ-operator *F: P(A)n → P(A)*. It is easy to show that in HF(M) any subset is a Σ∗-set.

Let M be a structure of relational computable signature ⟨Pn0, …, Pnk, …⟩ and let A be an admissible set. A structure M is called A-constructivizable, or Σ-definable in A, if there is an A-constructively generated isomorphic copy C ≃ M. Informally Ershov calls such M Σ-definable in A if there exists a computable sequence of Σ-formulas that can generate a chain sequence on formula definition the admissible set structures that have a quotient congruent model at the limit that is Σ-definable.

For arbitrary cardinal α, let Kα be the class of all structures (of computable signatures) of cardinality ≤α. We define on Kα an equivalence relation ≡Σ as follows: for M, N ∈ Kα, M≡ΣN if M ≤$_Σ$N and N ≤$_Σ$M. SΣ(α) = ⟨Kα/ ≡Σ, ≤$_Σ$⟩ structure is an upper semilattice with the least element, and, for any M, N ∈ Kα, where (M, N) denotes the model-theoretic pair of M and N.

Definition 8.9: Structure A is called Σ-definable in HF(B) if A is HF(B)-constructively generated, for example,

A ⊆ HF(B) is a Σ-subset of HF(B), and all the signature relations and functions of A are Δ-definable in HF(B).

For a countable structure M, the following are equivalent:
(i) M is constructivizable (computable);
(ii) M is Σ-definable in HF($\varnothing$).

For arbitrary structures M and N, we denote by M def$\leq_\Sigma$N the fact that M is Σ-definable in HF(N).

Proposition 8.2: For any structure A, there exists a graph (in fact, a lattice) GA such that A $\equiv_\Sigma$GA.

Structure A is called sΣ-definable in HF(B) (denoted as A def$\leq_s$B) if A is HF(B)-constructively generated, for example, A $\subseteq$ HF(B) is a Σ-subset of HF(B), and all the signature relations and functions of A are Δ-definable in HF(B). The development is further expanded to a Σ-jump notion.

Definition 8.10: *For a structure* A, *by a Σ-jump of* A *we mean the structure* A' = (HF(A), Σ-SatHF(A)), *where Σ-SatHF(A) is the satisfiability relation for Σ-formulas in* HF(A).

Correctness: for any structures A and B, A $\equiv\Sigma$B implies A' $\equiv\Sigma$B'.

Proposition 8.3: (Puzarenko, 2009)

Mappings i:D$\rightarrow$SΣ *and* j:De $\rightarrow$SΣ *are semilattice embedding's respecting the jump operation.*

Theorem 8.6: (Stukachev, 2010)

Let A be a structure such that $0' \leq_\Sigma$A. There exists a structure B such that

B' $\equiv_\Sigma$A.

There is for example, the following corollary.

Corollary 8.1: Let A be a countable structure such that $0' \leq_\Sigma$A. Then there is a structure B such that, B'$\equiv$m A, $\equiv$m denoting Muchnic reducible.

8.5   FIELDS, FRAGMENTS OF PEANO ARITHMETIC

8.5.1   FILTERS AND PRODUCTS

Let I be a nonempty set. Let S(I) be the set of all subsets of I. A filter D over I is defined to be a set D < S(I) such that I ε D; if X, Y ε D, hen X $\cap$ Y ε D,; if X ε D and X < Z < I, then Z ε D. Note that every filter D is a nonempty set since I ε D, for example, filters are the trivial filter D = {I}. The improper filter D = S(I). For each Y < I, the filter D = {X <I; Y <

X}; this filter is called the principal filter generated by Y. D is said to be a proper filter iff it is not the improper filter S(I). Let E be a subset of S(I). By the filter generated by E we mean the intersection D of all filters over I which include E: D = ∩ {F: E < F and F is a filer over I}. E is said to have the finite intersection property iff the intersection of any finite number if elements of E is nonempty. Can prove that the filter D generated by E, any subset E of S(I), is a filter over I.

Certain products on filters called reduced products are what we will apply to fields on what follows to create model factorizations over algebraically closed fields. The specific techniques and application areas, modula Vaught's obvious accomplishments, are new as far as we are aware. Suppose I is a nonempty set, D a proper filter over I, and for each $i \in$ I, Ai is nonempty. Let C = $\prod i \in$ I Ai be the Cartesian product of the sets. C is the set of all functions f with domain I s.t. for each $i \in$ I, f(I) Ai. For functions f and g $i \in$ C, say that f and g are D-equivalent, f =D g, iff {$i \in$ I: f(i)=g(i)} $\in$ D. The relation =D is easily proved an equivalence relation over C. Let fD. be the equivalence class of f.fD = {g $\in$ C: f =D g}.

Reduced product for sets Ai modula D is defined to be the set of all equivalence classes of =D, denoted by $\prod$D Ai. $\prod$D Ai.={fD: f$\in\prod$I$\in$I Ai}. On models reduced products are defined based on D, a proper filter over I, are defined as follows. For a language L, with usual conventions on functions and relation symbols, let Ri be a model for L.

Definition 8.11: The reduced product $\prod$D Ri. Is the model for L as follows:

(i)   the universe set is $\prod$D Ai.;
(ii) Let P be an n-placed relation symbol of L. The interpretation of P in $\prod$DRi.is the relation S s.t. S(f1D...fnD) iff {$i \in$ I, Ri ($f_{1(i)}$ ...$f_{n(i))}$ $\in$ D;
(iii) Let F be an n-placed function symbol of L. The F is interpreted in $\prod$DRi.by the function H, H (f_1D....f_nD) = <Gi (f_1 (i) ...fn(i): $i \in$ I> D Let c be a constant of L. Then c is interpreted by the element b $\in$ $\prod$D Ai, where b = <ai: $i \in$ I> D.

This section is a preliminary outline to how fragment consistent models can be applied to create models for algebraically closed fields from the authors 2005–2012 ASL-AMS briefs.

Starting with the unique factorization theorem on Galois fields. Consider polynomials f(D) over an algebraically closed Galois field of prime characteristic p. The above field is complete based on the stated theorems. Let us apply Vaught's theorem:.

Lemma 8.2: The model completion T* on the theory T of algebraic closed fields of characteristic (p 0 or prime) defines a proper filter on the set of T formulas.

Theorem 8.7: (Author, 2006) Prime Model Factorization consider polynomials factorization over the algebraic closed fields of characteristic p (p=0 or prime) on polynomials definable by T fragments. Let M be the prime model to T. Then there are prime models Mi, modeling the factors, respectively, such that there is a reduced product based on a proper filter D, defined on T, principal omitting type on fragments with ΠD Mi monomorphically embedded in M.

Theorem 8.8: $\Re$i is a prime model of T iff $\Re$i are elementarily embedded in every countable model of the fragment of T* that $\Re$i models.

The computational characterizations on omitting type, saturation and field extension were examined in the manuscript (Nourani, 2014). The following section introduces areas that are relevant to enumerability degrees.

8.5.2 FIELDS, ISOMORPHIC TYPES, AND TURING DEGREES

Here we take a glimpse on fields the isomorphism types of which have arbitrary Turing degrees. First, we state two definitions and prove two lemmas to describe sequences we will consider and their role in the construction of fields with isomorphism types of arbitrary Turing degrees.

From Rogers (1961), the following theorem that can be applied to the following products.

Theorem 8.9: (Rogers) Let T be a theory in a finite language L such that there is a computable sequence A0, A1, A2,... of finite structures for L, which are pairwise non-embeddable. Assume that for every set $X \subseteq \omega$, there is a countable model A_X of T such $AX \leq e X$, that and for every $j \in \omega$, Aj is embeddable in $AX \Leftrightarrow j \in X$. Then, for every Turing degree d, there is a countable model of T whose isomorphism type has degree d.

Richter applies the above to show that for every Turing degree d, there is an Abelian torsion group the isomorphism type of which has degree A generalization to this Theorem allows infinite structures in the computable sequence of structures.

Starting with Tarski(1931) K is real closed iff K is elementarily equivalent to $(R, +, \cdot, 0, 1, <)$. Next we review more applications to reducibility on real closed fields (Kuhlmann). The usual Peano arithmetic (PA) is the first-order theory, in the language $L := \{+, \cdot, <, 0, 1\}$, of discretely ordered commutative rings with 1 whose set of non-negative elements satisfies, for each formula $\Phi(x, y)$, the associated induction axiom:

$$\forall y \, [\Phi(0, y) \& \forall x \, [\Phi(x, y) \rightarrow \Phi(x + 1, y)] \rightarrow \forall x \Phi(x, y)].$$

Open Induction (OI) is the fragment of PA obtained by taking the induction axioms associated to open formulas only.

Theorem 8.10: (Shepherdson) IP's of real closed fields are precisely the models of OI.

This we can extraplate to construct non-standard models of fragments of arithmetic.

Note that Z is an IP of K iff K is Archimedean iff K is isomorphic to a subfield of R. Consider non-Archimedean fields. An ordered field K need not admit an IP. In general, different IP's need not be isomorphic, not even elementarily equivalent. Does every real closed field admit an IP? If yes, how to construct such? Let us first construct real closed fields.

- Let Γ be any ordered set, $\{A\gamma; \gamma \in \Gamma\}$ a family of divisible Archimedean groups (subgroups of R).
- For $g \in \Pi\Gamma A\gamma$, set support $g := \{\gamma \in \Gamma; g\gamma \neq 0\}$
- The Hahn group is the subgroup of $\Pi\Gamma A\gamma$ $H\Gamma A\gamma := \{g;$ support is well-ordered in $\Gamma\}$ ordered lexicographically by "first differences."
- The Hahn sum is the subgroup $\oplus\Gamma A\gamma := \{g;$ support g is finite$\}$

Theorem 8.11: (Hahn's embedding, 1907) Let G be a divisible ordered abelian group, with rank Γ and archimedean components $\{A_\gamma; \gamma \in \Gamma\}$. Then G is (isomorphic to) a subgroup of $H_\Gamma A_\gamma$.

- Let G be any divisible ordered abelian group, k a real closed archimedean field (a real closed subfield of R).
- The Hahn field is the field of generalized power series
  $$k((G)) = \{s = \sum_{g \in G} S_g t^g \; ; \text{support s is well-ordered in } G\} \; g \in G.$$

Does a RCF admit an IP which is a model of normal open induction? of full PA?

Let k be any real closed subfield of R. Let $G \neq \{0\}$ be any DOAG which is not an exponential group in k. Consider the Hahn field $k((G))$ and its subfield $k(G)$ generated by k and $\{tg: g \in G\}$. Let K be any real closed field satisfying $k(G)rc \subseteq K \subseteq k((G))$ where $k(G)rc$ is the real closure of $k(G)$. Any such K has G as value group and k as residue field. By Corollary above, K does not admit an IPA.

IPA real closed fields are recursively saturated, and the converse holds in the countable case of Real closed fields and models of Peano arithmetic; (D'Aquino-Knight –Starchenko, 2010). We now relate IPA real closed fields to the algebraic characterization of recursive saturation given in D'Aquino and Lange (2010).

- A subset $T \subset 2^{<\omega}$ is a tree if every substring of an element of T is also an element of T. If $\sigma, \tau \in 2^{<\omega}$, we let $\Sigma \prec \tau$ denote that Σ is a substring of τ. A sequence $f \in 2\omega$ is a path through a tree T if for all $\sigma \in 2^{<\omega}$ with $\sigma \prec f$, we have $\sigma \in T$. For any $\sigma \in 2^{<\omega}$, the length of σ, denoted by length(σ), is the unique $n \in \omega$ satisfying $\Sigma \in 2^n$.

A nonempty set $S \subset R$ is a Scott set if S is computably closed, for example, if $r1,... rn \in S$ and $r \in R$ is computable from $r1 \oplus ... \oplus rn$ (the Turing join of $r1, ..., rn$), then $r \in S$. If an infinite tree $T \subset 2^{<\omega}$ is computable in some $r \in S$, then T has a path that is computable in some $r' \in S$.

Definition 8.12: Let L be a computable language. An L-structure M is *recursively saturated* if for every computable set of L-formulas $\tau(x, y)$ and every tuple a in M (of the same length as $\bar{y}$) such that $\tau(x, \bar{a})$ is finitely satisfiable in M, then $\tau(x, \bar{a})$ is realized in M.

Proposition 8.4: A countable exponential group in C is recursively saturated if and only if C is a countable Scott set.

Proof: Follows from Valuation theoretic characterization of recursively saturated divisible ordered abelian groups] (Harnik-Ressayre and D'Aquino-S.K-Lange, 2010).

A conclusion on computability here is that R is an IPA real closed field then on the one hand the value group is exponential in the residue field, and on the other hand the value group is recursively saturated.

8.6 ARITHMETIC, BOREL HIERARCHY, AND TOPOLOGICAL STRUCTURES

Definition 8.13: Let E and F be equivalence relations on standard Borel spaces X and Y respectively. E is Borel reducible to F, written E $\leq$B F, if there is a Borel f:X $\to$Y such that xEy $\Leftarrow\Rightarrow$ f(x)Ff(y).

Definition 8.14: Let E and F be equivalence relations on standard Borel spaces X and Y respectively. E is Borel reducible to F, written E $\leq$B F, if there is a Borel f:X $\to$Y such that xEy $\Leftarrow\Rightarrow$ f(x)Ff(y).

This means that the points of X can be classified up to E-equivalence by a Borel assignment of invariants that are F-equivalence classes. This means that the points of X can be classified up to E-equivalence by a Borel assignment of invariants that are F-equivalence classes. f is required to be Borel to make sure that the invariant f (x) has a reasonable computation from x.

Definition 8.15: (H. Friedman–Stanley)

Let E, F be equivalence relations on standard Borel spaces X, Y. We say that E is Borel reducible to F (written E $\leq$B F) iff there exists a Borel function f: X $\to$ Y satisfying xEx' $\Leftarrow\Rightarrow$f(x)Ff(x'). A Borel reduction f from E to the Borel equivalence relation F is said to be countable iff every E-class is countable.

Definition 8.16: The equivalence relation E on X is called smooth (or completely classifiable) iff there exists a standard Borel space I of invariants and a Borel function f: X $\to$ I such that

$$xEx' \Leftarrow\Rightarrow f(x)=f(x').$$

The map f tells you how to find complete invariants for the classification problem up to E.

Example: The isomorphism problem for countable divisible groups is smooth.

Effros Borel Space: Let H be a separable Hilbert space and vN(H) the set of von Neumann algebras on H. Let H be a separable Hilbert space and vN(H) the set of von Neumann algebras on H. vN(H) can be given a standard Borel structure called the Effros Borel structure.

Let H be a separable Hilbert space and vN(H) the set of von Neumann algebras on H. vN(H) can be given a standard Borel structure called the Effros Borel structure.

This Borel structure is generated by the sets {M ∈ vN(H): M ∩ U≠ ∅} where U is a weakly open subset of B(H).

This Borel structure is generated by the sets {M ∈ vN(H): M ∩ U≠ ∅} where U is a weakly open subset of B(H).

Theorem 8.12: (Effros-Glimm dichotomy (Harrington, 1990))

E is a Borel equivalence relation. Either E is smooth or E0 ≤ E.

The equivalence relations that are classifiable by countable structures include all equivalence relations that can be classified (reasonably) using countable groups, graphs, fields, etc., as complete invariants. (x, y) = f2(φ(x), φ(y)). S∞ acts on GRAPHS as GRAPHS={f:N×N→{0,1}f(x, x)=0;f(x, y)=f(y, x))} f1 ~ f2 ⇔ ∃φ: N → N bijection s.t. f1 Θ∈S∞, Θf(x, y)=f(Θ−1(x), Θ−1(y))

Analogies between the hierarchy of Borel sets (of finite order) in Euclidean spaces and the arithmetical hierarchy of recursion theory are well-studied in descriptive set theory. The class

ΔB2 of Borel sets is again structured by F. Hausdorff's difference hierarchy of resolvable sets,

whose discrete and effective counterpart for ΔB2 was introduced by Yu. Ershov by means of constructive ordinal numbers.

The so-called topological arithmetical hierarchy consists of classes of subsets of Euclidean spaces which are effectively defined analogously to those of the discrete arithmetical hierarchy, using the (topological) base given by the rational open balls. The present talk deals with the counterpart of the HausdorffErshov classification within the class Δ2 of the topological arithmetical hierarchy. This class is large enough to include several types of point sets which are, for example, essential for computable analysis. Also some notions of effective decidability for subsets of Euclidean spaces are studied from this point of view, Iskandar Kalimulin and Damir Azaintdnov explore limit-wise monotonic reducibility of Σ02-sets.

Research in modern computability theory focus on studying properties of limit-wise monotonic functions and limit-wise monotonic sets. Kalimullin and Puzarenko introduced the concept of reducibility on families of subsets of natural numbers, which is consistent with Σ-definability on admissible sets. Let FA denote the families of initial segments {{x | x < n} | n ∈ A}. Accordingly to Cooper (2004), we define the notion of limit-wise monotonic reducibility of sets as a Σ-reducibility of the corresponding initial segments, namely $A <_{lm} B \iff FA <_{\Sigma} FB$.

Let $A \equiv_{lm} B$ if $A <_{lm} B$ and $B <_{lm} A$. The limit-wise monotonic degree (also called lm-degree) of A is $\deg(A) = \{B: B \equiv_{lm} A\}$. Let Slm denote the class of all lm-degrees of Σ^0_2 sets. The degrees Slm form a partially ordered set under the relation $\deg(A) < \deg(B)$ iff $A <_{lm} B$.

We prove the following theorems:

Theorem 8.13: There exist infinite Σ^0_2-sets A and B such that $A <_{lm} B$ and $B <_{lm} A$.

Theorem 8.14: Every countable partial order can be embedded into Slm.

Theorem 8.15: (jointly with M. Faizrahmanov) There is no maximal element in Slm.

8.7   INFINITARY THEORIES AND COUNTABLE N MODELS

8.7.1   INTUITIONIST TOPOS

Let us have a glimpse at the intuitionistic arithmetic hierarchy to have a feel for countable N Models. Every first-order set of sentences T, viewed as a set of axioms, gives rise to a classical theory T when T is closed under consequences of classical logic, or an intuitionistic one, T i, when T is closed under consequences of intuitionistic logic. Every sentence, which is intuitionistically provable, is also classically provable, the question when the converse holds leads us to the so-called conservativity problem. More precisely, given a class Γ of formulas, we say that the theory T is Γ-conservative over its intuitionistic counterpart T i iff for all $A \in \Gamma$ we have $T i \vdash A$ whenever $T c \vdash A$. A typical example is that of classical Peano Arithmetic PA and its intuitionistic counterpart, Heyting Arithmetic HA. The well-known result concerning these two theories states that PA is $\Pi 2$-conservative over HA. This fact is proven syntactically using Dialectica interpretation or by means of the so-called Friedman translations.

In the Effective Topos, there is exactly one model of intuitionistic $I\Sigma 1$ (the basic theory of the nonnegative integers with induction for $\Sigma 1$-formulas). This generalizes and reinterprets a similar theorem by Charles McCarty. We conclude that in the Effective Topos, first-order arithmetic is essentially finitely axiomatized. In 1983, McCarty showed that in the

Friedman-McCarty realizability model of intuitionistic set theory, there is only one model of Heyting Arithmetic. The present note strengthens this result and reinterprets it. Let IΣ be the theory in the language {0, S, +, ·, ≤} axiomatized by the axioms of Q ≤ (see, [1]) and induction for Σ1-formulas; but based on intuitionistic logic.

Theorem 8.16: In the effective topos Eff there exists (up to isomorphism) precisely one model of IΣ, namely the standard model N (the canonical structure on the natural numbers.

8.7.2 CATEGORICAL ENUMERATION MODEL THEORY

In the paper, Myhill advocates the use of Baire category methods to prove results in degree theory. Those results which do not have such proofs can be considered "truly 'recursive'" while those results with such proofs are "merely set-theoretic." Myhill proves Shoenfield's theorem (Shoenfield, 1960) that there is an uncountable collection of pairwise incomparable degrees using category methods. He also states that a Baire category proof of the Kleene-Post theorem that there are incomparable degrees below $0'$ will be given in another publication, but this never appeared.

Baire category methods in degree theory are also investigated in Myhill (1961), Sacks (1963b), Martin (1967), Stillwell (1972), and Yates (1976). Martin's remarkable accomplishment is note worthy here. On the Baire category theorem the collection is nonempty and a set with the property exists. This author's categorical projective sets (ASL Sofia) on projective sets and the following development from D.A. Martin's theorem in the theory of determinacy and Turing degrees says the following:

Suppose $A \subseteq \omega^\omega$ is Turing invariant and determined. If $\forall x \exists y(x \leq_T y \,\&\, y \in A)$ then A contains a cone.

A is Turing invariant iff $\forall x \in A \forall y(x \equiv_T y \Rightarrow y \in A)$. Here, $\leq_T$ is the relation of Turing reducibility and $\equiv_T$ is the corresponding equivalence relation. "Determined" is in the usual sense of infinite games on integers. A cone is a set of the form $K_y = \{x \mid y \leq_T x\}$. These cones are Turing invariant. We say that y is the base of the cone K. If $\forall x \exists y(x \leq_T y \,\&\, y \in A)$ we say that A is cofinal. When Martin proved his theorem, he was reaching for proofs that determinacy fails (in ZF), by considering explicit sets coming from recursion theory. Instead, he found several results in recursion theory as a consequence.

Here are some examples:

For every x we have x<Tx', where x' is the Turing jump of x. This means that the set of jumps is cofinal. By Borel determinacy, it follows that there is a y such that if y≤Tx, then x≡Tz' for some z. Well known recursion theoretic results show that in fact we can take y=0'.

Again by Borel determinacy, there is a real x such that any y with x≤Ty is a minimal cover above some z. Again, recursion theoretic arguments show that we can take x=0(ω).

To brief further on the newer research areas the abstracts at the Heidelberg workshop are summarized here with newer insights from the chapters in the present manuscript. (R. Solomon) notes that families of computable and computably enumerable sets play an important role in computable model theory. One technique is to show that there are computable structures that are computably categorical but not relatively so, is to prove that there is a family of computable sets which has a unique Friedberg enumeration and which is discrete but not effectively so, for example with graph encodings. Similar techniques are being explored at concrete descriptive complexity.

Solomon, et al. shows that for a computable ordinal alpha, is it the case that there is a computable structure which is $\Delta_0 \alpha$ categorical but not relatively so for computable successor ordinals? The authors examine a enumeration theorems in computable model theory and to explore how to lift these results to higher levels in the hyperarithmetic hierarchy.

A computable structure A is *computably categorical* if there is only one computable structure isomorphic to it, up to computable isomorphism. Algebraic characterizations of this property for such structures as Boolean algebras and linear orders have previously been found by Dzgoev, Goncharov, and Remmel. Lempp et al. presented an algebraic property equivalent to computable categoricity for trees of finite height. The language includes partial order but not meet, and our trees are allowed to be infinite branching.

The *computable dimension* of a structure is the number of computable structures isomorphic to it, up to computable isomorphism. Thus, all computably categorical structures have computable dimension 1, the authors present that all other trees of finite height have computable dimension γ. However, in certain cases it remains open whether one can diagonalizable effectively against finitely many computable presentations of the same tree, for example, whether the tree has *effectively infinite dimension*. The authors prove that no tree of infinite height can be computably categorical, either in the language of partial orders or in the language with an infimum function.

The general conjecture is that in each language, as long as Kruskal's Theorem holds for trees in that language, the criterion for computable categoricity is parallel to the one for partial orders. Similarly, joint work between Kogabaev, Kudinov, and the speaker suggests that the same principle holds for Itrees, for example, trees with a downward closed subset distinguished by a unary predicate I. The Itree can be generalized to the notion of a *labeled tree,* in which every node is labeled by an element of a quasiorder. Since Kruskal's Theorem holds for labeled trees, it is reasonable to conjecture that the principle described above extends to labeled trees as well. In practice, however, the question is more complicated, since we have sometimes encountered cases in which the quasiorder is not computable, although the labeled tree is. Eberhard Herrmann examines the notion of major subset was introduced in Lachlan (1968) on the lattice of recursively enumerable.

This notion is a basic notion for the lattice structure of the computably enumerable sets and was further investigated in several papers under different points of view, for example, their Degrees (Jockusch and Lerman), isomorphisms between them (Maass and Stob), index sets (Lempp), edominance (Robinson) and generalizations of major subset (Herrmann) to a new theorem concerning a generalization of the notion of major subset to the factor lattice by the immune sets is stated.

Serikzhan A. Badaev studies the elementary properties of the Rogers semilattices of computable numberings in the case of families of c.e. sets, d.c.e sets, and arithmetical sets. The results in the case of the arithmetical sets were obtained jointly with Goncharov and Sorbi during the last two years.

Yang studies the nonbounding phenomenon in the structure of enumeration degrees. In a poset with the least element 0, a pair of nonzero elements x and y is said to form a *minimal pair* if their greatest lower bound is 0; an element is said to be *bounding* if there is a minimal pair below it, and nonbounding otherwise. We investigate bounding and nonbounding problems in the poset of the enumeration degrees of Σ^{0-}_2 sets, which coincide with the enumeration degrees below $0e$,' and obtain the following results:

Theorem 8.17: Every nonzero Δ^0_2 enumeration degree bounds a minimal pair of Δ^0_2 enumeration degrees.

Theorem 8.18: There is a nonzero Σ^{0-}_2 enumeration degree which is non-bounding. That is based on joint work with S. B. Cooper, A. Li and A. Sorbi.

Marat Arslanov on *Relative enumerability and the relativized difference hierarchy.* two generalizations of the theorem of Soare and Stob that for each noncomputable low c. e. degree **a** there exists a nonc.e. degree **b** > **a** which is c. e. in **a** are examined. The first generalization extends the idea of the isolated dc. e. degrees, whereas the second can be considered as a generalization of Cooper's theorem on the existence of a properly d c. e. degree.

8.8   KPU ORDINAL MODELS AND AUTOMATA

8.8.1   *ORDINAL MODELS AND COMPUTABILITY*

In this section, we briefly review some basic concepts of finite automata and computability theory. Let Σ be a finite nonempty alphabet; that is, a finite set of symbols. The set of all finite strings over Σ will be denoted by $\Sigma*$. We shall use $s \cdot s'$ to denote concatenation of two strings s and s'. The empty string is denoted by ϱ. One commonly refers to subsets of $\Sigma*$ as languages.

A nondeterministic finite automaton is a tuple A = (Q, Σ, q0, F, δ) where Q is a finite set of states, Σ is a finite alphabet, q0 $\in$ Q is the initial state, F $\subseteq$ Q is the set of final states, and δ: Q×Σ→ 2Q is the transition function. An automaton is deterministic if $|\delta(q, a)| = 1$ for every q and a; that is, if δ can be viewed as a function Q × Σ→ Q.

We say that a run is accepting if r(n) $\in$ F, and that A accepts s if there is an accepting run. For the case of a deterministic automaton, there is exactly one run for each string). The set of all strings accepted by A is denoted by L(A).

A language L is called regular if there is a nondeterministic finite automata on A such that L = L(A). It is well known that for any regular language L, one can find a deterministic finite automaton A such that L = L(A).

The set of all strings accepted by M is denoted by L(M). We call a subset L of $\Sigma*$ recursively enumerable, or r.e. for short, if there is a Turing machine M such that L = L(M).

Notice that in general, there are three possibilities for computations by a Turing machine M on input s: M accepts s, or M eventually enters a rejecting state, or M loops; that is, it never enters a halting state. We call a Turing machine halting if the last outcome is impossible. In other words, on every input, M eventually enters a halting state.

We call a subset L of $\Sigma*$ recursive if there is a halting Turing machine M such that $L = L(M)$. Halting Turing machines can be seen as deciders for some sets L: for every string s, M eventually enters either an accepting or a rejecting state, which decides whether $s \in L$. For that reason, one sometimes uses decidable instead of recursive. When we speak of decidable problems, we mean that a suitable encoding of the problem as a subset of $\Sigma*$ for some finite Σ is decidable.

A canonical example of an undecidable problem is the halting problem: given a Turing machine M and an input w, does M halt on w (i.e., eventually enters a halting state)? In general, any nontrivial property of recursively enumerable sets is undecidable. One result we shall use later is that it is undecidable whether a given Turing machine halts on the empty input.

Let L be a language accepted by a halting Turing machine M. Assume that for some function $f: N \rightarrow N$, it is the case that the number of transitions M makes before accepting or rejecting a string s is at most $f(|s|)$, where $|s|$ is the length of s. If M is deterministic, then we write $L \in DTIME(f)$; if M is nondeterministic, then we write $L \in NTIME(f)$.

Finally, we define the polynomial hierarchy. Let $\Sigma_0 p = \Pi_0 p = Ptime$. i−1 define inductively $\Sigma p = NP\Sigma p$, for $i \geq 1$. That is, languages in Σp are those accepted by a nondeterministic Turing machine running in polynomial time such that this machine can make "calls" to another machine that computes a language in Σp. Such a call is assumed to have unit cost. We define the class Πi as the class of languages whose complements are in Σi.

Considering addressing the "proof theoretic" or syntactic views to computability further let us examine ordinal computability here, from the predicative to impredicative. In informal proof-theoretic language a set of objects is said to be impredicative if it makes reference to a collection of sets that includes the set being defined. A classic example arises if one takes the real numbers to be lower Dedekind cuts of rationals, and then defines the least upper bound of a bounded set of reals to be the intersection of all the upper bounds. A theory is said to be impredicative if its intended interpretation depends on such a definition.

The metacircularity in an impredicative theory poses problems for its ordinal analysis, since the goal of ordinal analysis is to measure the theory's strength in terms of well-founded ordinal notations (Avigad, 2014). For that reason, the first ordinal analyzes of impredicative theories, due to Takeuti, Buchholz, and Pohlers were a landmark. Another important step was the move to studying fragments of set theory instead of second-order arithmetic, providing a more natural framework for the analysis of impredicativity.

The notion of a function from N to N defined by recursion on ordinal notations is fundamental in proof theory. Here this notion is generalized to functions on the universe of sets, using notations for well-orderings longer than the class of ordinals. The generalization is used to bound the rate of growth of any function on the universe of sets that is $\Sigma 1$-definable in Kripke-Platek admissible set theory with an axiom of infinity. Formalizing the argument provides an ordinal analysis. To that end let us examine a Kripke Platek primitive recursive fragment.

Rathjen has stated an axiomatic theory (PRS) that characterizes the primitive recursive set functions. In addition to the $\in$ symbol, the language of PRS has function symbols corresponding to the inductive definition of Prim. The axioms of PRS are extensionality, pair, union, the foundation axiom for sets, and the schema of $\Delta 0$ separation, together with the natural rendering of the defining equations above in the language of set theory. Note that $\Delta 0$ collection is not one of the axioms of PRS.

Of course, the axiomatic theory can be relativized to arbitrarily many free function variables f1, ..., fk, of various arities; I will denote the resulting theory by PRS[f1, ..., fk]. We will mostly be interested in the primitive recursive set functions relativized to the constant ω. That is, we will consider theories with an additional constant symbol, ω, described by a defining axiom, (ω), which asserts that: ω is transitive, and linearly ordered by $\in$; ω contains $\varnothing$, and is closed under the successor function and no element of ω contains $\varnothing$ and is so closed.

Here we will need a universally axiomatized theory that includes PRS[ω]. To that end, let us add a function symbol μ, with defining axiom $y \in x \rightarrow \mu(x) \in x \circ y \notin \mu(x)$. ($\mu$). The axiom states that if x is a nonempty set, $\mu(x)$ returns an $\in$-least element of x. For example, if one restricts one's attention to the constructible hierarchy, one can interpret $\mu(x)$ as returning the least element of x in the standard ordering of L. The following definition will be notationally convenient:

Definition 8.17: Let PRS ω denote the theory PRS [ω, μ] + (ω) + (μ). The fact we need is the following:

Proposition 8.6: PRSω has a set of universal axioms.

The proposition follows from the Lemmas 8.4 and 8.5.

Lemma 8.4: PRSω has a set of Π1 axioms.

Proof: (Exercises)

8.8.2 MORE ON COMPUTABILITY AND ADMISSIBLE SETS

Definition 8.18: A set of the form Lα is admissible if α is a limit and Lα satisfies Σ1 collection. We then also say that α is an admissible ordinal.

Definition 8.19: For M a model with of a binary relation, $\in$M, we let O(M) be the sup of the ordinals in the well-founded part. More precisely, O(M) is the sup of those ordinals α such that there exists some a $\in$ M, (α, $\in$) ~= ({b: b $\in$M a}, $\in$M).

Lemma 8.5: If M is a model of ZFC, then O(M) is admissible.

Proof: (Exercises)

Theorem 8.19: Every countable Σ1 set is included in $L\omega_1^{ck}$

Proof: In the case that the Σ1 set is countable, admissibility allows us to complete the derivation process of 3.2 inside $L\omega_1^{ck}$.

Theorem 8.20: Any two disjoint Σ1 sets can be separated by a Δ1 set. In other words, if we have A, B Σ1 subsets of a space such as ωω, then there is a Δ_1 D which includes and avoids B.

Proof: This follows from Spector-Gandy, using the uniformization principles, which hold for Σ1 inside the constructible and relatively constructible universes.

Let φ be a Σ1 formula such that x$\in$/A if and only if there is some $\alpha < \omega_1 x$ such Lα[x] $\models$ φ(x).

Let ψ be a similar formula for x $\in$/B. The assumption of disjointness imply that the complements of A and B cover ωω, and thus for each x that there is some $\alpha < \omega_1 x$ with Now let C be the set of pairs (x, e) such that the eth recursive in x linear order is a recursive well order of order type α with α least such that Exactly the same calculation as in the proof of 5.2 shows that this is a Σ_1 set. But then likewise appealing $L\alpha[x] \models \psi(x) \vee \varphi(x)$ to the proof of the effective version of Kunen-Martin shows that there is a bound $\delta < \omega_{ck}$ such that for each 1 (x, e) in C the order type of the eth recursive in x linear order is equal to some ordinal less than δ. Thus, for each x we have Then we let D be the set of x such that

$$L\delta\,[x] \models \psi$$

Definition 8.20: $(x) \vee \varphi(x)$. $L\delta[x] \models \neg\psi(x)$. We equip $\omega\omega$ with the product topology obtained by its natural identification with $n\omega\omega$. We then equip the product spaces of the form $(\omega\omega)n$ with the resulting product topology. We say that a set in one of these spaces is Borel if it appears in the smallest σ-algebra containing the open sets. Thus, the proof of 6.1 shows a bit more.

Theorem 8.20: Disjoint $\Sigma1$ sets can be separated by Borel sets, and hence every $\Delta1$ set is Borel.

8.8.3   AUTOMATA AND ORDINAL COMPUTING

The fundamental results of Büchi (1962) and Rabin (1969) presented that the monadic second-order (mso) theory of the ω-chain $(\omega, \leq)$ and of the complete binary tree $\{0, 1\}*, \preccurlyeq, \leq$ is decidable. In both cases the proof relies on a class of finite automata with expressive power equivalent to mso. Because of effective closure properties and decidability of the emptiness problem, the languages of ω-words and infinite trees definable in mso are called regular. For a broad introduction to the field of regular languages of infinite objects. While the emptiness problem is decidable for regular languages, there are some more subtle properties of a given regular language that one may want to decide. One of such questions asks how complicated a given language is. The thesis is devoted to studying such questions for regular languages of ω-words and infinite trees.

Out of the many acceptance conditions that were proposed for automata on infinite objects (e.g., Büchi, Rabin, Muller, Street) the parity condition is the most convenient. To implement this condition, each state q of an automaton should be equipped with a priority $\Omega(q) \in N$. Every regular language of infinite trees can be recognized by an alternating top-down automaton with the parity acceptance condition. However, it was shown that some languages require big indices: for every pair (i, j) there is a regular language of infinite trees that cannot be recognized by any alternating automaton of index (i, j). It means that the index hierarchy is strict.

The index of a language turns out to be a good measure of its complexity. For instance, in the case of languages of infinite trees definable in μ-calculus, the index corresponds precisely to the alternation of fix points used in the definition of a language (Niw, 1997). Also, parity condition of index (0,1) is equivalent to the Büchi acceptance condition and to languages definable in existential fragment of mso. One of the fundamental hierarchy-type problems in automata theory asks is it possible to compute what is the index required to recognize a given regular language of infinite trees. Since every regular language of infinite trees can also be recognized by a non-deterministic parity automaton, there is a non-deterministic variant of the index problem, one asks is there some non-deterministic automaton of index (i, j) recognizing the language of A. Since the translations between the non-deterministic and alternating automata modify the index, the two index problems are independent.

There is no effective procedure solving any of the index problems known, only some partial results have been obtained: both problems are solvable for deterministic languages and for Büchi automata. Additionally, a reduction of the general non-deterministic index problem to boundedness of certain counter automata was given by Colcombet and Löding (2008).

The results presented in the thesis of Skrzypczak (2014) gives wider classes of automata for which the index problems are shown to be decidable: provides an alternative proof of decidability in the case of Büchi automata while an automaton has index (i, j) if are decidable for game automata, a class of automata containing deterministic ones and closed under complement and certain kind of substitution. These automata are expressive enough to recognize languages lying arbitrarily high in both index hierarchies, therefore it is the first class for which the alternating index problem is solved in its full generality.

Rabin (1970) characterized the class of regular languages of infinite trees that are definable in weak monadic-second order logic (wmso)—a variant of mso where set quantifiers are restricted to finite sets. He proved that L is definable in wmso if and only if both L and the complement Lc can be recognized by non-deterministic automata of index (0, 1). Therefore, definability in wmso can be seen as a special case of the non-deterministic index problem. Even this restricted version of the problem seems to be out of reach of the currently known methods.

That thesis presents characterization of regular languages of thin trees (i.e., trees which have countably many branches) that are wmso-definable among all infinite trees is given.

The index hierarchy for automata on infinite trees turns out to be closely related to topological hierarchies from descriptive set theory. These relations motivate a number of interesting questions, one of them is the following conjecture, stated over 20 years ago.

Conjecture 1 (Skurczyński, 1993). If a regular language of infinite trees is Borel then it is definable in weak monadic second-order logic (wmso).

The converse implication is known to be true: every wmso-definable language is Borel. Therefore, the conjecture says in fact that a regular language of infinite trees is Borel if and only if it is wmso-definable. The conjecture has been proved only in the special case of deterministic languages (Niwiński and Walukiewiczm, 2003).

In general, there is no direct relationship between decidability of a logic and topological complexity t of languages it defines. For instance, the FO theory of the structure of arithmetic (ω, $\leq$, $+$, $*$) is undecidable, while it defines only Borel languages of ω-words. On the other hand one can construct a trivial logic that defines some particular language of very high topological complexity. However, as observed by Shelah (using Rabin (1969) for decidability), in the case of mso, the topological complexity and decidability are strongly related.

Theorem 8.21: (Shelah, 1975, Gurevich Shelah, 1982, Rabin, 1969). The mso theory of the real line (R, $\leq$) is undecidable. However, it becomes decidable if we restrict set quantification to $\Sigma02$-sets.

Among the ideas that are used in this study to decidability of the mso logic equipped with an additional quantifier U (as introduced by Bojańczyk and denoted mso+u). For example, the topological complexity of languages of ω-words definable in mso+u. It is shown that the topological complexity of these languages is as high as possible: examples of languages lying arbitrarily high in the projective hierarchy are given.

This observation implies that there is no simple model of automata capturing mso+u on ω-words. Basing on the topological hardness of mso+u, it is shown in Chapter 8 that the mso+u theory of the complete binary tree is unexpectedly expressive — it is possible to write mso+u sentences that express certain subtle properties of the universe of set theory (e.g., analytic determinacy). This observation was later used in has been applied to prove that the mso+u theory of the complete binary tree is not decidable in the standard sense. The proof is an adaptation of the method of Shelah (1975).

8.9   GENERIC COMPUTABILITY AND FILTERS

8.9.1   BASIC GENERIC COMPUTABILITY AND DEGREES

Let us start with basic areas on computability here.

The Pairing Function:

Theorem 8.22: There is a 1–1 and onto computable function $<, >: \Sigma^* \times \Sigma^* \rightarrow \Sigma^*$ and computable functions π_1 and $\pi_2: \Sigma^* \rightarrow \Sigma^*$ such that: $z=<w, t>$ iff $\pi_1(z)=w$ and $\pi_2(z)=t$.

Let us consider partial functions on the set of the natural numbers N.

Let $\{\varphi(n)\}i \in \omega, \}$ be the standard listings of the Turing computable i functions on n arguments. Here i is the code of the Turing machine Mi which computes $\varphi(n)$.

Π^{0-}_2 = languages of the form $\{ x \mid \forall y \exists z\, R(x, y, z) \}$

Theorem 8.23: TOTAL = $\{ M \mid M$ halts on all inputs $\}$

Is in Π^{0-}_2 TOTAL = $\{ M \mid \forall w\, \exists t\, [M$ halts on w in t steps] $\}$

The Turing computable functions coincide with the μ-recursive ones.

Definition 8.21: A set B is said to be A-Turing computable, or Turing reducible to A (B $\leq$T A) if B is A-Turing computable, for example, the characteristic function cB is A-Turing computable.

Σ^{0-}_2 = languages of the form $\{ x \mid \exists y \forall y\, z\, R(x, y, z) \}$

Theorem 8.24: F = $\{M \mid L(M)$ is finite $\}$ is in Σ^{0-}_2.

F= $\{M \mid \exists n \forall w \forall t\,$ [Either $|w| < n$, or M doesn't accept w in t steps]$\}$.

dT(A)=$\{ B \equiv B \mid$ T A$\}$.

Restating the degrees:

Definition 8.22: The Turing degree of the set A is the equivalence relation defined as follows:

A set X is said to be Turing reducible to a set Y if there is an oracle Turing machine that decides membership in X when given an oracle for membership in Y. The notation X $\leq$T Y indicates that X is Turing reducible to Y.

Two sets X and Y are defined to be Turing equivalent if X is Turing reducible to Y and Y is Turing reducible to X. The notation $X \equiv_T Y$ indicates that X and Y are Turing equivalent. The relation $\equiv_T$ can be seen to be an equivalence relation, which means that for all sets X, Y, and Z:

$X \equiv_T X$

$X \equiv_T Y$ implies $Y \equiv_T X$

If $X \equiv_T Y$ and $Y \equiv_T Z$ then $X \equiv_T Z$.

A Turing degree is an equivalence class of the relation $\equiv_T$.

The relation $\equiv_T$ is an equivalence relation; class containing A:

Note that: $dT(A) \leq dT(B) \rightleftharpoons A \leq_T B$, $\rightleftharpoons$ is "if and only if." Recall from a preceding chapter that DT, the set of all Turing degrees, has $(DT, \leq)$ a partial order.

Definition: $A \equiv_T B \rightleftharpoons (A \leq_T B \,\&\, B \leq_T A)$

Proposition 8.6: $(\forall A \subseteq N)(A \equiv_T A)$.

$A \leq_e B$ if there exists an effective procedure that, given any enumeration of B, computes an enumeration A.

Theorem 8.25: (Selman) $A \leq_e B \Leftrightarrow (\forall X - \text{total})(B \leq_e X \Rightarrow A \leq_e X)$.

Proof: ($\Rightarrow$) From the transitivity of $\leq_e$.

($\Leftarrow$) Suppose that $A \leq_e B$. Then we construct a B-regular enumeration η, s.t. $A \leq_e \langle\eta\rangle$, but $\langle\eta\rangle$ is total and $B \leq_e \langle\eta\rangle$, a contradiction.

Now, let us consider what in complexity is called computational genericity. The basic unsolvable problems "outside" of computability theory are computably enumerable (c.e.). The c.e. sets can intuitively be viewed as unbounded search problems, a typical example being those formulas provable in some effectively given formal system. The concept first appeared in print in Kleene's 1936 article and his definition is equivalent to the modern one except that he does not allow the empty set as computably enumerable. He used the term recursively enumerable instead of computably enumerable.)

Post in 1921 invented an equivalent concept, which he called generated set. This work was not submitted for publication until 1941 and did not appear until 1965 (Post, 1965). The final concept whose origins we wish to comment on is the jump operator. In 1936, Kleene showed in Kleene (1936) that $K = \{x: \varphi x(x) \downarrow\}$ (or precisely, the predicate $\exists y T(x, x, y)$) is computably enumerable but not computable. Not having a definition of reducibility at this point, Kleene could not show that K was complete (i.e., that every computably enumerable set is reducible to K).

In his 1943 paper, Kleene again shows that K is c.e. but not computable and here he has a definition of reducibility, but the completeness of K is not shown. Thus, it was Post in his 1944 paper, who first showed the completeness of K, in fact he showed that every c.e. set is 1-reducible to K. Now let us examine how we ever can deploy genericity to computability. The author's beginnings were at the 1984 logic colloquium to present genetic sets to nonmonotic jump computability. At the time the authors first brief on computational complexity was written. Comparing to the concept in the (Ambos-Spies, 1996) 1984 ICALP paper, the accomplishments are bit unusual to this author, since it was a genericity concept for tally (i.e., unary) sets and not with a relation to a specific forcing property or mentioning a relation to forcing.

The classes of p-generic sets form comeager classes (with respect to characteristic sequences and the standard topology on the Cantor space. An example of an application of p-genericity: P- generic sets are P-immune, for example, do not contain infinite polynomial time computable sets. Baire category has been considered in terms of forcing notions and, the particular sets here obtained are called generic sets. This author had written considerably, for example, the preceding chapters on proper forcing computability notions. Therefore, this blurred conceptualization must be treated with care.

According to these authors since private discussions, we have statements that Feferman (1965) has introduced arithmetically generic sets and Hinman (1969) has further developed with n-generic sets, related to the nth level $\Sigma 0n$ of the arithmetical hierarchy. Paraphrasing, an n-generic set has all properties that can be forced by a $\Sigma 0n$-extension strategy. The former is carrying on proof theory and predicative techniques and the latter on recursion at the arithmetic hierarchy. These notions, are however, not with specific forcing properties to address computability instantiations with specific models, or forcing companions for the genercity notions.

Since the class of n-generic sets is comeager, advantages of the Baire category approach are preserved but since there are n-generic sets computable in the nth jump $\varnothing(n)$, at the same time one can obtain results on initial segments. We begin with the first computational genericity notion bearing in mind that there is a confusion with using these terms in a loose sense outside a prespecified forcing property with respect to a genericity notion can be contemplated.

8.9.2 GENERICITY AND REDUCIBILITY

With the preceding proviso with begin to examine specific genericity notions, for example, computational enumerable genercity that are being contemplated at the contemporary r.e. degrees. From Jockusch et.al. (2014), we have

Definition 8.25: Y is enumeration reducible to X (written Y $\leq$e X) if Y = W(X) for some enumeration operator W.

It is well known that the enumeration operators are closed under composition and hence that enumeration reducibility is transitive. Also, each enumeration operator W is obviously $\subseteq$-monotone in the sense that if U $\subseteq$ V then W (U) $\subseteq$ W (V). We are now ready to define generic reducibility.

A generic description of a set A is a partial function Ψ which agrees with the characteristic function of A on its domain and which has a domain of density 1. If Ψ is a partial function, let $\gamma(\Psi) = \{\langle a, b \rangle \colon \Psi(a) = b\}$, so that $\gamma(\Psi)$ is a set of natural numbers coding the graph of Ψ. A listing of the graph of a generic description of a set A is called a generic listing for A. Intuitively, the idea is that A is generically reducible to B if there is a fixed oracle Turing machine M which, given any generic listing for B on its oracle tape, generically computes A. It is again convenient to use enumeration operators in the formal definition.

Definition 8.26: A is generically reducible to B (written A $\leq$g B) if there is an enumeration operator W such that, for every generic description Ψ of B, W($\gamma(\Psi)$) = $\gamma(\Theta)$ for some generic description Θ of A. Note that $\leq$g is transitive because enumeration operators are closed under composition. (It is also easy to check transitivity from the intuitive definition.) Thus generic reducibility leads to a degree structure as usual.

Definition 8.27: The sets A and B are generically interreducible, written A $\equiv$g B, if A $\leq$g B and B $\leq$g A.

The generic degree of A, written deg(A), is $\{C \colon C \equiv_g A\}$. Of course, the generic degrees are partially ordered by the ordering induced by $\leq$g.

The generic degrees have a least element 0g, and the elements of 0g are exactly the generically computable sets. The generic degrees form an upper semi-lattice, with join operation induced by $\oplus$ where $A \oplus B = \{2n: n \in A\} \cup \{2n + 1: n \in B\}$.

Definition 8.28: Let us call a set A c.e. generic, if for every c.e. set S of finite parts:

$(\exists \alpha \subseteq A) (\alpha \in S \vee (\forall \beta \supseteq \alpha)(\beta \in / S))$.

Proposition 8.7: Let S be dense in A, for example, $(\forall \alpha \subseteq A)(\exists \beta \in S)$ $(\alpha \subseteq \beta)$. Then A is c.e. generic, iff A meets S just in case S is dense in A, for example, $(\exists \alpha \subseteq A)(\alpha \in S)$.

Proposition 8.8: Suppose A is a c.e. generic set. If $V \subseteq A$ is c.e., then V is finite.

Proof: Let $S=\{\alpha|(\exists x)(\alpha(x)\simeq 0 \& x \in V)\}$-c.e. Since A is c.e. generic then $\exists \alpha s.t. \alpha \in S \vee (\forall \beta \supseteq \alpha)(\beta \in /S)$. $\alpha \in /S$. Therefore, $(\forall \beta \supseteq \alpha)(\forall x)$ $(\beta(x)\simeq 0 \Rightarrow x \in /V)$. Let $n \geq |\alpha|$, then for every $\beta \supseteq \alpha$, with $|\beta|=n$, $\beta(n)=0$ hence $n \in /V$. Thus, V is finite.

Now we can state the jump inversion theorem for c.e. sets.

Theorem 8.26: (Friedberg) Let $\varnothing' \leq T B$. There exists a c.e. generic A, s.t. $A' \equiv T B$.

Proof: Construct A piecemeal as follows: so that $A \leq_T B$ and A – generic. Then $A' \equiv T \varnothing \oplus A \Rightarrow A' \leq_T B$. For the reverse direction code B in $A \oplus \varnothing'$. On each step n define a finite part αn of cA. Let $\alpha 0 = \varnothing$. If αn is constructed then we ask: Is it true that: $(\exists \beta \supseteq \alpha n)(\beta \in Sn)?$." Since the set $V=\{(\alpha, n)|(\exists \beta \supseteq \alpha)(\beta \in Sn)\}$ is c.e, then $V \leq T K=\varnothing'$.If yes, set $\alpha n*$ will be the minimal such β, if no, then $\alpha n* = \alpha n$. Thus, assures that A is generic. Set $\alpha n+1 = \alpha n* * cB (n)$. 1 $A \leq T B$. Since $|\alpha x+1| \geq x$, $x \in A \rightleftharpoons x \in \alpha x+1$.But $\alpha n \leq T B \oplus \varnothing' \leq T B$.

A is generic, since $\alpha n*$ assures genericity with respect to Sn. 3 $B \leq T A \oplus \varnothing'$. We have $k \in B \Leftrightarrow \Rightarrow \alpha k+1(|\alpha k*|)=1$.Wecan construct B repeating the construction, changing cB(n) with cA($|\alpha n*|$). So, using oracle A and $\varnothing'$ we have $B \leq T A \oplus \varnothing'$. Thus, A is c.e. generic and $A' \equiv T B$.

An important result in degree theory after the Jump is the Density Theorem of Sacks (1964).

Theorem 8.27: (Sacks) The c.e. degrees are dense.

Given two sets C, D with $C<T D$, it is necessary to construct a c.e. set A with $C <_T A <_T D$. $C \leq_T A$ is obtained by directly coding C into A. $D \underset{\leq}{/T}$ A is obtained by the preservation method which is used to ensure that if $D = \{e\}A$ then D would be computable in C. $A \leq_T D$ is not obtained by permitting, but rather because D can compute all of the numbers that have to be put into A to meet the other requirements. The key new idea in the density proof is the method to obtain $A \leq_T C$. As long as $\{e\}C$ looks like A, more and more of D is put into A. Because $D \not\leq_T C$, eventually a difference between A and $\{e\}C$ must appear. Thus, diagonalization is obtained by coding. The density theorem and the techniques used in its proof were very influential in the study of the c.e. degrees.

To compare to standard forcing genercity let us state our definition from a preceding chapter. This notion, is for example, based on this author's positive forcing and Keisler's model-theoretic forcing.

Definition 8.29: A forcing property for a language L is a triple F = (S, <, f) such that

   (i) (S, <) is a partially ordered structure with a least element, for example, 0,

   Thus, < is not always proper and could be read as < or =;

   (ii) f is a function which associates with each p in S a set f(p) of atomic sentences of L[C];

   (iii) whenever p < q f(p) is subset of f(q)

   (iv) let l and t be terms of L[C] without free variables and p in S, φ a formula of L[C] with one free variable. Then if (l-=t) is in f(p) then (t=l) is in f(q) for some q > p. φ (t) in f(p) implies φ (l) is in f(q) for some q > p. For some c in C and q > p, (c=l) is in f(q). The elements of S are called conditions for F.

The well-known genericity notions are defined with respect to a forcing property, for example, Chapter 4. Based on these standard structures we can define genercity notions as follows.

Proposition 8.9: Say that a set A is c.e. generic only iff every subset G of A is generic with respect to the above forcing property.

P-genericity was encoded already in Chapter 7 with Σ_1 compact admissible sets for standard forcing properties (Author, 2003).

For example, in the preceding sections, Chapter 7, we have seen that for $L_{\omega 1}$ computability if R is a relation of countable arity α. R is c.e. if the set of ordinal codes for sequences in R is definable by a Σ_1 formula in $(L_{\omega 1}, \in)$. R is computable if it is both c.e. and co-ce. These and admissible sets languages are all compatible with the forcing and genericity notions developed here.

Now let us consider the following set-theoretic development in this authors publications (Nourani, 2014). In axiomatic set theory, the Rasiowa–Sikorski lemma, named after Roman Sikorski and Helena Rasiowa, an important fundamental facts used in the technique of forcing. In the area of forcing, a subset D of a forcing notion $(P, \leq)$ is called dense in P if for any $p \in P$ there is $d \in D$ with $d \leq p$. A filter F in P is called D-generic if $F \cap E \neq \varnothing$ for all $E \in D$.

Now, we state the Rasiowa–Sikorski lemma:

Theorem 8.28: Let $(P, \leq)$ be a poset and $p \in P$. If D is a countable family of dense subsets of P then there exists a D-generic filter F in P such that $p \in F$.

Proof: Since D is countable, one can enumerate the dense subsets of P as $D_1, D_2, \ldots$. By assumption, there exists $p \in P$. Then by density, there exists $p_1 \leq p$ with $p_1 \in D_1$. Repeating, one gets $\ldots \leq p_2 \leq p_1 \leq p$ with $pi \in Di$. Then $G = \{ q \in P : \exists\, i, q \geq pi \}$ is a D-generic filter.

The Rasiowa–Sikorski lemma can be viewed is a weaker form of an equivalent to Martin's axiom. More specifically, it is equivalent to $MA(2^{\aleph_0})$.

For $(P, \geq) = (\text{Func}(X, Y), \subset)$, the poset of partial functions from X to Y, define $Dx = \{ s \in P : x \in \text{dom}(s) \}$. If X is countable, the Rasiowa–Sikorski lemma yields a $\{Dx : x \in X\}$-generic filter F and thus a function $\cup\, F : X \to Y$.

If D is uncountable, but of cardinality strictly smaller than $2^{\aleph_0}$ and the poset has the countable chain condition, we can instead use Martin's axiom.

The following theorem from the an author's ASL publication (Nourani 2005) was prompted by a comment-question from Dana Scott on Boolean and Heyting models to the author.

Theorem 8.29: There is a Horn dense corollary to the Rasiowa-Sikorski lemma.

Proof: Form the above and preceding chapter on forcing, there are generic Horn filters with a Horn-dense base.

Proposition 8.10: (Author, 2014) There are Horn definable Kalimullin pairs.

8.10   MORE ON THE GENERICITY, PRIORITIES AND R.E. SETS

Not all finite lattices can be embedded in the r.e. degrees, that preserves suprema and infima (Lachlan and Soare, 1980). Post's Problem was solved independently by Friedberg (1957) and Mucnik (1956). Both show that there are incomparable c.e. degrees and therefore that incomplete, non-computable c.e. sets exist. The new technique introduced by both papers to solve the problem has come to be known as the *priority method*. The version used in these papers is specifically known as the finite injury priority method. In the priority method, one has again requirements or conditions which the sets being constructed must meet, as in the finite extension method.

Usually when the priority method is used, the set to be constructed must be c.e., so it is constructed as the union of a uniformly computable increasing sequence of finite sets, the i-th finite set consisting of those elements enumerated into the set by the end of stage i of the construction. The requirements are listed in some order with requirements earlier in the order having higher priority than ones later in the order. In a coinfinite extension argument, at stage n action is taken to meet requirement n.

This action consists of specifying that certain numbers are in the set being constructed and others are not in the set. The status of infinitely many numbers is left unspecified. Action at all future stages obeys these restrictions. Because the determination of what action to take at a given stage cannot be made effectively, the set constructed by this method is not c.e. In the priority method, at stage n action is taken for whichever is the highest priority requirement $R_i n$ that appears to need attention at the stage.

Action consists of adding numbers into the set (which cannot be undone later) and wanting to keep other numbers out of the set. If at a later stage a higher priority requirement acts and wants to put a number into the set which $R_i n$ wanted to keep out, then this number is added and Rin is injured and must begin again. On the other hand, no lower priority requirement can injure $R_i n$.

In a finite injury priority argument, each requirement only needs to act finitely often to be met, once it is no longer injured. For the solution to Post's problem, each requirement needs to act at most twice after it is no longer injured. By induction, it follows that each requirement is injured only finitely often, is met, and acts only finitely often. The priority method is fundamental for computably enumerable degrees and has further applications in other areas of computability. Friedberg (1958) contains three further applications of the finite injury method.

He shows that every noncomputable c.e. set is the union of two disjoint noncomputable c.e. sets (the Friedberg Splitting Theorem), that maximal sets exist, and that there is an effective numbering of the c.e. sets such that each c.e. set occurs exactly once in the numbering. The Friedberg Splitting Theorem is a very simple priority argument as there are no injuries. Priority is just used to decide which requirement to satisfy at a given stage when there is more than one requirement that can be satisfied.

8.10.1 *SYNTACTIC COMPUTATIONAL GENERICITY*

To follow on the newer development on more syntactic genericity areas (Jockusch et al., 2014), let us preview what is being accomplished to address computability with specific structures, for example, groups since (Schlupp, 2010). Generic-case complexity was introduced by Kapovich et al. (2003) as a complexity measure, which is much easier to work with. The basic idea is that one considers partial algorithms, which give no incorrect answers and fail to converge only on a "negligible" set of inputs as defined below.

Definition 8.30: (Jockusch et al., 2014) Let S be a subset of $\Sigma*$ with characteristic function χS. A partial function Φ from $\Sigma*$ to $\{0, 1\}$ is called a generic description of S if $\Phi(x) = \chi S(x)$ whenever $\Phi(x)$ is defined (written $\Phi(x) \downarrow$) and the domain of Φ is generic in $\Sigma*$. A set S is called generically computable if there exists a partial computable function Φ which is a generic description of S. We stress that all answers given by Φ must be correct even though Φ need not be everywhere defined, and, indeed, we do not require the domain of Φ to be computable.

Magnus (1930) proves that one-relator groups have solvable word problem, but we do not know any precise bound on complexity over the entire class of one-relator groups. We do not even know whether or not the isomorphism problem restricted to one-relator presentations is solvable, but the problem is strongly generically linear time. Even undecidable problems can be generically easy. For example, in Boone's group with unsolvable word problem, the word problem is strongly generically linear time. Miasnikov and Osin have constructed finitely generated recursively presented groups whose word problem is not generically computable, but it is not known if there is a finitely presented group whose word problem is not generically computable.

The idea of generic computability is very different from that of Turing reducibility, since generic computability depends on how information is distributed in a given set.

Amongst the syntactic genericity observations is that:

Every Turing degree contains a set which is strongly generically computable in linear time.

Let A be an arbitrary subset of ω and let $S \subseteq \{0,1\}*$ be the set $\{0n:n \in A\}$. Now S is Turing equivalent to A and is strongly generically computable in linear time by the algorithm Φ which, on input w, answers "No" if w contains a 1 and does not answer otherwise. Here all computational difficulty is concentrated in a negligible set, namely the set of words containing only 0's. Note that since the algorithm given is independent of the set A, the observation shows that one algorithm can generically decide uncountably many sets. The next observation is a general abstract version of Miasnikov's and Rybalov's proof that there is a finitely presented semigroup whose word problem is not generically computable.

Every nonzero Turing degree contains a set which is not generically computable. Let A be any noncomputable subset of ω and let $T \subseteq \{0,1\}*$ be the set $\{0n1w:n \in A, w \in \{0,1\}*\}$. Clearly A and T are Turing equivalent. For a fixed n0, $\rho(\{0n01w: w \in \{0,1\}*\}) = 2-(n0+1) > 0$. A generic algorithm for a set must give an answer on some members of any set of positive density. Thus, T cannot be generically computable since if Φ were a generic algorithm for T we could just run bounded simulation of Φ on the set $\{0n1w: w \in \{0,1\}*\}$ until Φ gave an answer, thus deciding whether or not $n \in A$. Here the idea is that the single bit of information $\chi A(n)$ is "spread out" to a set of positive density in the definition of T. Also note that if A is c.e. then T is also c.e. and thus every nonzero c.e. Turing degree contains a c.e. set which is not generically computable.

Observation 1.6 above, is that every nonzero Turing degree contains a set of natural numbers which is not generically computable.

Jockusch et.al. (2014) define the notion of being densely approximable by a class C of sets and observe that a set A is generically computable if and only if it is densely approximable by the class of c.e. (computably enumerable) sets. We call a set A of natural numbers coarsely computable if there is a computable set B such that the symmetric difference of A and B has density 0. We show that there are c.e. sets which are coarsely computable but not generically computable and c.e. sets which are generically computable but not coarsely computable. We also prove that every nonzero Turing degree contains a set which is not coarsely computable.

Considering a relativized notion of generic computability and also introduce a notion of generic reducibility which gives a degree structure and which is related to enumeration reducibility. We identify the set N = {0,1, ...} of natural numbers with the set ω of finite ordinals, In this article we focus on generic computability properties of subsets of ω to see how density interacts with some classic concepts of computability theory.

Our notation for computability is mostly standard, except that we use Φe for the unary partial function computed by the e-th Turing machine, and we let Φe, s be the part of Φe computed in at most s steps. Let We be the domain of Φe. We identify a set $A \subseteq \omega$ with its characteristic function χA.

Definition 8.31: (Jockusch et al., 2014) Let C be a family of subsets of ω. A set $A \subseteq \omega$ is densely C-approximable if there exist sets C_0, $C_1 \in C$ such that $C_0 \subseteq A$, $C_1 \subseteq A$ and $C_0 \cup C_1$ has density 1.

The following proposition corresponds to the basic fact that a set A is computable if and only if both A and its complement A are computably enumerable.

Proposition 8.10: A set A is generically computable if and only if A is densely CE-approximable where CE is the class of c.e. sets.

Proof: If A is densely CE-approximable, then there exist c.e. sets $C0 \subseteq A$ and $C1 \subseteq A$ such that $C0 \cup C1$ has density 1. For a given x, start enumerating both C0 and C1 and if x appears, answer accordingly.

If A is generically computable by a partial computable function Φ, then the sets C0 and C1 on which Φ respectively answers "No" and "Yes" are the desired c.e. sets.

8.10.2   CONCLUDING COMMENTS

To conclude this section with thought experiments on what Myhill proved on Shoenfield's 1969 theorem that there is an uncountable collection of pairwise incomparable degrees using category methods. Baire category methods in degree theory are also investigated in Myhill (1961), Sacks (1963b), Martin (1967), and Yates (1976). If the collection of all sets with a certain property is a comeager subset of 2ω (under the usual topology) then by the Baire category theorem the collection is nonempty and a set with the property exists. Martin showed the existence of a noncomputable set whose degree has no minimal predecessors using this method. Baire category can also shed light on the finite extension method of Kleene and Post. In this method, one shows that the collection of sets meeting each requirement contains a dense open set. Thus, the collection of sets meeting all the requirements is comeager and so nonempty. It follows that if the collection of all sets meeting all requirements is not comeager, then the finite extension method cannot be used to produce a set meeting all the requirements.

EXERCISES

1. Turing degrees D a linear order? Are there f and g such that $f \not\leq_T g$ and g $\not\leq_T$ f?
2. Any infinite cardinal is admissible.
3. If α is a countable ordinal and ψ is a formula in the language of set theory, then the set of x for which $L\alpha[x] \models \psi(x)$ is Borel.
4. For M a model with of a binary relation, $\in M$, we let O(M) be the sup of the ordinals in the well-founded part. Show that if M is a model of ZFC, then O(M) is admissible.
5. Show that HA proves that addition is commutative: $\forall x \forall y\ x + y = y + x$. Hint: It suffices to show $\forall y\ 0+y = y+0$ and $\forall y(x+y = y + x) \rightarrow \forall y(Sx + y = y + Sx)$.
6. Let $\oplus$ be the semi-lattice join operation. Prove that $dT\ (A \oplus B)$ is the least upper bound of $dT\ (A)$ and $dT(B)$.
7. If α is a countable ordinal and ψ is a formula in the language of set theory, then the set of x for which $L\alpha[x] \models \psi(x)$ is Borel.
8. Prove that the Turing jump has the monotonicity property $A \leq_T B \Rightarrow A' \leq_T B'$, by first proving that $A \leq_T B \vdash A' \leq_m B'$. (Hint: Apply the Post-Kleene Theorem).
9. Let A be a c.e. generic set. If $V \subseteq A$ is c.e., then V is finite.
10. The generic degrees have a least element 0g, and the elements of 0g are exactly the generically computable sets. The generic degrees form an upper semi-lattice, with join operation induced by $\oplus$ where $A \oplus B = \{2n: n \in A\} \cup \{2n + 1: n \in B\}$.
11. $PRS\omega$ has a set of $\Pi 1$ axioms.
   Hint: (Use the explicit function symbols for pairing and union to eliminate the existential quantifiers in the pairing and union axioms),

KEYWORDS

- admissible finite models
- Borel hierarchy
- fields
- Gödel hierarchy
- Peano arithmetic fragments
- recursion on arithmetic fragments
- topological structures

REFERENCES

Barry S. Cooper, (2004). Computability Theory, Chapman and Hall/CRC.

Benno van den Berg, Jaap van Oosten (2014). Department of Mathematics Utrecht University Arithmetic is Categorical. Armin Hemmerling *The topological Ershov hierarchy, Heideger workshop.* Workshop on, Computability and Logic Heidelberg, Germany, June 23–27, 2003.

Carl, M., D'Aquino, P., Kuhlmann, S. (2012). Value groups of real closed fields and fragments of Peano arithmetic (arXiv: 1205.2254).

Phillp W. Carruth 1947 , Generlized Power Series Fields, AMS, April 1947.

Cooper, S. B. (1990). The jump is definable in the structure of the degrees of unsolvability. Bull. Amer. Math. Soc. (N. S.), 23(1):151–158.

Cooper, S. B. (1997a). Beyond Godel's theorem: the failure to capture information content. In Complexity, Logic, and Recursion Theory, volume 187 of Lecture Notes in Pure and Appl. Math., pages 93–122. Dekker, New York.

Cooper, S. B. (1997b). The Turing universe is not rigid. Preprint Series 16, University of Leeds, Department of Pure Mathematics.

Cooper, S. B. (1999a). Local degree theory. In Handbook of Computability Theory, volume 140 of Stud. Logic Found. Math., pages 121–153. North- Holland, Amsterdam.

Cooper, S. B. (1999b). Upper cones as automorphism bases. Siberian Adv. Math., 9(3):17–77.

Cooper, S. B. (2001). On a conjecture of Kleene and Post. Math. Log. Q., 47(1):3–33.

Cooper, S. B. *Computability theory.* Chapman & Hall/CRC, Boca Raton, FL, 2004. ISBN 1–58488–237–9.

Cooper, S. B., Harrington, L., Lachlan, A. H., Lempp, S., and Soare, R. I. (1991). The d.r.e. degrees are not dense. Ann. Pure Appl. Logic, 55(2), 125–151.

Crossley, J. N. (1975). Reminiscences of logicians. With Contributions by C. C. Chang, John Crossley, Jerry Keisler, Steve Kleene, Mike Morley, Vivienne Morley, Andrzej Mostowski, Anil Nerode, Gerald Sacks, Peter Hilton and David Lucy. In Algebra and logic (Fourteenth Summer Res. Inst., Austral. Math. Soc., Monash Univ., Clayton, 1974), volume 450 of Lecture Notes in Math., pages 1–62. Springer, Berlin.

D'Aquino, P., Knight, J. F., Kuhlmann, S., Lange, K. (2012). Real closed exponential fields, Fund. Math. 219, 163–190.

Damian Niwiński, and Igor Walukiewicz (2003). A gap property of deterministic tree languages. *Theor. Comput. Sci.,* 1(303), 215–231.

Degtev, A. N. (1973). tt- and m-degrees. Algebra i Logika, 12:143–161, 243. Dekker, J. C. E. (1954). A theorem on hyper-simple sets. Proc. Amer. Math. Soc., 5, 791–796.

Dekker, J. C. E., Myhill, J. (1958). Some theorems on classes of recursively enumerable sets. Trans. Amer. Math. Soc., 89, 25–59.

Downey, R. G., Lempp, S. (1997). Contiguity and distributivity in the enumerable Turing degrees. J. Symbolic Logic, 62(4), 1215–1240.

Epstein, R. L. (1979). Degrees of unsolvability: structure and theory, volume 759 of Lecture Notes in Mathematics. Springer, Berlin.

Feferman, S. (1957). Degrees of unsolvability associated with classes of formalized theories. J. Symbolic Logic, 22, 161–175.

Feferman, S. (1965). Some applications of the notions of forcing and generic sets: Summary. In Theory of Models (Proc. 1963 Internat. Sympos. Berkeley), pages 89–95. North-Holland, Amsterdam.

Feiner, L. (1970). The strong homogeneity conjecture. J. Symbolic Logic, 35:375–377.

Fejer, P. A. (1983). The density of the nonbranching degrees. Ann. Pure Appl. Logic, 24(2), 113–130.

Fejer, P. A., Soare, R. I. (1981). The plus-cupping theorem for the recursively enumerable degrees. In Logic Year 1979–80 (Proc. Seminars and Conf. Math. Logic, Univ. Connecticut, Storrs, Conn., 1979/80), volume 859 of Lecture Notes in Math., pages 49–62. Springer, Berlin.

Friedberg, R. M. (1956). The solution of Post's problem by the construction of two recursively enumerable sets of incomparable degrees of unsolvability (abstract). Bull. Amer. Math. Soc., 62, 260.

Friedberg, R. M. (1957a). A criterion for completeness of degrees of unsolvability. J. Symbolic Logic, 22, 159–160.

Friedberg, R. M. (1957b). The fine structure of degrees of Unsolvability of recursively enumerable sets. In Summaries of Cornell University Summer Institute for Symbolic Logic, Communications Research Division, Inst. for Def. Anal., Princeton., pages 404–406.

Friedberg, R. M. (1957c). Two recursively enumerable sets of incomparable degrees of unsolvability (solution of Post's problem, 1944). Proc. Nat. Acad. Sci. USA, 43, 236–238.

Friedberg, R. M. (1958). Three theorems on recursive enumeration. I. Decomposition. II. Maximal set. III. Enumeration without duplication. J. Symbolic Logic, 23, 309–316.

Greenberg, N., Shore, R. A., Slaman, T. A. (2015). The theory of the metarecursively enumerable degrees. To appear.

Griffor, E. R., editor. (1999). Handbook of Computability Theory, volume 140 of Studies in Logic and the Foundations of Mathematics. North-Holland Publishing Co., Amsterdam.

Grohe, M. (2014). Definable Graph Structure Theory. Technical Report.

Groszek, M. J., Slaman, T. A. (1983). Independence results on the global structure of the Turing degrees. Trans. Amer. Math. Soc., 277(2), 579–588.

Grzegorczyk, A. (1951). Undecidability of some topological theories. Fund. Math., 38, 137–152.

Hajek, P., Pudlak, P. (1988).Metamathematics of First-Order Arithmetic, Perspectives in Mathematical Logic, Springer-Verlag.

Harrington, L. (1978). Plus cupping in the recursively enumerable degrees. Handwritten notes.

Harrington, L. (1980). Understanding Lachlan's monster paper. Handwritten notes.

Harrington, L. (1982). A gentle approach to priority arguments. Handwritten notes for a talk at AMS-ASL Summer Institute on Recursion Theory, Cornell University.

Harrington, L., Kechris, A. S. (1975). A basis result for $\Sigma 03$ sets of reals with an application to minimal covers. Proc. Amer. Math. Soc., 53(2):445–448.

Harrington, L., Shelah, S. (1982). The undecidability of the recursively enumerable degrees. Bull. Amer. Math. Soc. (N. S.), 6(1):79–80.

Harrington, L., Slaman, T. A. (2015). Interpreting arithmetic in the Turing degrees of the recursively enumerable sets.

Harrington, L., Soare, R. I. (1991). Post's program and incomplete recursively enumerable sets. Proc. Nat. Acad. Sci. USA, 88(22), 10242–10246.

Harrington, L., Soare, R. I. (1996). Dynamic properties of computably enumerable sets. In Computability, Enumerability, Unsolvability, volume 224 of London Math. Soc. Lecture Note Ser., pages 105–121. Cambridge Univ. Press, Cambridge.

Hinman, P. G. (1969). Some applications of forcing to hierarchy problems in arithmetic. Z. Math. Logik Grundlagen Math., 15, 341–352.

Hugill, D. F. (1969). Initial segments of Turing degrees. Proc. London Math. Soc. (3), 19, 1–16.

Ilya Kapovich, Alexei Miasnikov, Paul Schupp, Vladimir Shpilrain, (2003), Generic-case complexity, decision problems in group theory and random walks, J. Algebra 264, 665–694.

Ilya Kapovich, Paul Schupp, Vladimir Shpilrain, (2006), Generic properties of Whitehead's Algorithm and isomorphism rigidity of one-relator groups, Pacific Journal of Mathematics, 223, 113–140.

Institüt für Informatik, theorems, incompleteness theorems and arithmetic, Transactions of the American Mathematical models of Society, vol. 239 (1978).

Jockusch, Jr., C. G. (1980). Degrees of generic sets. In Recursion Theory: Its Generalization and Applications (Proc. Logic Colloq., Univ. Leeds, Leeds, 1979), volume 45 of London Math. Soc. Lecture Note Ser., pages 110–139. Cambridge Univ. Press, Cambridge.

Jockusch, Jr., C. G. (1981). Three easy constructions of recursively enumerable sets. In Logic Year 1979–80 (Proc. Seminars and Conf. Math. Logic, Univ. Connecticut, Storrs, Conn., 1979/80), volume 859 of Lecture Notes in Math., pages 83–91. Springer, Berlin.

Jockusch, Jr., C. G. (1985). Genericity for recursively enumerable sets. In Recursion Theory Week (Oberwolfach, 1984), volume 1141 of Lecture Notes in Math., pages 203–232. Springer, Berlin.

Jockusch, Jr., C. G. (2002). Review of On a conjecture of Kleene and Post by S. B. Cooper. Mathematical Reviews, MR1808943 (2002m:03061).

Jockusch, Jr., C. G., Posner, D. B. (1981). Automorphism bases for degrees of unsolvability. Israel J. Math., 40(2), 150–164.

Jockusch, Jr., C. G., Shore, R. A. (1984). Pseudojump operators. II. Transfinite iterations, hierarchies and minimal covers. J. Symbolic Logic, 49(4):1205–1236.

Jockusch, Jr., C. G., Simpson, S. G. (1976). A degree-theoretic definition of the ramified analytical hierarchy. Ann. Math. Logic, 10(1), 1–32.

Jockusch, Jr., C. G., Slaman, T. A. (1993). On the $\Sigma 2$-theory of the upper semilattice of Turing degrees. J. Symbolic Logic, 58(1), 193–204.

Kalimullin, I. Sh. (2003). Definability of the jump operator in the enumeration degrees, Journal of Mathematical Logic, vol. 3, pp. 257–267.

Kjos-Hanssen, B. (2002). Lattice Initial Segments of the Turing Degrees. PhD thesis, University of California at Berkeley.

Kjos-Hanssen, B. (2003). Local initial segments of the Turing degrees. Bull. Symbolic Logic, 9(1), 26–36.

Kleene, S. C. (1936). General recursive functions of natural numbers. Math. Ann., 112, 727–742.

Kleene, S. C. (1943). Recursive predicates and quantifiers. Trans. Amer. Math. Soc., 53, 41–73.

Kleene, S. C. (1952). Introduction to Metamathematics. D. Van Nostrand Co., Inc., New York, NY.

Kleene, S. C. (1981). Origins of recursive function theory. Ann. Hist. Comput., 3(1), 52–67.

Kleene, S. C., Post, E. L. (1954). The upper semi-lattice of degrees of recursive unsolvability. Ann. of Math. (2), 59, 379–407.

Knight J., 2014, Computable structure theory and formulas of special forms, VSL, Vienna, July, 2014.

Kuhlmann, S. (2000). Ordered Exponential Fields, Fields Institute Monographs 12

Kumabe, M. (1996). Degrees of generic sets. In Computability, Enumerability, Unsolvability, volume 224 of London Math. Soc. Lecture Note Ser., pages 167–183. Cambridge Univ. Press, Cambridge.

Kurt Gödel (1929). Thesis (1929): Gödel Completeness Theorem. Vienna. (1931). "Über *formal unentscheidbare Sätze der* Principia Mathematica *und verwandter Systeme I."* Gödel's First Incompleteness Theorem and a proof sketch of the Second Incompleteness Theorem.

Kurt Gödel (1931). "Über *formal unentscheidbare Sätze der* Principia Mathematica *und verwandter Systeme I."* Gödel's First Incompleteness Theorem and a proof sketch of the Second Incompleteness Theorem.

Lachlan, A. H. (1966a). The impossibility of finding relative complements for recursively enumerable degrees. J. Symbolic Logic, 31, 434–454.

Lachlan, A. H. (1966b). Lower bounds for pairs of recursively enumerable degrees. Proc. London Math. Soc. (3), 16, 537–569.

Lachlan, A. H. (1968). On the lattice of recursively enumerable sets", Trans. Amer. Math. Soc. 130, pp. 1–37.

Lachlan, A. H. (1968b). Distributive initial segments of the degrees of unsolvability. Z. Math. Logik Grundlagen Math., 14, 457–472.

Lachlan, A. H. (1972). Embedding nondistributive lattices in the recursively enumerable degrees. In Conference in Mathematical Logic—London '70 (Proc. Conf., Bedford Coll., London, 1970), pages 149–177. Lecture Notes in Math., Vol. 255. Springer, Berlin.

Lachlan, A. H. (1975a). A recursively enumerable degree which will not split over all lesser ones. Ann. Math. Logic, 9:307–365.

Lachlan, A. H. (1975b). Uniform enumeration operations. J. Symbolic Logic, 40(3), 401–409.

Lachlan, A. H. (1979). Bounding minimal pairs. J. Symbolic Logic, 44(4), 626–642.

Lachlan, A. H., Lebeuf, R. (1976). Countable initial segments of the degrees of unsolvability. J. Symbolic Logic, 41(2), 289–300.

Lachlan, A. H., Soare, R. I. (1980). Not every finite lattice is embeddable in the recursively enumerable degrees. Adv. in Math., 37(1), 74–82.

Lacombe, D. (1954). Sur le semi-reseau constitue par les degres d'indecidabilite recursive. C. R. Acad. Sci. Paris, 239, 1108–1109.

Ladner, R. E., Sasso, Jr., L. P. (1975). The weak truth table degrees of recursively enumerable sets. Ann. Math. Logic, 8(4), 429–448.

Lempp, S., Nies, A., and Slaman, T. A. (1998). The Π3-theory of the computably enumerable Turing degrees is undecidable. Trans. Amer. Math. Soc., 350(7), 2719–2736.

Lerman, M. (1971). Initial segments of the degrees of unsolvability. Ann. of Math. (2), 93:365–389.

Lerman, M. (1973). Admissible ordinals and priority arguments. In Cambridge Summer School in Mathematical Logic (Cambridge, 1971), pages 311–344. Lecture Notes in Math., Vol. 337. Springer, Berlin.

Lerman, M. (1977). Automorphism bases for the semilattice of recursively enumerable degrees. J. Symbolic Logic, 24:A–251.

Lerman, M. (1983). Degrees of Unsolvability. Perspectives in Mathematical Logic. Springer-Verlag, Berlin.

Lerman, M. (1996). Embedding's into the recursively enumerable degrees. In Computability, Enumerability, Unsolvability, volume 224 of London Math. Soc.

Lerman, M. Degrees of unsolvability. Perspectives in Mathematical Logic. Springer-Verlag, Berlin, 1983. ISBN 3–540–12155–2.

Lyndon, R. C., Schupp, P. E. Combinatorial Group Theory, Ergebnisse der Mathematik, Band 89, Springer 1977. Reprinted in the Springer Classics in Mathematics series, 2000.

Maass, W. (1982). Recursively enumerable generic sets. J. Symbolic Logic, 47(4), 809–823.

Magnus, Das W. (1932). Identitatsproblem fur Gruppen mit einer definierenden Relation, Math. Ann., 106, 295–307. A. Miasnikov and D. Osin, Algorithmically finite groups, preprint.

Marchenkov, S. S. (1976). A class of incomplete sets. Math. Notes, 20, 823–825. Martin, D. A. (1966). Classes of recursively enumerable sets and degrees of unsolvability. Z. Math. Logik Grundlagen Math., 12, 295–310.

Martin, D. A. (1968). The axiom of determinateness and reduction principles in the analytical hierarchy. Bull. Amer. Math. Soc., 74, 687–689.

Martin, D. A. (1975). Borel determinacy. Ann. of Math. (2), 102(2), 363–371.

McCarty, D. C. (1983). Realizability and recursive mathematics Technical Report CMUCS. Department of Computer Science CarnegieMellon University Report version of the author. PhD thesis Oxford University.

Miasnikov, A., Rybalov, A. (2008), Generic complexity of undecidable problems, Journal of Symbolic Logic, 73, 656–673.

Mucnik, A. A. (1956). On the unsolvability of the problem of reducibility in the theory of algorithms. Dokl. Akad. Nauk SSSR (N. S.), 108, 194–197.

Mucnik, A. A. (1958). Solution of Post's reduction problem and of certain other problems in the theory of algorithms. Trudy Moskov. Mat. Obsc., 7, 391–405.

Myhill, J. (1956). The lattice of recursively enumerable sets (abstract). J. Symbolic Logic, 21:220.

Myhill, J. (1961). Category methods in recursion theory. Pacific J. Math., 11, 1479–1486.

Nerode, A. and Remmel, J. B. (1986). Generic objects in recursion theory. II. Operations on recursive approximation spaces. Ann. Pure Appl. Logic, 31(2–3), 257–288.

Nerode, A., Shore, R. A. (1980). Second order logic and first order theories of reducibility orderings. In The Kleene Symposium (Proc. Sympos., Univ. Wisconsin, Madison, Wis., 1978), volume 101 of Stud. Logic Foundations Math., pages 181–200. North-Holland, Amsterdam.

Nies, A., Shore, R. A., Slaman, T. A. (1998). Interpretability and definability in the recursively enumerable degrees. Proc. London Math. Soc. (3), 77(2), 241–291.

Nourani, C. F. (1980). On Induction for Programming Logic, University of Michigan, Ann Arbor, Eurpean Association for TCS (EATCS 1981).

Nourani, C. F. (1997). Functorial Computability and Initial Computable Models, May 1997, AMS 927 Milwakee, Wisconsin, 1997. Abstract number 97T-68–191. Volume 18, No. 4, p. 624.

Nourani, C. F. (1997). Functorial Models, Admissible Sets, and Generic Rudimentary Fragments, March 1997, Summer Logic Colloquium, Leeds, July 1997. BSL, vol. 4, no.1, March 1998. www.amsta.leeds.ac.uk/events/logic97/con.html

Nourani, C. F. (1998). Functorial Computability and Generic Definable Models, International Congress Mathematicians, Berlin, August 18–27, 1998.

Nourani, C. F. (1998). Functorial Metamathematics, Maltsev Meeting, Novosibirsk, Russia, November 1998. Positive Forcing—The Unrecorded Specifics, ASL January 1998, San Antonio.

Nourani, C. F. (1999). Functorial Models and Infinitary Godel Consistency, March 4, 1999 Goedel Conference 5th Barcelona Logic Meeting and 6th Kurt Godel Colloquium.

Nourani, C. F. (2000). Generic Limit Functorial Models and Toposes, TOPO2000, August 2000, AMCA: 2000 Summer Conference on Topology and its Applications, Oxford, Ohio atlas-conferences.com/c/a/e/u/04.htm

Nourani, C. F. (2002). Functorial Models and Implicit Complexity International Congress of Mathematicians ICM2002, Beijing, China, August 20–28, 2002.

Nourani, C. F. (2002). Functorial Models and Implicit Hierarchy Degrees October 2002. Algebra and Discrete Mathematics Under the Influence of Models, 26–31 July, Hattingen-www. esf.org/euresco/03/pc03101 Abstracts http://www.esf.org/generic/1850/Truss03101.pdf

Nourani, C. F. (2003). Higher Stratified Consistency and Completeness Proofs, April 2003 SLK 2003, Helsinki August 14–20 http://www.math.helsinki.fi/logic/LC2003/abstracts/

Nourani, C. F. (2005). Functorial String Models, ERLOGOL-2005: Intermediate Problems of Model Theory and Universal Algebra, June 26–July 1, State Technical University/Mathematics Institute, Novosibirsk, Russia. www.nstu.ru/science/conf/erlogol-2005.phtml; www.ams. org/mathcal/info/ 2005_jun26-jul1_novosibirsk.html

Nourani, C. F. (2013). Product Models and Term Algebras, Brief draft, Berlin.

Nourani, C. F. (2014). More on Completion with Horn Filters, VSL, Vienna, July 2014.

Nourani, C. F., (1998). Admissible Models and Peano Arithmetic, ASL, March 1998, Los Angeles, CA. BSL, vol.4, no.2, June 1998.

Nourani. C. F. (2008). Positive Realizability on Horn Filters, Logic Colloquium 2008.

Odifreddi, P. G. (1989). *Classical Recursion Theory*, Studies in Logic and the Foundations of Mathematics 125, Amsterdam: North-Holland, ISBN 978–0–444–87295–1, MR 982269.

Paris, J. B. (1972). ZF ⊢ Σ04 determinateness. J. Symbolic Logic, 37, 661–667.

Post, E. L. (1965). Absolutely unsolvable problems and relatively undecidable propositions. In Davis, M., editor, The Undecidable. Basic Papers on Undecidable Propositions, Unsolvable Problems and Computable Functions, pages 338–433. Raven Press, Hewlett, N. Y.

Post, E. L. (1994). Solvability, Provability, Definability: The Collected Works of Emil L Post. (Davis, M., editor). Contemporary Mathematicians. Birkhauser Boston Inc., Boston, MA.

Richter, L. J. (1979). On automorphisms of the degrees that preserve jumps. Israel J. Math., 32(1), 27–31.

Robinson, R. W. (1971a). Interpolation and embedding in the recursively enumerable degrees. Ann. of Math. (2), 93, 285–314.

Robinson, R. W. (1971b). Jump restricted interpolation in the recursively enumerable degrees. Ann. of Math. (2), 93, 586–596.

Rogers, H. *The Theory of Recursive Functions and Effective Computability*, MIT Press. ISBN 0–262–68052–1; ISBN 0–07–053522–1.

Rogers, Jr., H. (1959). Computing degrees of unsolvability. Math. Ann., 138, 125–140.

Sacks, G. E. (1961). A minimal degree less than 0′. Bull. Amer. Math. Soc., 67, 416–419.

Sacks, G. E. (1963a). Degrees of Unsolvability, volume 55 of Ann. of Math. Studies. Princeton University Press, Princeton, NJ.

Sacks, G. E. (1963b). On the degrees less than 0′. Ann. of Math. (2), 77:211–231. Sacks, G. E. (1963c). Recursive enumerability and the jump operator. Trans. Amer. Math. Soc., 108, 223–239.

Sacks, G. E. (1964). The recursively enumerable degrees are dense. Ann. of Math. (2), 80, 300–312.

Sacks, G. E. (1966). Degrees of Unsolvability, second edition, volume 55 of Ann. of Math. Studies. Princeton University Press, Princeton, NJ.

Sacks, G. E. (1999). Selected Logic Papers, volume 6 of World Scientific Series in 20th Century Mathematics. World Scientific Publishing Co. Inc., River Edge, NJ.

Sacks, Gerald E. *Degrees of Unsolvability* (Annals of Mathematics Studies), Princeton University Press. ISBN 978–0691079417.

Shoenfield, J. R. (1959). On degrees of unsolvability. Ann. of Math. (2), 69, 644–653.

Shoenfield, J. R. (1960). Degrees of models. J. Symbolic Logic, 25, 233–237. Shoenfield, J. R. (1960/1961). Undecidable and creative theories. Fund. Math., 49, 171–179.

Shoenfield, J. R. (1965). Applications of model theory to degrees of unsolvability. In Theory of Models (Proc. 1963 Internat. Sympos. Berkeley), pp. 359–363. North-Holland, Amsterdam.

Shoenfield, J. R. (1966). A theorem on minimal degrees. J. Symbolic Logic, 31:539–544.

Shoenfield, J. R. (1971). Degrees of Unsolvability. North-Holland Publishing Co., Amsterdam.

Shoenfield, J. R. (1975). The decision problem for recursively enumerable degrees. Bull. Amer. Math. Soc., 81(6), 973–977.

Shoenfield, J. R. (1976). Degrees of classes of RE sets. J. Symbolic Logic, 41(3), 695–696.

Shoenfield, Joseph R. *Degrees of Unsolvability*, North-Holland/Elsevier, ISBN 978–0720420616.

Shore, R. (1993). The theories of the T, tt, and wtt r.e. degrees: undecidability and beyond. Proceedings of the IX Latin American Symposium on Mathematical Logic, Part 1 (Bahía Blanca, 1992), 61–70, Notas Lógica Mat., 38, Univ. Nac. del Sur, Bahía Blanca.

Shore, R. A. (1978). On the ∀∃-sentences of α-recursion theory. In Generalized Recursion Theory, II (Proc. Second Sympos., Univ. Oslo, Oslo, 1977), volume 94 of Stud. Logic Foundations Math., pages 331–353. North-Holland, Amsterdam.

Shore, R. A. (1999). The recursively enumerable degrees. In Handbook of Computability Theory, volume 140 of Stud. Logic Found. Math., pp. 169–197. North-Holland, Amsterdam.

Simpson, S. Degrees of unsolvability: a survey of results. *Handbook of Mathematical Logic*, North-Holland, 1977, pp. 631–652.

Slaman, T. A., Woodin, W. H. (1986). Definability in the Turing degrees. Illinois J. Math., 30(2), 320–334.

Soare, R. (1987). *Recursively enumerable sets and degrees.* Perspectives in Mathematical Logic. Springer-Verlag, Berlin, ISBN 3–540–15299–7.

Soare, R. I. (1996). Computability and recursion, Bulletin of Symbolic Logic 2 (1996), p.284–321.

Spector, C. (1956). On degrees of recursive unsolvability. Ann. of Math. (2), 64, 581–592.

Thomason, S. K. (1970). A theorem on initial segments of degrees. J. Symbolic Logic, 35, 41–45.

Titgemeyer, D. (1962). Untersuchungen uber die Struktur des Kleene-Postschen Halbverbandes der Grade der rekursiven Unlosbarkeit. PhD thesis, University of Munster.

Titgemeyer, D. (1965). Untersuchungen uber die Struktur des Kleene-Postschen Halbverbandes der Grade der rekursiven Unlosbarkeit. Arch. Math. Logik Grundlagenforsch., 8, 45–62.

Towsner, H. (2014). Peano and Heyting Arithmetic, UPenn Technical report.

Turing, A. M. (1936). On computable numbers, with an application to the Entscheidungsproblem. Proc. Lond. Math. Soc., II. Ser., 42, 230–265.

Turing, A. M. (1939). Systems of logic based on ordinals. Proc. Lond. Math. Soc., II. Ser., 45, 161–228.

CHAPTER 9

REALIZABILITY AND COMPUTABILITY

CONTENTS

9.1   INTRODUCTION

There are at least five realizability areas that are not all mutually disjoint on thought precepts:

  (i)   realizing a type with closures omitting extras;
  (ii) realizing a model or type with morphism and filters;
  (iii realizability on topos, triposes, or sheaves;
  (iv) realizing universal arrows in categories; and
  (v) Kleene and intuitionistic realizability.

From this authors' perspective the topics were presented in part in his publications including the 2014 book on functorial model theory during contemporary times. Since the present volume is more on mathematics from the computational perspective the author has minimized the exposure to pure foundational areas from mathematics. However, we add what has been contributed on all the five areas with respect to the new development from the authors on the field. Historic notions to the present on what the realizability schools are with this author merged only at the Savoie meeting 2012. A major reason for there not being mergers is that the recent realizability schools are essentially lifts on combinatory algebras or Hyting algebras to topos characterizations in the making even as we speak. The overview is as follows and reconciled to a degree on the ending sections. Computability with the intuitionistic schools this author we began to address on model computability on initial models during the past decade or more with a newer look was at FMCS Vancouver 5–6 years ago, but dates 20 years. Newer areas are treated in the author 2014 book on realizability on generic filters, ultrafiters and ultraproduct projective computability on saturated types, Gödel sets, and presheaves.

From Longley (2012) we have the glimpse that amongst the preceding schools there is no general definition of what realizability is. Realizability interpretations for logics such as Heyting calculus as realizability applications, quintessential applicative programming models, arithmetic type systems are the trends. The idea of realizability was first introduced by Kleene in a paper of 1945. One can see Kleene's work as an attempt to recasts independent views from Brouwer's thought in more accessible terms. More on that is presented in the last section.

Universal realizations were goals in ADJ's IBM accomplishments at the start in the 1970's of course. structures. The treatments are on further on filters and ultraproduct models, beginning with Hyting and Kleene on intuitionistic logic. This chapter is a brief overview including the newer developments from authors during the past two years.

This author's (Nourani, 2014) volume expounds on new realizability areas based on omitting types, forcing, and adjoint functors, with applications to Kleene.

9.2   CATEGORICAL MODELS AND REALIZABILITY

There is some confusion between realizability to models and model categories vs. topos realizability, for example on second order types, Martin Löf types, being an example. The alternate track realizability authors have being focusing on combinatory types and not treated what was in this author's 2014 on homotopy realizations on Martin Löf with model categories. Where models for set theory are being examined at pure mathematics is addressed on the last section. Since the techniques here have already addressed forcing on topos and model categories at algebraic set theory (Nourani 2014, Blass, 2014).

9.2.1   CATEGORICAL INTUITIONISTIC MODELS

Lawvere recognized the Grothendieck topos, introduced in algebraic topology as a generalized space, as a generalization of the category of sets, on mathematical logic on *quantifiers and sheaves* (1970). With Myles Tierney, Lawvere then developed the notion of elementary topos, thus establishing the fruitful field of topos theory, which provides a unified categorical treatment of the syntax and semantics of higher-order predicate logic. The resulting logic is formally intuitionistic. Andre Joyal is credited, in the term Kripke–Joyal semantics, with the observation that the sheaf models for predicate logic, provided by topos theory, generalize Kripke semantics. Joyal and others applied these models to study higher-order concepts such as the real numbers in the intuitionistic setting.

An analogous development was the link between the simply typed lambda calculus and Cartesian-closed categories (Lawvere, Lambek, Scott), that were for domain theory developments. The realizability concept of Kleene is also a serious route to intuitionistic logic towards categorical logic. From Kolmogorov there is one way to explain the protean Curry–Howard isomorphism. The Curry-Howard-Lambek correspondence provides a deep isomorphism between intuitionistic logic, simply-typed lambda calculus and Cartesian closed categories. Intuitionistic arithmetic hierarchy is an area that the developments in the next few sections might be applied to a primitive realizability. In what follows we state specific areas with the Kiesler fragment since this author's decade or two example developed areas, but the developments are for the most part are all applicable to infinitary admissible languages in general, for example, on Chapter 10 (Nourani, 2014).

9.2.2   INFINITARY LANGUAGE PRODUCT MODELS

Starting with Infinite language categories (Nourani, 1995) specific computing models based on fragments are presented. Positive categories and Horn categories are new fragment categories are presented as example with positive process algebraic computing that were presented in the preceding chapters since Nourani (2005). For example, the author defined the category $L_{P,\omega}$ to be the category with objects positive fragments and arrows the subformula preorder on formulas to present models. The model bases are Fragment Consistency Models, where new techniques for creating generic models are defined. Infinitary positive language categories are briefed and infinitary complements to Robinson consistency from the author's preceding volume and publication is summarized to present new positive omitting types techniques for realizability with infinitary positive fragment higher stratified computing categories.

9.2.3   POSITIVE GENERIC MODELS

Definition 9.1: Let M be a structure for a language L, call a subset X of M a generating set for M if no proper substructure of M contains X, for example, if M is the closure of X U {c(M): c is a constant symbol of L}. An assignment of constants to M is a pair <A, G>, where A is an infinite set of constant symbols in L and G: A $\rightarrow$ M such that {G(a): a in A} is a set of generators for M. Interpreting a by g(a), every element of M is denoted by at least one closed term of L(A). For a fixed assignment <A, G> of constants to M, the diagram of M, D<A, G>(M) is the set of basic (atomic and negated atomic) sentences of L(A) true in M. (Note that L(A) is L enriched with set A of constant symbols.).

Definition 9.2: A generic diagram for a structure M is a diagram D<A, G>, such that the G in definition above has a proper definition by a specific function set.

Starting with $L_{\omega1,K}$ language fragments, define ω-chain models, and back and forth to a limit model.

Remember we had defined the fragment by letting C be a countable set of new constants symbols and form the first order language K by adding to L the constants c in C. Let K<A> be the set of formulas obtained from formulas φ in L<A> by replacing finitely many free variables by constants c in C.

Let $L_{\omega1,K}$ be the least fragment of $K_{\omega1, \omega}$ which contains L<A>. Each formula φ in K<A> contains only finitely many c in C. This implies when raking leaves on the trees, there are only finite number of named branches claimed by constant names. The infinite trees are defined by function names, however. The functions define the model with schemas on the generic diagram functions.

From chapters before on a theory T for $L_{\omega1,K}$ fragments, let T* be T augmented with induction the constants. From the functorial view what follows resembles to a Cosmic Archeology, scooping out the model theoretic specifics for a functorial model theory.

For example, Σ_1 Skolem functions are a specific function set. We abbreviate generic diagrams by G-diagrams. From chapters before on a theory T for $L_{\omega1,K}$-fragments, let T* be T augmented with induction schemas on the generic diagram functions. Like onto skating on a thin functorial linguistics layer, a 1990's project by the author, c.f. Chapter 3, let us define functorial string models. We start to define functorial models by infinite chains, back and forth designs and elementary diagrams. We start to define functorial models by infinite chains, back and forth designs, and elementary diagrams. A chain of models is an increasing sequence of models A_0 $\subset A_1 \subset A_3 \subset A_4 \subset ... A_\beta$ $\beta<\alpha$, whose length is an ordinal α. The union of chains is the model A= $\cup A\beta$, defined as follows.

Defining ω-chain models, we take the union of chains to be the model A = $\cup A_\beta$ as follows. The universe of A is the set A = $\cup A_\beta$. Each relation R of A is the union of the corresponding relations of $A_\beta R = \cup R_\beta$. Similarly, each function G of A is the union of the corresponding functions of Aβ, G= $\cup G_\beta$, $\alpha < \beta$. The models Aβ and A all have the same constants. An elementary chain is A chain of models $A0 \subset A1 \subset A3 \subset A4 \subset ...A\beta$ such that $A\gamma < A\beta$ whenever $\gamma<\beta < \alpha$, < is the elementary extension relation.

Elementary chains are applied to define Robinson's consistency theorem and we plan to define similar techniques for defining String Models on the infinitary $L_{1,K}$ fragments. Of course the task is quite difficult and intricate functorial models on language fragments have to be designed. From basic model theory we have the following theorem and the chain lemma on the following section.

Theorem 4.1: (Robinson's Consistency Theorem)

Let L1, L2 be two languages. Let $L = L1 \cap L2$. Suppose T is a complete theory in L and $T1 \supset T$, $T2 \supset T$ are consistent in L1, L2, respectively. Then $T1 \cup T2$ is consistent in the language $L1 \cup L2$.

We want to define functorial models piecemeal from language fragments by an infinite limit. There are two ways to view it.

(a) Take $L_{\omega1,K}$ language fragments, define ω-chain models, and back and forth to a limit diagram model.

(b) Define models for the Fi from elementary diagrams, that is, define a limit model by embedding from a D<A, G> model.

What complete theory can we fall onto? It has to be the theory Th(F), where F is the generic model functor defined to Set, that is,

Th (F: $\mathbf{L}^{op}_{\omega1,K} \to \mathbf{Set}$).

9.2.4 OMITTING TYPES REALIZABILITY

Definition 9.3: Let Σ be a set of formulas in the variables $x1, \ldots, xn$. Let R be a model for L. We say that R realizes Σ iff some n-tuple of elements of A satisfies Σ in R. R omits Σ iff R does not realize Σ.

Let $\Sigma(x1, \ldots, xn)$ be a set of formulas of L. A theory T in L is said to locally realize Σ iff there is a formula $\varphi (x1, \ldots, xn)$ in L s.t. (i) φ is consistent with T; (ii) for all $\sigma \in \Sigma$, $T \models \varphi \to \sigma$. That is every n-tuple of T which realizes φ satisfies Σ. We say that T locally omits Σ iff T does not locally realize Σ.

For our purposes we define a new realizability basis (Nourani, 2006).

Definition 9.4: Let $\Sigma(x1, \ldots, xn)$ be a set of formulas of L. Say that a positive theory T in L positively locally realize Σ iff there is a formula $\varphi (x1, \ldots, xn)$ in L s.t. (i) φ is consistent with T: (ii) for all $\sigma \in \Sigma$, $T \models \varphi$ or $T \cup \sigma$ is not consistent.

Definition 9.5: Given models A and B, with generic diagrams DA and DB we say that DA homomorphically extends DB iff there is a homomorphic embedding f: A → B.

Consider a complete theory T in L. A formula φ (x1, ..., xn) is said to be complete (in T) iff for every formula ψ(x1, ..., xn), exactly one of T ⊨ φ → ψ or T ⊨ φ → ¬ ψ. A formula θ (x1, ..., xn) is said to be completable (in T) iff there is a complete formula φ (x1, ..., xn) with T models φ → θ. If that can't be done θ is said to be incompletable.

Proposition 9.1: Let R and |B be models for L. Then R is isomorphically embedded in B iff B can be expanded to a model of the diagram of R.

Proposition 9.2: Let R and |B be models for L. Then R is homomorphically embedded in B iff B can be expanded to a model of the positive diagram of R.

Theorem 9.2: Let L1, L2 be two positive languages. Let L = L1 ∩ L2. Suppose T is a complete theory in L and T1 ⊃ T, T2 ⊃ T are consistent in L1, L2, respectively. Suppose there is model M definable from a positive diagram in the language L1 ∪ L2 such that there are models M1 and M2 for T1 and T2 where M can be homomorphically embedded in M1 and M2. Then (i) T1 ∪ T2 is consistent. (ii) There is model N for T1 ∪ T2 definable from a positive diagram that homomorphically extends that of M1 and M2.

Theorem 9.3: Let L1, L2 be two positive languages. Let L = L1 ∩ L2. Suppose T is a complete theory in L and T1 ⊃ T, T2 ⊃ T are consistent in L1, L2, respectively. Then,

(i) T1 ∪ T2 has a model M, that is a positive end extension on Models M1 and M2 for T1, and T2, respectively.

(ii) M is definable from a positive diagram in the language L1 ∪ L2. Recall that a companion closure generic filter was characterized with a set T*.

Omitting types is an important technique on model-theoretic forcing and set-theoretic considerations. On the preceding section we have examined positive omitting types that this author developed on the decades publications towards a basis for topos model characterizations, natural for algebraic theories. The more general techniques are presented here that has not had direct application to Topos.

Given a theory T and a nonnegative integer n, let n(T) be the set of all nature (φ) $\subseteq$ signature (T). Say that $\subseteq$ n(T) is consistent over T if there exist A s.t. A $\models$ T and a1, ..., an $\in$ |A| such that A $\models \varphi$(a1, ..., an) for all $\varphi \in Y$. Note that by compactness, if each finite subset of is consistent over T then so is. formulas φ(x, ..., x) with no free variables other than x1, ..., xn, such that sig- 1n.

A realization of $\subseteq$ n (T) is an n-tuple a1, ..., an $\in$ |A|, A s.t. A $\models$ T, and A $\models \varphi$(a1, ..., an) for all $\varphi \in$. We then say that is realized in the model A.

We say that $\psi \in$ n (T) is a logical consequence of Y $\subseteq$ n (T) over T if every realization of Y is a realization of ψ.

Definition 9.6: An n-type over T is a set p $\subseteq$ n(T) which is consistent over T and closed under logical consequence over T.

Definition 9.7: An n-type p over T is said to be complete if, for all $\varphi \in$ n(T) either $\varphi \in$ p or $\neg \varphi \in$ p. The set of all complete n-types over T is denoted n(T).

Definition 9.8: A type p(x) is said to be isolated by j if there is a formula j (x) with the property that $\forall y(x) \in$ p(x), j(x) $\rightarrow$ y(x). Since finite subsets of a type are always realized in M, there is always an element b $\in$ Mn such that j (b) is true in M; that is, M models j, thus b realizes the entire isolated type. So isolated types will be realized in every elementary substructure or extension. So isolated types can never be omitted.

A direct statement for omitting types is as follows:

Theorem 9.4: Omitting Types Theorem: Let L be a countable language. Let T be an L-theory. Let {pi: i $\in$ N } be a countable set of nonisolated n-types of T. There is a countable L-structure M such that M models T and M omits each pi.

Proof: This proof is by the well-known Henkin techniques, which results in a model all of whose elements are the interpretations of constant symbols from some language. The construction for this proof in particular is done so that the theory this model satisfies asserts that each tuple of constants fails to satisfy at least one f from each type pi. As a result, the model will not realize any of the types.

Definition 9.9: Let p be an n-type over a theory T. A model A of T is said to omit p if there is no n-tuple a1, ..., an $\in$ |A| which realizes p.

We have already seen how to use the compactness theorem to construct a model, which realizes p. It is somewhat more difficult to construct a model, which omits p. Indeed, such a model may not even exist. The omitting types theorem gives a sufficient condition for the existence of a model of T, which omits p. (If T is complete, this sufficient condition is also necessary.)

Definition 9.10: We say that p is principal over T if it is generated by a single formula, for example, there exists $\varphi \in \Phi n$ (T) such that p = {ψ $\in \Phi n$ (T): T $\models \forall v1 \cdots \forall vn(\varphi(v1, ..., vn) \rightarrow \psi(v1, ..., vn))$. Such a φ is called a generator of p.

We say that p is essentially nonprincipal over T if p is not included in any principal n-type over T.

Note that the hypothesis that T is countable cannot be omitted from the theorem.

Standard example is considering a language with one binary connective, $\in$. Let M be the model $\omega \in \omega$, the ordinal standard well ordering. Let T denote the theory of this model. The set of formulas p(x): {$n \in x$| $n \in \omega$} is a type. Let p0 $\in$ p(x). Take the successor of the largest ordinal mentioned in the set of formulas p0(x). Then this will clearly contain all the ordinals mentioned in p0(x).Thus, we have that p(x) is a type. Now, note that p(x) is not realized in M. Since that would imply there is some n $\in \omega$ that contains every element of w.

9.3   POSITIVE REALIZABILITY MORPHISMS AND MODELS

Definition 9.11: A formula is said to be positive iff it is built from atomic formulas using only the connectives &, v and the quantifiers $\forall$, $\exists$.

Definition 9.12: A formula $\varphi(x1, x2, ..., xn)$ is preserved under homomorphisms iff for any homomorphisms f of a model A onto a model B and all a1, ..., an in A if A $\models \varphi$ [a1, ..., an] then B $\models \varphi$ [fa1, ..., fan].

Theorem 9.6: A consistent theory is preserved under homorphisms iff T has a set of positive axioms.

Lemma 9.1: For a chain A_β, $\beta < \alpha$, of models, $\cup A\beta$ is the unique model with universe $\cup A_\beta$ which contains each Aβ as a submodel.

A functor F: $\mathbf{L}^{op}_{\omega 1, K} \rightarrow$ Set can be defined by sets Fi, where the Fi's are defining a free structure on some subfragment of $L_{\omega 1, K}$. To be specific we can define the subfragment models A(Fi) straight from the ω-inductive definition of the Infinitary fragment. F0 assigns names to the Set members, for example, F1 can define 1-place functions and relations, so on and so forth.

Taking the languages defined above on Kiesler fragments and the Robinson's consistency theorem define Functorial Limit Chain models as follows. We shall refer to it by FLC-models (Nourani, 1997).

Let A and B, be models for Fi and Fi+1, respectively. Let $A \equiv_L B$, and f: $A <_L B$ mean the L-reduct of A and B are elementarily equivalent and that f is an elementary embedding of A|L into B|L. Let A FLC model be the limit model defined by the elementary chain on the L-reducts of the models defined by the Fi's. A specific FLC model is defined by theorem 4.4's proof.

9.3.1   FRAGMENT PRODUCT ALGEBRA REALIZABILITY

The propositions 9.2–9.4 and theorem are from basic model theory. Proposition 9.4 and Theorem 9.1 are from Nourani (2005).

Proposition 9.3: Let φ be a universal sentence. Then φ is a (finite) direct product sentence iff φ is equivalent to a universal Horn sentence.

Proposition 9.4: Let φ be a universal sentence. Then φ is a (finite) direct product sentence if and only if φ is equivalent to a universal Horn sentence.

Proposition 9.5: (Author, 2006) Let I be the set T*. Let $\varphi (x1...xn)$ be a Horn formula and let $\Re I$, $i \in I$ be models for language L. let $a1....an \in \Pi i \in I \, Ai$. The $\Re i$ are fragment Horn models.

If $\{i \in I: \Re i \models \varphi[(a1(i) ...an(i)]\}$ then the direct Π_D on $\Re i \models \varphi[a1D ...anD]$, where D is the generic filter on T*.

We can prove the following as a lemma and a consequent theorem.

Theorem 9.7: (Author, 1997) The embedding to form elementary chains on Fi's can be defined by a back and forth model design from the strings in $L_{\omega 1, B}$ language fragments.

Theorem 9.8: (Author, 1997) By defining models corresponding to the Fi on the fragments as an ω-chain from the elementary diagrams on the Th (A(Fi)) a generic model is defined by the limit.

9.3.2 POSITIVE REALIZABILITY ON HORN FILTERS

During 2005–2007, the author explored positive and Horn fragment categories based on his 1981 on positive forcing on Kiesler fragments. Positive forcing had defined T^* on a theory T to be T augmented with induction schemas on the generic diagram functions, for a model M fo T. Let P be a poset, F a family of sets, G subset of P. The author defined positive local realizability on 2005–2007 publications (ASL).

Theorem 9.9: (Author, 1981). T^* is a F-generic filter.

Lemma 9.2: (Author, 2006) T^* is a principal proper filter.

Proposition 9.6: (Author, 2006) Let I be the set T^*. Let $\phi(x1 \ ...xn)$ be a Horn formula and let Ri $i\varepsilon I$ be models for language L, let $a1 \ ...an \ \varepsilon \ \Pi i \ \varepsilon I$ Ai. The Ri are fragment Horn models.

> If $\{i\varepsilon I$: Ri
> $\phi[(a1(i) \ ...an \ (i)]\}$ then the direct product ΠD Ri
> $\phi[a1D...an \ D]$, where D is the generic filter on T^*.
> Applying positive local realizability from above sections:

Theorem 9.9: (Author, 2012) Let $(P, \leq)$ be a positive Horn Poset and $p\in P$. If D is a countable family of dense subsets of P then $T^*(P)$ is D-generic filter F in P such that $p\in F$ and every $p\in F$ has a positive local realization.

Proof: (e.g., Nourani, 2014).

The above theorem is a Horn density counterpart to the important set foundational Rasiowa–Sikorski lemma.

9.3.3 COMPUTABILITY AND POSITIVE REALIZABILITY

Let us start from certain model-theoretic premises with propositions known form basic model theory.

Proposition 9.7: Let $\Re$ and $|B$ be models for L. Then $\Re$ is isomorphically embedded in $|B$ iff $|B$ can be expanded to a model of the diagram of $\Re$.

Proposition 9.8: Let $\mathfrak{R}$ and $|B$ be models for L. Then $\mathfrak{R}$ is homomorphically embedded in $|B$ iff $|B$ can be expanded to a model of the positive diagram of $\mathfrak{R}$.

Definition 9.13: Given models A and B, with generic diagrams D_A and D_B we say that D_A homomorphically extends D_B iff there is a homomorphic embedding f: $A \rightharpoonup B$.

Consider a complete theory T in L. A formula $\varphi\,(x1,..., xn)$ is said to be complete (in T) iff for every formula $\psi(x1,..., xn)$, exactly one of T $\models \varphi \lozenge$ ψ or T $\models \varphi \lozenge \neg \psi$. A formula $\theta\,(x1, ..., xn)$ is said to be completable (in T) iff there is a complete formula $\varphi\,(x1,..., xn)$ with T models $\varphi \lozenge \theta$. If that can't be done θ is said to be incompletable.

Example realizability based on Horn filters is the following.

Lemma 9.3: (author ASL, 2007) Every formula on the presentation P is completable in T*.

Proof: New proof ASL, Vienna (2014). Applies Theorem 9.10.

Let $L_{P, \omega}$ be the positive fragment obtained from the Kiesler fragment.

Define the category $L_{P, \omega}$ to be the category with objects positive fragments and arrows the subformula preorder on formulas.

Define a functor F: $L^{op}_{P, \omega} \rightarrow$Set by a list of sets Mn and functions fn. The functor F is a list of sets Fn, consisting of:

(a) the sets corresponding to an initial structure on $L_{P, \omega}$, for example the free syntax tree structure, where to f(t1, t2,.. tn) in $L_{P, \omega}$ there corresponds the equality relation f(t1, ..., tn)=ft1...tn in Set;

(b) the functions fi: Fi+1 $\rightarrow$ Fi.

Proposition 9.9: (author, 2010—positive morphic extensions) The following are equivalent:

Every positive sentence holding on $\eta\mathfrak{R}$ also holds on B. There are elementary extensions $\eta\mathfrak{R} < \eta\mathfrak{R}$', B < B' such that B' is a homomorphic image of $\eta\mathfrak{R}$'

The ordering we call morphic might further be computationally appealing ever since computable functors were defined by us. We define initial ordered structures with a slight change in terminology, since we have not yet defined the relation between preorder algebras and admissible sets. Admissible sets are as Barwise (1968). Functorial admissible models in the admissible set sense are due to Nourani (1997).

Definition 9.14: A preorder $<<$ on a $\sum$-algebra A is said to be morphic iff for every s in s, s1...sn and ai, Ki in Asi, Ksi, respectively, if ai, bi for *i* in [n] then, sA (a1, ..., an) $<<$ sB (b1, ..., Kn).

What I have called a morphic order is called an admissible order by (Thatcher Wagner, et.al., 1979) to save confusion from ambiguity to admissible sets. The ordering's relation to admissible sets and functores are explored at (Nourani, 1997).

Proposition 9.10: If $<<$ is a morphic preorder on a $\sum$-algebra A, then the equivalence relation $\sim$ determined by $<<$ is a congruence relation.

Proposition 9.11: The morphic quotient structure for an ordered $\sum$-algebra A, is an ordered $\sum$-algebra.

We apply function and relation definably in the following theorem. To have a glimpse as to what it implies let us see an example. Let A be a model with built-in Skolem functions. A function $f \in A$, for A the model's universe, is said to be definable iff there exits a formula j (xyz1...zn) in the language and elements a1, ..., an $\in$ A such that for all a, K $\in$ A, A j [aba1, ..., an] iff f(a)= b. We have defined generic Functors on the category Lω1, K with computable hom sets (Nourani, 1996).

Definition 9.15: For functions f and g in a structure for a language L, define the morphic *preorder* f $<<$ g iff there are formulas j and y from L such that the formulas define f and g respectively, and j is a subformula of y in the sense of definition 3.1.

9.3.4 *MORPHIC REALIZATION FUNCTORS*

Define a functor F: $\mathbf{L}_{\omega1,K} \rightarrow$ Set by a with he free syntax tree structure, where to f(t1, t2,.. tn) in $L_{\omega1,K}$ there corresponds the equality relation f(t1, ..., tn)=ft1...tn in Set; (b) the functions fi:Fi+1 $\rightarrow$ Fi. The functor F is a list of sets Fn, consisting of (a) the sets corresponding to a morphic preorder on $L_{\omega1,K}$ defined by the deniability order from the subformula Preorder on $L_{\omega1,K}$.

Theorem 9.11: The morphic preorder functor defines initial models for Lω1, K.

To prove the above there are two routes. Route A applies Keisler, and author 1997 on with new techniques similar to Robinson's consistency theorem, however, with varying languages and fragments called functorial consistency (Nourani, 1997). Route B is outlined on the following paragraph from authors' publications where functorial morphic models are defined. One consequence is a theorem, which can get us Theorem 9.12's proof. This is obtained by techniques, which are by no means obvious applying admissible sets and Urelements.

Theorem 9.12: The morphic preorders on initial structures are definable by formulas, which are preserved by end extensions on fragment consistent models.

Towards the proof: Since Nourani (1997) European Research council brief as follows. At the Summer 1996 European logic colloquium (Nourani, 1996) had put forth descriptive computing principles, defining descriptive computable functions. Amongst the theorems at is that for A an admissible computable set, A is descriptive computable.

Generic diagrams were applied to define admissible computable sets and models.

Definition 9.16: A model is admissible iff its universe, functions, and relations are defined with or on admissible sets.

Theorem 9.13: (author, 1990) Admissible models are obtained by taking a reduct from the admissible hull to the Skolem hull definable by a generic diagram.

The proof for the above applied a theorem from Ferferman and Kriesel that might have applications to predicative realizability models.

ADJ (1979) defines ordered initial MacLane (1971) categories important for computing. The ordering's significance is its being operation preserving. The author defined a model-theoretic ordering for the initial ordered structures to reach for models for which operation preserving orderings are definable. The author's dissertation was in part on a model theory on many-sorted categorical logic and its definability with finite similarity type. That proved to be a newer look at intuitionistic forcing a year or two later and announced at ASL, Boston. The ordering we call *morphic* might further be computationally appealing ever since computable functors were defined by Nourani (1996, 1998).

Definition 9.17: For functions f and g in a structure for a language L, define the morphic *preorder* f << g iff there are formulas j and y from L such that the formulas define f and g respectively, and j is a subformula of y in the sense of definition of the fragments for infinitary logic (Keisler 1973).

Theorem 9.14: (Ferferman-Kriesel, 1966) If a formula q is preserved under end extensions then it is equivalent to a $_1$ formula.

We can apply our generic diagrams and sets from + forcing Nourani (1982) to the Σ_1 formulas (Barwsie, 1972; Kripke-Platek, 1966) by applying a theorem form (Feferman-Kriesel, 1966.) It might also put our mind at serenity with the preorder functors and initial structures on the forthcoming chapters. We apply the Ferferman-Kriesel theorem to check for conditions on admissible sets starting with what we call *generic rudimentary sets*. With formula belongs to a condition for fragment consistent Functorial models. Kiesler (1971) and Barwise (1969, 1972) admissible fragments are also pertaining. We had defined functors from the language category positive forcing (Nourani, 1983) and fragment consistency models (Nourani, 1997), we can check if the $L\omega1$, K defined on the Keisler (1973) $L_{\omega1,\omega}$ fragment to the category *Set*. Its arrows are the preorder arrows and its objects the fragments. Its properties and further areas called functorial model theory, admissible models, and ordered structures are defined by the author 1994 on. We conclude with the morphic order functor theorem. The functor creates limits on the category *Set* applying the morphic preorder to $L_{\omega1,K}$.

Theorem 9.15: (Author 1999)The morphic preorder functor defines initial models for $L_{\omega1,K}$. The model has the morphic preorder property.

9.4 POSITIVE CATEGORIES AND CONSISTENCY MODELS

Let $L_{P,\omega}$ be the positive fragment obtained from Kiesler fragment.

Define the category $L_{P,\omega}$ to be the category with objects positive fragments and arrows the subfoumual preorder on formulas.

Define a functor F: $\mathbf{L}^{op}_{P,\omega} \to$ Set by a list of sets Mn and functions fn. The functor F is a list of sets Fn, consisting of

(a) the sets corresponding to an initial structure on $L_{P,\omega}$, for example the free syntax tree structure, where to f(t1, t2,.. tn) in $L_{P,\omega}$ there corresponds the equality relation f(t1, ..., tn)=ft1...tn in Set;

(b) the functions fi:Fi+1 $\to$ Fi.

Theorem 9.16: (author, 1997) Infinitary Fragment Consistency on Algebras.

Let T<i, i+1> be complete theories for L<i, i+1> = Li intersect Li+1. Let Ti and Ti+1 be arbitrary consistent positive theories for the subfragments Li and Li+1, respectively, satisfying Ti contains T<i, i+1> and Ti+1 contains T<i, i+1>. Let Ai be A(Fi) and Bi be A(Fi+1).

(i) There are iterated elementary extensions Bi+1> Bi and an embedding fi: Ai < Bi;

(ii) Slalom between the language pairs Li Li+1 gates to a limit model for L, a model M for $L_{P,\omega}$.

Proof: Starting with a basis model $\Re_{<0,1>}$ and $B_{<0,1>}$, models of T_0 and T_1. $\Re_0|L<0,1>$ and $B_0|L<0,1>$ are models of a complete theory, therefore $\Re_0|L<0,1> \equiv B_0|L<0,1>$. It follows that the elementary diagram of $A_0|l<0,1>$ is consistent with the elementary diagram of $B_0|L<0,1>$. Let T<i, i+1> be the positive theory in L<i, i+1> = Li intersect Li+1 defined by T* (A(Fi)) intersect A(Fi+1)), where T* is the theory obtained from the positive theory T augment with inductive consequences. We shall construct a limit model M realizing (i) enroute. Every $\Sigma(x1...xn)$ a set of formulas of L<0,1> that is provable in T, $T*(A_0|L<0,1>)$ positively locally realizes Σ, that is there is a formula $\varphi (x1...xn)$ in L<01> s.t. φ is consistent with $T*(A_0|L<0,1>)$ and for all $\Sigma\varepsilon \Sigma$, $T* \models \varphi$ or $T* \cup \{\sigma\}$ is not consistent. Every finite subset Σ of $T*(A_0|L<0,1>)$ has a model, therefore $T*(A0|L<0,1>)$ has a model M_0, where for every $\varphi \varepsilon L <0,1> T*(A0|L<0,1>) \models \varphi$ iff M0 $\models \varphi$. Starting with a basis models $\Re_0$ and B_0, we construct a positive tower. On the iteration we realize that by s preceding proposition at each stage there are elementary extensions to $\Re_i$ and B_i there are elementary extensions $\Re < \Re,$' $|B < |B$' such that $|B$' is a homomorphic image of $\Re.$' Therefore, there are elementary extensions B1> B0 and an embedding f1: A0 < B1 at L<0,1>. Passing to the expanded language L<0,1>A0, we have (A0, a)aε A0 $\equiv$ L<0,1> A0 (B1, fa) a ε A0. g1 inverse is an extension of f1.

Iterating, we obtain the tower depicted sideways.

```
AO   <     A1  <    A2  < .........
   \  f1   | g1   \  f2   | g2
BO  <      B1  <    b2  < .....
```

Slalom between the language pairs Li Li+1 gates to a limit model for L. For each m, fm $\subset$ inverse(gm) $\subset$ fm+1, fm: Am-1 < Bm at L<m-1, m>.

Let $A = \cup Am$, $m < \omega$, $B = \cup Bm$, $m < \omega$. B is isomorphic to a model B' such that $A|L = B'|L$. Piecing A and B' together we obtain a model M for $L_{P,\omega}$. ?

9.4.1 HORN COMPUTABILITY AND REALIZABILITY

Define the category $L_{H,\omega}$ to be the category with objects Horn fragments and arrows the subfoumual preorder on formulas.

Define a functor $F: L_{H,\omega}{}^{op} \to$ Set by a list of sets Mn and functions fn. The functor F is a list of sets Fn, consisting of

(a) the sets corresponding to an initial structure on $L_{P,\omega}$, for example the free syntax tree structure, where to f(t1, t2,.. tn) in $L_{P,\omega}$ there corresponds the equality relation f(t1, ..., tn)=ft1...tn in Set;

(b) the functions fi:Fi+1 $\to$ Fi. $_\subseteq$

Proposition 9.12: Infinitary Horn Fragment Consistency

Let T<i, i+1> be the complete theory in L<i, i+1> = Li intersect Li+1, defined by Th (A(Fi) intersect A(Fi+1)), Let Ti and Ti+1 be arbitrary consistent positive Horn theories for the subfragments Li and Li+1, respectively, satisfying Ti contains T<i, i+1> and Ti+1 contains T<i, i+1>. Let Ai be A(Fi) and Bi be A(Fi+1). Starting with a basis model A0, A0|L<0,1> and B0|L<0,1> are models of a complete theory, hence, A0|L<0,1> $\equiv$ B0|l<0,1>. It follows that the elementary diagram of A0|l<0,1> is consistent with the elementary diagram of B0|L<0,1>.

(i) There are iterated elementary extensions Bi+1> Bi and an embedding fi: Ai < Bi

(ii) Slalom between the language pairs Li, Li+1 gates to a limit model for L, a model M for $L_{H,\omega}$.

Define the category $L_{H,\omega}$ to be the category with objects Horn fragments and arrows the subfoumual preorder on formulas.

Op

Define a functor $F: L_{H,\omega} \to$ Set by a list of sets Mn and functions fn. The functor F is a list of sets Fn, consisting of

(a) the sets corresponding to an initial structure on $L_{P,\omega}$, for example the free syntax tree structure, where to f(t1, t2,.. tn) in $L_{P,\omega}$ there corresponds the equality relation f(t1, ..., tn)=ft1...tn in Set;

(b) the functions fi:Fi+1 $\to$ Fi. $_\subseteq$

9.4.2 INTUITIONISTIC TYPES AND REALIZABILITY

The idea of realizability was first introduced by Kleene in a paper of 1945. One can see Kleene's work as an attempt to recast (some aspects of) Brouwer's thought in more accessible terms. Starting with the language of first order logic, for example N, in intuitionism, to say what a (closed) formula 'means' is to say what counts as a 'proof' of it: An atomic formula has a trivial 'proof' iff it's (verifiably) true. If p, q are proofs of P, Q respectively, the pair (p, q) is a proof of P $\circ$ Q. If p is a proof of P then (0, p) is a proof of P $\lor$ Q. Likewise for (1, q). A proof of P $\Rightarrow$ Q is a constructive operation that transforms any proof of P into a proof of Q (...). A proof of $\exists$x.P is a pair (n, p) where p is a proof of P[n/x]. A proof of $\forall$x.P is a constructive operation that transforms a value n into a proof of P[n/x] (...). • There's no proof of $\perp$ (falsity). For the above example, Kleene's definition can be viewed as making the Brouwer-Heyting-Kolmogorov 'interpretation' precise. The construction is to be Church-Turing computable. Proofs were thus elementarily arithmatized. Kleen's notions: n |~ P was the number n "realizes the formula P." Kleene relies on the pairing and application operations on N computable on a Turing machine. For example, For P atomic, 0 |~ P iff P is true. • n |~-P $\circ$ Q iff n = $\langle$p, q$\rangle$ n |~ P $\Rightarrow$Q iff for all m |~ P, n•m is defined and n•m |~ Q. Say P is realizable if some n |~ P.

Kleene realizability is based on a definition of |||— that is pending on a real "computability" notion. So is useful for questions of intuitionistic provability. For example. a sentence provable in Heyting Arithmetic is realizable by induction on structure of HA proofs. The converse is unrealizable, because the halting problem is undecidable. From that is the insight that a consistency proof is ultimately a relative consistency proof. A typical feature of intuitionistic systems like HA is the existence property: if $\vdash$ $\exists$x.P, then there's some m such that P [m/x]. Realizability gives a nice proof of this for HA.

The definition and applications of Kleene '1945' realizability are the prototype for all other realizability interpretations of logic. The general pattern is that one defines a relation n |||— P, where P is a formula in some logic (e.g., HA), n is an entity with some computational or algorithmic content (e.g., a natural number), ~|— is some relation of 'providing constructive evidence for.'

Considering the intuitionistic hierarchy on Heyting arithmetic one might examine parallers on realizability with formulas from Section 9.4.

Proposition 9.12: Positive formula is Kleene realized only if is locally realized as in section 9.4.

To state a more general theorem here is deferred to intuitionistic type theorist. This author is not certain that there is a very specific notion for Kleene realizability for first or higher order logics that can be a state on a direct theorem here.

For example to apply positive realizability to Heyting arithmetic and consider the intuitionistic fragments Φn (Burr, 2000) on arithmetic hierarchy are prenex normal form realizable in the sense of Section 9.4.

In view of the above, however, considering more contemporary realizability, one considers closed terms in an extension of the untyped λ-calculus for realizers. For realizing classical logic that will not be enough and some typing controls must be there. Intuitionistic second order logic is a realizability model based on closed λ-terms (Krivine & Parigot, 1990; Krivine, 1990a). Based on a Griffin (1990) paper proof term assignment for classical logic via a λ-calculus with control operators as realizers for classical principles like *reductio ad absurdum* were developed (Streicher & Reus, 1998). Krivine developed his theory of Classical Realizability for extensions of classical second order logic and Zermelo-Fraenkel set theory. The comparison to realizability as initiated by M. Hyland in (Hyland 1982), further expanded in van Oosten (2008) to the above evades most except when one considers how forcing on realizability topos was already on this authors over a decade publications (Nourani, 2014).

Triposes and toposes are two classes of categorical models of higher order intuitionistic logic, of slightly different kinds: triposes are indexed preorders P: Cop $\rightarrow$ Ord where logical formulas are interpreted as elements $\varphi \in P(I)$ (called predicates) of some fiber, whereas toposes are categories and predicates are interpreted by monomorphisms (also called subobjects). The class of triposes is more general since the subobjects in any topos form a tripos. For any tripos P, the tripos-to-topos construction produces a topos TP whose objects are partial equivalence relations internal to P, and whose morphisms are (external) equivalence classes of internal functional relations. A way of understanding the construction is that it 'freely' internalizes the 15 abstract predicates of a tripos as subobjects in a topos, and it is the word 'freely' that a universal characterization has to make precise. To formulate a universal property for the tripos-to-topos construction, we have to place triposes and toposes as objects in a categorical framework, and the natural choice is to consider 2-categories of toposes and triposes where 1-cells of toposes are functors, and the 1-cells of triposes are natural transformations relative to a 'change of base' functor between the base categories. A logical intuition on these 1-cells is that they map the interpretation of types, terms, and formulas from one model into another, in a way that commutes with the interpretation of logical connectives. However, the 25 requirement to preserve all logical connectives is too restrictive so one rather focus on some fragment of first order logic whose interpretation we want to preserve.

A pleasant aspect of triposes is that they give rise to a conceptually clear account of iteration of model constructions which is also explained in van Oosten (2008). This framework we use for explaining the iterated model construction of (Krivine, 2008). Moving onto Hyland et.al. 2009, that work motivated by strongly normalizing untyped λ*-terms (where * is just a formal constant) forms a partial applicative structure with the inherent application operation. The quotient structure satisfies all but one of the axioms of a partial combinatory algebra (PCA). Such applicative structures are called *conditionally partial combinatory algebras (c-pca)*.

Applying a notation form Behtke, Klop, et. al. (1999), consider a structure $e = \langle A, s, k, \cdot \rangle$, where A is some set containing the distinguished elements s, k, equipped with a binary operation $\cdot$ on A, called application, which may be partial.

- Instead of a·b we write ab; and in writing applicative expressions, the usual convention of association to the left is employed. So for elements a, b, c $\in$ A, the expression aba(ac) is short for $((a \cdot b) \cdot a) \cdot (a \cdot c)$.
- 2 ab $\downarrow$ will mean that ab is defined; ab $\uparrow$ means that ab is not defined. Obviously, an applicative expression can only be defined if all its subexpressions are.
- 3 If t1, t2 are applicative expressions, t1 $\sim$= t2 means that either both t1 $\uparrow$ and t2 $\uparrow$, or t1 $\downarrow$ andt2 $\downarrow$andt1 =t2.

Definition 9.18: A structure e as indicated above is a partial combinatory algebra (pca) if for all a, b, c $\in$ A: ka$\downarrow$, sa$\downarrow$, sab$\downarrow$, 2 kab $\sim$= a and sabc $\sim$= ac(bc).

An example of a pca is Kleene's u = $\langle x, s, k, app \rangle$, where app is defined as the Kleene-bracket application from recursion theory: app(m, n) $\sim$= {m}(n), and s and k are so chosen to satisfy the characterizing axioms.

There are (infinitely) many possible choices for s, k, yielding an u. Another example is on Uniformly Reflexive Structures, for example, Wagner (1969).

John Longley's dissertation (Longley, 2011) develops the groundwork for the study of the dynamics of PCAs is developed with a useful 2-category structure on the class of pcas. Applictive morphims with a category **Ass** for assemblies are presented in view of Hyland's realizability.

Definition 9.19: Let A and B be pcas. An applicative morphism A → B is a total (or, as some people prefer, 'entire') relation from A to B, which we see as a map γ from A to the collection of nonempty subsets of B, which has a realizer, that is: an element $r \in B$ satisfying the following condition: whenever a, a' $\in$ A are such that aa'↓, and b $\in \gamma(a)$, b' $\in \gamma(a')$, then rbb'↓ and rbb' $\in \gamma(aa')$.

Given two applicative morphisms γ, δ: A → B we say $\gamma \le \delta$ if some element s of B satisfies: for every a $\in$ A and b $\in \gamma(a)$, sb↓ and sb $\in \delta(a)$.

Pcas, applicative morphisms and inequalities between them form a preorder-enriched category. Applicative morphisms have both good mathematical properties and a computational intuition: if a pca is thought of as a model of computation, then an applicative morphism is a simulation of one model into another.

To study geometric morphisms RT(B) → RT(A) one looks at those applicative morphisms γ:A→B for which $\gamma*$ has aright adjoint.

The following definition is from (? [4]). Extending our notational conventions about application a bit: for a $\in$ A, $\alpha \subseteq$ A we write aα↓ if ax↓ for every x $\in \alpha$, and in this case we write aα for the set {ax|x $\in \alpha$}.

Definition 9.20: An applicative morphism γ: A → B is computationally dense if there is an element m $\in$ B such that the following holds:

For every b$\in$B there is an a$\in$A such that for all a' $\in$A: if bγ(a')↓, then aa'↓ and mγ(aa')↓ and mγ(aa') $\subseteq$ bγ(a').

Theorem 9.17: (Johnstone 2014) An applicative morphism γ: A → B induces a geometric morphism RT(B) → RT(A) precisely when it is computationally dense.

Obvious drawbacks of this theorem are the logical complexity of the definition of 'computationally dense' and the fact that, prima facie, the theorem only says something about geometric morphisms which are induced by a Γ— functor between categories of assemblies, in other words: geometric morphisms RT(B) → RT(A) for which the inverse image functor maps assemblies to assemblies. Both these issues were successfully addressed in Peter Johnston's paper.

Theorem 9.18: (Johnstone) An applicative morphism γ: A $\rightarrow$ B is computationally dense if and only if there exist an element r $\in$ B and a function g: B $\rightarrow$ A satisfying: for all b $\in$ B and all b' $\in$ γ(g(b)), rb' = b.

We might, extending the notation for inequalities between applicative morphisms, express the last property as: γg $\leq$ idB. An arbitrary *right absorptive c-pca* gives rise to a *tripos* provided the underlying intuitionistic predicate logic is given an interpretation in the style of Kreisel's *modified realizability*, as opposed to the standard Kleene-style realizability. Starting from an arbitrary right-absorptive *C-PCA U*, the tripos-to-topos construction due to Hyland et al. can then be carried out to build a *modified realizability topos* TOPm(*U*) of non-standard sets equipped with an equality predicate. Church's Thesis is internally valid in *TOP m (K* 1) (where the *pca k* 1 is "Kleene's first model" of natural numbers) but not Markov's Principle. There is a topos inclusion of SET-the "classical" topos of sets-into **TOP** m(*U*); the image of the inclusion is just sheaves for the ¬¬-topology. Separated objects of the ¬¬-topology are characterized.

We identify the appropriate notion of PER's (partial equivalence relations) in the modified realizability setting and state its completeness properties. The topos *TOP m (U)* has enough completeness property to provide a category-theoretic semantics for a family of higher type theories which include Girard's System F and the Calculus of Constructions due to Coquand and Huet. As an important application, by interpreting type theories in the topos *TOP m (SN.)*, a clean semantic explanation of the Tait-Girard style strong normalization argument is obtained. Hyland et.al. illustrate how a strong normalization proof for an impredicative and dependent type theory may be assembled from two general "stripping arguments" in the framework of the topos *TOP m (SN.)*. This opens up the possibility of a "generic" strong normalization argument for an interesting class of type theories.

9.5 REALIZABILITY ON ULTRAFILTERS

From the author's volume on functorial model theory 2014 we have the following developments on topological realizability models and their computational properties.

Lemma 9.4: (author, 2013) There is a topological space on $L\omega 1$, K fragments, with set of points X and a system $\Omega(X)$ of open sets of elements from X, that is, a subset of the powerset of X.

Proof: (c.f. Nourani 2014)

Lemma 9.5: (author, 209) Starting with the discrete topology on the Keisler fragment K, on L1.. Let M be the infinite product copies on K. Give K the product topology. Let $F = \{K, K2, ...\}$. Observations: (i)A subset of elements of F from a topological space with a pointset topology. (ii) M is homomorphic to its product with itself.

Proof: (c.f. Nourani, 2014).

There is a functor from the K--fragment category to the category of Boolean models with a Stone representation. Recall that: Let I be a non--empty set. A proper filter U over I is a set of subsets of I such that: (i) U is closed under supersets; if $X \in U$ and $X \subseteq Y \subseteq I$ then $Y \in U$. (ii) U is closed under finite intersections; if $X \in U$ and $Y \in U$ then $X \cap Y \in U$. (iii)$I \in U$ but $\varnothing \in /U$. An ultrafilter over I is a proper filter U over I such that: (iv) For each $X \subseteq I$, exactly one of the sets X, I\X belongs to U. We take the fragment models and projection pointsets on the product topology above. The following proposition shows that the points ultrafilters.

Proposition 9.14: (author, 2014) There is a natural transformation that maps stone representation category for the Boolean models above to the $L_{\omega 1, K}$ fragment models.

We can apply the sheaf topos on authors ASL with forgetful functors on fragments. Forgetful functors to **Set** are often representable. In particular, a forgetful functor is represented by (A, u) whenever A is a free object over a singleton set with generator u. The forgetful functor **Top** $\rightarrow$ **Set** on the category of topological spaces is represented by any singleton topological space with its unique element. To prove proposition 7.5 define two functors: F1: $L\omega 1$, $K \rightarrow$ Set F2: a forget functor: Top $\rightarrow$ ΠK where the singleton element is the discrete product topology on the Keisler fragment K. h1 is the natural transformation functor from K products to $L\omega 1$, K; h2 is projections to fragment sets. From the above developments the Topos Upward Lowenhiem Skolem realization is as follows.

Theorem 9.21: (author, 2014) κ-cardinal realization models are at the κ-saturation point set models from the above proposition.

Proof: (c.f. Nourani, 2014).

9.5.1 COMPUTING MORPHISMS ON TOPOS

Geometric morphisms between realizability toposes are studied in terms of morphisms between partial combinatory algebras (PCAs). The morphisms inducing geometric morphisms (the computationally dense ones) are seen to be the ones whose 'lifts' to a kind of completion have right adjoints. We characterize topos inclusions corresponding to a general form of relative computability. We characterize PCAs whose realizability topos admits a geometric morphism to the effective topos.

The study of geometric morphisms between realizability toposes was initiated by John Longley in his thesis (Longley, 2012). He began an analysis of partial combinatory algebras, the structures underlying realizability toposes by defining a 2-categorical structure on them. Longley's "applicative morphisms" characterize regular functors between categories that commute with the global sections functors to Set. A partial combinatory algebra or, as Johnstone calls them, Schonfinkel algebra is a structure with a set A and a partial binary function on it, called application such that every element of A encodes a partial function on A, and ab is the result of the function encoded by a applied to b.

9.5.2 RELATIVE REALIZABILITY ON TOPOS

Hyland showed how to treat realizability in terms of categorical models. This isolates a rich structure that can be studied in advance of choosing a logic. It also turns out that this structure provides a natural home for many other things besides logics. Hyland's intuitionistic models are essentially applicative realizations that do not have a direct correspondence with model categories or the newer homotopy models with applications to Martin Löf types in this authors publications. An interpretation of a logic ('P is satisfied if...') can often be cast as a model: a mathematical structure given independently of the logic in which formulae can be assigned denotations: $P \mapsto [[\,P]]$. Example: Interpreting formulae P with one free variable as predicates on a set A.

Interpretation: Define a relation a $\models$ P for a $\in$ A. Model: Define a mapping P $\mapsto$ [[P]] $\in$ P(A) (a Boolean algebra). Working with arbitrary PCA (A, •). Hyland defined a realizability topos RT(A), a universe for 'intuitionistic set theory.' RT(K1) is known as the effective topos. For now, we'll work with a simpler category PER(A) $\rightsquigarrow$ RT(A). • Objects: PERs (i.e., symmetric transitive relations) on A. • Morphisms R $\rightarrow$ S: define a PER SR by aSRa′ $\Leftrightarrow$ ($\forall$b, b′.bRb′$\Rightarrow$a•bSa′•b′) A morphism R $\rightarrow$ S is an equivalence class for SR. Intuition: PERs are 'data types' implementable on the 'abstract machine' A. Elements a with aRa are 'machine representations' ('realizers') of data values. Elements a, b with aRb realize the same data value. Morphisms are machine-computable functions.

Structure in PER(A): Any PCA admits a representation of natural numbers: n $\mapsto$ n. So in PER(A) we have a natural number object N: n Nn⁻ for every n and that's all. PER(A) is Cartesian closed (exponentials SR as on previous slide). The finite types over N are exactly those we saw earlier. PER(A) is locally Cartesian closed and regular. In any such category, one can interpret first order logic over whatever types are around, using standard ideas from categorical logic. In the case of PER(A), this agrees precisely with the standard realizability interpretation (e.g., for HA$^\omega$). A predicate P on type Σ is modeled as a subobject [[P]] of [[σ]]. Languages like System F can only express total functions (and our PER semantics reflects this). However, most programming languages allow partial functions to be defined using iteration and/or general recursion. Consider the simple types over ι, interpreted in PER(K1) by setting [[ι]] = N$\perp$, [[$\Sigma \rightarrow \tau$]] = [[τ]][σ], where mN$\perp$n $\Leftrightarrow$ m•0$\simeq$n•0 It turns out that every [[σ]] admits a fixed point operator: a morphism Yσ: [[σ]][σ] $\rightarrow$ [[σ]]. (Cf. Myhill-Shepherdson theorem.) This means one interpret Plotkin's language PCF (simply typed λ-calculus with arithmetic and general recursion).

9.5.3 REALIZABILITY TRIPOSES

It is more or less straightforward how to interpret intuitionistic second order logic in a realizability model based on closed λ-terms. This was studied in detail by J.-L. Krivine and M. Parigot in the late 1980s, see (Krivine & Parigot, 1990; Krivine, 1990a). Around 1990 due to the seminal paper (Griffin 1990) it got clear to many researchers how to give a proof term assignment for classical logic via a λ-calculus with control operators which serve as realizers for classical principles like *reductio ad absurdum* or Peirce's law (Streicher & Reus, 1998). Krivine developed his theory of Classical Realizability for extensions of classical second order logic and Zermelo-Fraenkel set theory. In more recent yet unpublished work (Krivine, 2008) Krivine has embarked on the long-term project of providing a realizability interpretation for full ZFC, for example, Zermelo-Fraenkel set theory with the full Axiom of Choice. This is to be achieved by considering forcing interpretations inside classical realizability models. In Krivine (2008) he has shown how to contract this two-step model construction into one step. Reading through Krivine's papers introducing classical realizability is not clear that has relations to the structural (i.e., semantic) approach to realizability as initiated by Hyland (1982) and fully described in van Oosten (2008).

9.5.4   MORE ON TOPOS REALIZABILITY

Longley's "applicative morphisms" characterize regular functors between categories of assemblies *Ass* that commute with the global sections functors to Set. Longley was thus able to identify a class of geometric morphisms with adjunctions between partial combinatory algebras. In these papers the computationally dense applicative morphisms are studied as those which, when "lifted" to the level of order-PCAs, do have a right adjoint. One can prove that every realizability topos which is a subtopos of Hyland's effective topos is on a partial combinatory algebra of computations with an "oracle" for a partial function on the natural numbers. Authors employ "oracles" to realize such computational characterizations.

Hyland et.al. motivated by the discovery that an appropriate quotient SN of the strongly normalizing untyped λ *terms (where * is just a formal constant) forms a partial applicative structure with the inherent application operation. The quotient structure satisfies all but one of the axioms of a partial combinatory algebra (*PCA*). The authors call such structures *conditionally partial combinatory algebras (c-PCA)*. An arbitrary *right absorptive c-PCA* gives rise to a *tripos* provided the underlying intuitionistic predicate logic is given an interpretation in the style of Kreisel's *modified realizability*, as opposed to the standard Kleene-style realizability. Starting from a tripos-to-topos construction due to Hyland et al. There is a topos inclusion of SET-the "classical" topos of sets-into image Sheaves on topologies. The authors apply an appropriate notion of PER's (partial equivalence relations) in the modified realizability setting and state its completeness properties. The topos inclusion is from the classical set topos to a *TOP m (U)* has enough completeness property to provide a category-theoretic semantics for a family of higher type theories, including Girard's System F and the Calculus of Constructions due to Coquand and Huet.

As an important application a clean semantic explanation of the Tait-Girard style strong normalization argument is accomplished. The relation to realizability as initiated by Hyland (1982) and fully described in van Oosten (2008) is not all explored. That has to consider triposes is that they give rise to a conceptually clear account of iteration of model constructions which is also explained in van Oosten (2008). Hyland is essentially applicative morphisms on fibers but is not direct model categories as developed in Nourani (2014) and Arnds (2014). The interesting part on Hyland is the applications for Cartesian closed categories with separable complete and co-complete diagrams to have a realization for the propositional exiom of choice. On set theory models there are direct realizations from algebraic set theory in the functorial model theory book (Nourani, 2014) with the more recent on Blass and Dubrinen (2014).

9.6   ON PRESHEAVES TOPOS REALIZABILITY

The author had defined infinite language categories the on the Keisler $L\omega1,\omega$ fragments to present a functorial model theory since 1996. Here a basic Grothendeick topology is defined on the fragments and further new categorical areas are presented. Generic functors and functorial model theory are applied to presheaves providing a glimpse onto the functorial models on the topologies. The techniques a new glimpse on embedding's with natural transformations, where further generalizations to Yoneda and applications to Grothendeick topology can be explored. Let us define a small-complete category $\mathbf{L}_{\omega1,K}$ from $L\omega1,\omega$. The category is the preorder category defined by the formula ordering defining the language fragment. The objects are small set fragments. There are three categories at play— the category $\mathbf{L}_{\omega1,K}$, the category Set, and the category D<A, G>. The D<A, G> category is the category for models definable with D<A, G> and their morphisms. The techniques from Nourani (1995) are a three categories play without having to present a categorical interpretation for $\mathbf{L}_{\omega1,K}$, in categorical logic as in Lawvere's (1967). A notaion point A-diagrams are the usual arrow diagrams. Model diagrams are at times referred to by M-diagrams.

A site I given by a category C (the underlying category of the site) together with Grothendeick topology on C given by a class Cov (A) for each object A of C.

The elements of Cov (A) are families (sets) (Ai fi → A) i∈ I of morphisms with domain A. An element of Cov(A) is called a covering family of A, A ∈ C.

A site I given by a category C (the underlying category of the site) together with Grothendeick topology on C given by a class Cov (A) for each object A of C.

The elements of Cov (A) are families (sets) $(A_i f_i A)_{i\,I}$ of morphisms with domain A. An element of Cov(A) is called a covering family of A, A C.

The topology has to satisfy the following conditions (for example, note the mapping arrows are prefixed with the mapping name, f_i):

(i) Every isomorphism A' f →A gives an elementary covering family {A' f → A} Cov (A).

(ii) (Satisfiability under pull-backs), Whenever $(A_i f_i A)_{i\,I}$ Cov (A) and B g A is an isomorphism in C then $(A_i \times B f_i B)_{i\,I}$ Cov (B).

$$A_i \mathbf{\;—\;} f_i\, A$$

$$\uparrow \uparrow g$$

$$A_i \times_A B - f_i'\, B$$

Is any pullback diagram for each i I.

(iii) "Clousure under composition" whenever $(A_i f_i A)_{i\,I}$ Cov (A) and $(A_{ij} g_{ij} Ai)_{j\,Ji}$ Cov (Ai) for every i I? we have that $(A_{ij} f_i \circ g_{ij} A)_{j\,Ji,\,i}$ belongs to Cov(A).

(iv) "Montonicity" If $(A_i f_i A)$ Cov (A) and $(B_j g_j A)_{j\,J}$ is such that from any j J there is an I and a morphism $B_j A_i$ with $B_j g_j A$ and A_i Ai

Theorem 9.22: (Author, 2000) There is a natural Gorothedieck topology on the category $\mathbf{L}_{\omega 1, K}$.

Generic Limit Functors: As an example a theorem might be proved with intermediate difficulty as an alternate way to define functorial models.

Definition 9.18: A functor $V: A \to X$ creates limits for a functor $F: J \to A$ if

  (i) to every limiting cone tt: $x -^* \to VF$ in X there is exactly one pair $<a, \sigma>$ consisting of an object a in A with $Va = x$ and a cone σ:a $-^* \to F$ with $V\sigma = tt$ and if moreover

  (ii) This cone σ:a $-^* \to F$ is a limiting cone in A. Op

Define a functor F: $\mathbf{L\omega 1, K} \to$ Set to be the generic functor.

Define a functor V: D<A, G> $\to \mathbf{L\omega 1, K}$ by universal embedding from the diagram functions.

Theorem 9.23: (author, 1997) V Creates a limit for F.

From the above we note that there is a contravariant functors F: $\mathbf{L}_{\omega 1, K}$ $\to$ **Set** to form a presheave.

Proposition 9.16: There are limiting cones in F that define example realization presheaves for the objects in $\mathbf{L}_{\omega 1, K}$.

Proof: Follows form small completeness at $\mathbf{L}_{\omega 1, K}$ with the generic filter on T*.

Theorem 9.24: The natural transformations on functors sending arbitrary objects to sets on fragment string sets, arrowed by preorder functions, with cones to base F from {}, on the above on the functor F: $\mathbf{L}_{\omega 1, K} \to$ **Set,** are filter creating realizaion presheaves for the objects in $\mathbf{L}_{\omega 1, K}$.

Proof: The preceding lemma and theorem and Nourani (2014).

Remark: The obvious direction from here is what comma category realizes that above.

That is a route to explore a new glimpse on a Yoneda-like embedding and the Grothendeick topology. Newer applications were (Nourani 2011) were generic functors and functorial model theory are applied to presheaves providing a glimpse onto the functorial models on the topologies. Further applications to filtering on Joyal Simplicity can be developed on higher stratified functorial models. The authors 2004 book furthermore paper presents generic functors that present model categories based on fiberations. Example applications to Martin Löf type systems are stated. A functortial model theory defines positive omitting types topos that create models. The above is the basis for the natural transformations on functors sending arbitrary objects to sets on fragment string sets, arrowed by pre-order functions.

EXERCISES

1. A typical feature of intuitionistic systems like HA is the existence property: if ⊢ ∃x.P, then there's some m such that P [m/x].
    What is a Kleene Realizability proof of this for HA.
2. For example to apply positive realizability to Heyting arithmetic and consider the positive intuitionistic fragments Φn on arithmetic hierarchy are prenex normal form and show that the fragment is realizable as in Section 9.4.
3. Prove that a positive formula is Kleene realized only if is locally realized as in Section 9.4.
4. Prove that two-categories realize triposes.
5. Can you apply two-categories to provide an explication for the remark following the last theorem above with comma categories.
6. Prove Theorem 9.15.
7. Prove Proposition 9.16.

KEYWORDS

- **categorical models**
- **positive categories and consistency models**
- **PreSheaves topos realizability**
- **realizability on ultrafilters**
- **realizability, positive realizability morphisms**

REFERENCES

Abraham, H. Taub (editor) (1961). John von Neumann. Collected works, vol. 1, Pergamon Press, New York.

Addisson, J. (1960). "The Theory of Hierarchies," in Proceedings of the International Congress of Logic, Methodology, and Philosophy of Science, Stanford, 1960, pp. 26–37.

ADJ-Goguen, J. A., Thatcher, J. W., Wagner, E. G., Wright, J. B. "An Introduction to Categories, Algebraic Theories and Algebras," IBM Research Report, RC5369, Yorktown Heights, NY, April 1975.

ADJ-Goguen, J. A., Thatcher, J. W., Wagner, E. G., Wright, J. B. "Initial Algebra Semantics and Continuous Algebras," IBM Research Report RC5701, November 1975, JACM24, 1977, 68–95.

ADJ-Goguen, J. A., Thatcher, J. W., Wagner, E. G., Wright, J. B. "A Junction Between Computer Science and Category Theory," (parts I and II), IBM T. J. Watson Research Center, Yorktown Heights, N. Y. Research Report, RC4526,1973.

ADJ-Thather, J. W., Wagner, E. G., Wright, J. B. (1979). "Notes On Algebraic Fundamentals For Theoretical Computer Science," IBM T. J. Watson Research Center, Yorktown Heights, NY, Reprint from Foundations Computer Science III, part 2, Languages, Logic, Semantics, J. deBakker, J. van Leeuwen, editors, Mathematical Center Tract 109.

Alfred Tarski (1983). Logic, semantics, metamathematics. Papers from 1923 to 1938, second ed., Hackett, Indianapolis, translations by J. H. Woodger.

Alfred Tarski (1986). Collected papers, (Steven, R. Givant and Ralph, N. McKenzie, editors), Birkhauser, Basel.

Alfred Tarski, Robert, L. Vaught (1957). Arithmetic extensions of relational systems, Composition Mathematical, vol. 13, pp. 81–102, reprinted in Tarski (1986), vol. 3, pp. 653–674.

Andrej Bauer. The Realizability Approach to Computable Analysis and Topology, thesis. On the Failure of Fixed-Point Theorems for Chain-complete Lattices in the Effective Topos.

Andrews, Peter, B. (2002). *An Introduction to Mathematical Logic and Type Theory: To Truth Through Proof*, 2nd ed, Kluwer Academic Publishers, ISBN 1–4020–0763–9

Armstrong, M. A., *Basic Topology*, Springer; 1st edition (May 1, 1997). ISBN 0–387–90839–0.

Artin, E., O. Schreier, 1926, "Algebraische Konstruktion reeller Korper", Hamb. Abh. 5 (1926), 85–99.

Artin, M., Grothedieck, A., Vedier, J. L. Therore des Topos et Cohomolgie Etale des Schemas. Lecture Notes in Mathematics, vol. 1, no. 269, vol.2. No. 270, Springer Verlag.

Artin, M., Grothendieck, A., Verdier, J. L. *Séminaire de Géométrie Algébrique du Bois Marie (SGA 4)*. Lecture Notes in Mathematics 269, Springer Verlag, 1972. Exposé I, 2.7.

Artin, Michael, Mazur, Barry (1969), *Etale homotopy*, Lecture Notes in Mathematics, No. 100, Berlin, New York: Springer-Verlag

Awodey, S., Warren, A. M. Homotopy theoretic models of identity types, Math. Proc. of the Cam. Phil. Soc. (2009).

Barwick, C. On left and right model categories and left and right bousfield localizations, Homology, Homotopy and Applications, Vol. 12 (2010), No. 2,

Barwise, J., "Syntax and Semantics of Infinitary Languages," Springer-Verlag Lecture Notes in Mathematics, vol. 72, 1968, Berlin-Heidelberg-NY.

Barwise, J., "Implicit Definability and Compactness in Infinitary Languages," in The Syntax and Semitics of Infinitary Languages, Edited by, J. Barwise, Springer-Verlag LNM, vol.72, Berlin-Heidelberg, NY.

Barwise, J. "Infinitary logic and admissible sets," JSL, 34, 226–252. 1969.

Barwise, J. (1978). Handbook of Mathematical Logic, North-Holland, second edition.

Barwise, J., Robinson, A. "Completing Theories By Forcing," Annals of Mathematical Logic, vol.2, no.2, 1970, 119–142.

Batanin, M. A., Monoidal globular categories as a natural environment for the theory of weak n-categories, Adv. Math. 136 (1998), no. 1, 39–103.

Baues, Hans Joachim, Algebraic homotopy, Cambridge Studies in Advanced Mathematics, vol. 15, Cambridge University Press, Cambridge, 1989.

Benno van den Berg, Ieke Moerdijk. Aspects of Predicative Algebraic Set Theory II: Realizability, to appear in TCS.

Bethke, I., Jan. W. Klop, Vrijer, R. D. (1999). Extending partial combinatory algebras Math. Struct. in Comp. Science, vol. 9, pp. 483–505. Cambridge University Press.

Bethke, I. (1987) On the existence of extensional partial combinatory algebras. Journal of Symbolic Logic 52 (3) 819–833.

Bethke, I., Klop, J. W. (1996) Collapsing partial combinatory algebras. In: Dowek, G., Heering, J., Meinke, K. and Moller, B. (eds.) Higher-Order Algebra, Logic, and Term Rewriting (HOA '95). Springer-Verlag Lecture Notes in Computer Science 1074–57–73.

Blass, A., "A model-theoretic view of some special ultrafilters," pp. 79–90 in Logic Colloquium 77, ed. by, A. Macintire et al., North-Holland 1978.

Blass, A., F-generic ultra filters, Mathematics Department, University of Michigan, 2010 North American Annual Meeting In Washington, DC.

Booth, D., "Ultrafilters on a countable set," Annals of Math. Logic 2 (1971), 1.

Borceux, F. Handbook of Categorical Algebra: vol 1 Basic category theory (1994) Cambridge University Press, (Encyclopedia of Mathematics and its Applications) ISBN 0–521–44178–1.

Bourbaki, Nicolas; *Elements of Mathematics: General Topology*, Addison-Wesley (1966).

Bredon, Glen, E., *Topology and Geometry* (Graduate Texts in Mathematics), Springer; 1st edition (October 17, 1997). ISBN 0–387–97926–3.

Brown, K. S., Abstract homotopy theory and generalized sheaf cohomology, Transactions of the American Mathematical Society 186 (1973), 419–458.

Burris, Stanley, N., Sankappanavar, H. P., 1981. *A Course in Universal Algebra.* Springer-Verlag. ISBN 3–540–90578–2.

Carboni, A. Some free constructions in realizability and proof theory. Journal of Pure and Applied Algebra, 103, 117–148, 1995.

Carnap, R., Abriss der Logistik, Wien, Austria, Springer.

Carnap, R., Bachman, F. (1936) Uber Extremalaxiome, Erkenntnis, 6, 166–188. 1929.

Čech, Eduard; *Point Sets*, Academic Press (1969).

Chang, C. C., Keisler, H. Jerome, Model Theory, Studies in Logic and the Foundations of Mathematics, vol. 73, North Holland, 1977, 2nd edition.

Colin McLarty, *Elementary Categories, Elementary Toposes.* Oxford Univ. Press. A nice introduction to the basics of category theory, topos theory, and topos logic. Assumes very few prerequisites. 1992.

Cyrus F. Nourani, Positive realizability on Horn filters.

Dana, S. Scott, *Definitions by abstraction in axiomatic set theory, Bulletin of the American Mathematical Society,* vol. 61, p. 442, 1955.

David MacIver, (2004) *Filters in Analysis and Topology (Provides an introductory review of filters in topology and in metric spaces.)*

Ehrig, H. "Embedding Theorems in the Algebraic Theory of Graph Grammars." FCT 1977: 245–255.

Eilenberg, S., MacLane, S. (1950). Relations between homology and homotopy groups of spaces. II Ann. of Math. 51, pp. 514–533

Eilenberg, S., MacLane, S., (1945). Relations between homology and homotopy groups of spaces Ann. of Math. 46, pp. 480–509

Fachbereich Mathematik, TU Darmstadt, Schlossgartenstrasse 7, Darmstadt, Germany, 2102.

Feferman, S., Kreisel-66, G. Persistent and invariant formulas relatives to theories of higher types, Bull Amer. Math. Soc., 72, 1966, 480–485.

Fragment Consistent Kleene Models, Fragment. Topologies, and Positive Process Algebras. Algebraic Topological Methods In Computer Science (ATMCS) III, Paris, France.

Frayne, T. E., A. C. Morel, and, D. S. Scott, Reduced direct products, Fund. Math. 51 (1962), 195–228 (Abstract: Notices Amer. Math. Soc. 5 (1958), p. 674).

Friedlander, Eric, M. (1982), *Étale homotopy of simplicial schemes*, Annals of Mathematics Studies, 104, Princeton University Press, ISBN 978–0-691–08288–2; 978–0-691–08317–9

Fulton, William, *Algebraic Topology*, (Graduate Texts in Mathematics), Springer; 1st edition (September 5, 1997). ISBN 0–387-94327–7.

Gallager, R. G. 1968, Information Theory and Reliable Communication, Wiley, SBN 471–29048–3. May 03, 2008, 1984.

Ghilardi, S. Free Heyting algebras as bi-Heyting algebras, Math. Rep. Acad. Sci. Canada XVI., 6, 240–244, 1992.

Gray, J. W. ed. Categories in Computer Science and Logic, Contemporary Math. 92 (1989), 1–7, Amer. Math. Soc.2000 Mathematics Subject Classification: 18 C10,03C65. Amer. Math. Soc., 1989.

Grothendieck and Verdier: *Théorie des topos et cohomologie étale des schémas* (known as SGA4)". New York/Berlin: Springer. (Lecture notes in mathematics, 269–270).

Hartmut Ehrig, (1977). Embedding Theorem in the Algebraic Theory of Graph Grammars. FCT, 245–255

Henkin, L. "On Mahematical Induction," American Mathematical Monthly, 67, 1960.

Henkin, L., "The Completeness of First Order Functional Calculus," Journal of Symbolic Logic," vol. 14, 1949.

Hinman, P. G. Recursion Theoretic Hierarchies, Spring-Verlag, 1980.

Hofstra, P., Oosten, van J. Ordered partial combinatory algebras. Math. Proc. Camb. Phil. Soc., 134, 445–463, 2003.

Huber, P. J. Homotopical cohomology and Čech cohomology, *Mathematische Annalen* 144, 73–76, 1961.

Hyland, J. M. E. The effective topos. In, A. S. Troelstra, D. Van Dalen, editors, The, L. E. J. Brouwer Centenary Symposium, pages 165–216. North Holland Publishing Company, 1982.

John Lane Bell (2005) *The Development of Categorical Logic*. Handbook of Philosophical Logic, Volume 12. Springer. Version available online at John Bell's homepage.

John von Neumann, Zur Einfuʾhrung der transfiniten Zahlen, Acta Litterarum ac Scientiarum Regiae Universitatis Hungaricae Francisco-Josephinae (Szeged), sectio scientiarum mathematicarum, vol. 1, pp. 199–208, 1923, reprinted in 1961 below, vol. 1, pp. 24–33; translated in van Heijenoort (1967), pp. 346–354.

John, R. Myhill, Dana, S. Scott (1971). *Ordinal definability, Proceedings of symposia in pure mathematics* (Dana, S. Scott, editor), Axiomatic Set Theory, vol. 13, part 1, American Mathematical Society, Providence, pp. 271–278.

Johnstone, P. T. Geometric Morphisms of Realizability Toposes. Theory and Applications of Categories, 28(9), 241–249, 2013.

Johnstone, P. T. *Topos Theory*, L. M. S. Monographs no. 10. Academic Press. ISBN 0–12–387850–0. 1997.

Johnstone, P. T., *"Sketches of an Elephant: A Topos Theory Compendium."* Oxford Science Publications. 2002.

Johnstone, Peter, T. (1982). *Stone Spaces*. Cambridge University Press. ISBN 0–521–23893–5.

Joyal, A., Moerdijk, I. *Algebraic set theory*, London Mathematical Society Lecture. Note Series, vol. 220, Cambridge University Press, 1995.

Keisler, H. J. Forcing and The Omitting Types Theorem, Studies in Model Theory, Mathematical Association of America, 96–133, M. D. Morley editor.

Keisler, H. J. Some applications of the theory of models to set theory, pp. 80–86 in Logic, Methodology and Philosophy of Science, ed. by, E. Nagel et. al., Stanford Univ. Press 1962.

Keisler, H. J. Ultraproducts and elementary classes, Koninkl. Ned. Akad. Wetensch. Proc. Ser. A 64 (= Indag. Math. 23) (1961), 477–495.

Keisler, H. J., (1967). Forcing and the Omitting Types Theorem, Studies in Model Theory, Mathematical Association of America, 96–133, M. D. Morley editor.

Keisler, H. J., "Limit ultrapowers," Trans. Amer. Math. Soc. 107 (1963), 383–408.

Keisler, H. J., Model Theory for Infinitary Logic, North Holland, Amsterdam, 1971.

Keisler, H. J., On cardinalities of ultraproducts, Bull. Amer. Math. Soc. 70 (1964), 644–647.

Keisler, H. J., Ultraproducts and elementary classes, Koninkl. Ned. Akad. Wetensch. Proc. Ser. A 64 (= Indag. Math. 23) (1961), 477.495..

Knight, J., "Generic Expansions of Structures," JSL, 38, 1973, 561–570.

Kripke, S, "Transfinite recursion on admissible ordinals," I, II, JSL, 29, 161–162, abstract. 1964.

Kripke, S. A., 1965, 'Semantical analysis of intuitionistic logic, I.' In: J. N. Crossley and, M. A. E. Dummett (eds.): Formal Systems and Recursive Functions. Amsterdam: North- Holland, pp. 92–130.

Kunen, Kenneth (1980). Set Theory: An Introduction to Independence Proofs. North-Holland. ISBN 0–444–85401–0.

Kusraev, A. G. Samson Semenovich Kutateladze (1999). Boolean valued analysis. Springer. p. 12. ISBN 978–0–7923–5921–0.

Lambek, J., Scott, P. J., 1986. Introduction to Higher Order Categorical Logic, Cambridge University Press, ISBN 0–521–35653–9

Lambek, J., Scott, O. J. Introduction to Higher-Order Categorical Logic Cambridge Studies in Advanced Mathematics, 1988.

Lambekt, J. The mathematics of sentence structure, American Mathematical Monthly, vol. 65, pp. 154–170.

Lars Birkedal, Jaap van Oosten. (2002). Relative and Modified Relative Realizability, Annals of Pure and Applied Logic 118, 115–132

Lawvere, F. W., "Functorial Semantics of Algebraic Theories," Proc. National Academy of Sciences, USA, 1963.

Lawvere, F. W., 1963 Functorial Semantics of Algebraic Theories, Proceedings of the National Academy of Science 50, No. 5 (November 1963), 869–872.

Lawvere, F. W., Elementary Theory of the Category of Sets. In Proceedings of the National Academy of Science 52, No. 6 (December 1964), 1506–1511.

Lawvere, F. W., Quantifiers and Sheaves. In Proceedings of the International Congress on Mathematics (Nice 1970), Gauthier-Villars (1971) 329–334.

Lipschutz, Seymour; *Schaum's Outline of General Topology*, McGraw-Hill; 1st edition (June 1, 1968). ISBN 0–07–037988–2.

Longley, J. Realizability Toposes and Language Semantics. PhD thesis, Edinburgh University, 1995.

Mac Lane, S. Categories for The Working Mathematician, GTM Vol. 5, Springer-Verlag, NY Heidelberg Berlin, 1971.

Mac Lane, Saunders (1998), *Categories for the Working Mathematician* (2nd ed.), Berlin, New York: Springer-Verlag, ISBN 978–0-387–98403–2, section IX.1.

Macintire, A, "Twenty years of p-adic model theory", pp. 121–153. in Logic colloquium '84 (Manchester, 1984), edited by, J. B. Paris et al., Stud. Logic Found. Math. 120, North-Holland, Amsterdam, 1986

Macintye, A. "Model Completeness," Handbook Mathematical Logic, J. Barwise, editor, 1978, North-Holland.

Makaai, M. 1981 "Admissible Sets and Infinitary Logics" Handbook Chapter A7, (Barwise, editor), Studies in Logic and Foundations, Vol. 90, 1981.

Marshall, H. Stone (1936) "The Theory of Representations of Boolean Algebras," *Transactions of the American Mathematical Society 40*: 37–111.

Martin, D. A. "Descriptive Set Theory," in Hand Book of Mathematical Logic, Barwise, J. Editor, North Holland. 1978.

McLane, S. Categories For the Working Mathematician, GTM, Springer-Verlag, Berlin-NY-Heidelberg, 1971.

Michael Barr and Charles Wells (1985) *Toposes, Triples and Theories*. Springer Verlag. Corrected online version at http://www.cwru.edu/artsci/math/wells/pub/ttt.html. More concise than *Sheaves in Geometry and Logic*, but hard on beginners.

Milies, César Polcino; Sehgal, Sudarshan, K. An introduction to group rings. Algebras and applications, Volume 1. Springer, 2002. ISBN 978–1-4020–0238–0; Jacobson. Basic Algebra II. Dover. 2009. ISBN 0–486–47187-X.

Moredij, I., Palmgren, E. "Minimal models of Heyting arithmetic, Journal of Symbolic Logic, vol. 62, pp. 1448–1460, 1997.

Morley, M., Vaught, R. Homogeneous universal models, Math. Scand. 11 (1962), 37–57.

Munkres, James; *Topology*, Prentice Hall; 2nd edition (December 28, 1999). ISBN 0–13–181629–2.

Neumann, B. H., "The Isomorphism Problem for Algebraically Closed Groups," in Word Problems, Boone et.al. editors, North-Holland, 1973, 553–562.

Nourani, C. F. "Filters, Fragment Constructible Models, and Sets," www.ams.org/meetings/sectional/mtgs-2164–1051–03–12.pdf

Nourani, C. F. "Forcing with Universal Sentences and Initial Models, Annual Meeting of the Association for Symbolic Logic, Boston, MA., December 1983, Journal of Symbolic Logic, vol. 49 (1984), p. 1444.

Nourani, C. F. "Functorial Computability and Generic Definable Models," International Congress, Mathematicians, Berlin, August 18–27, 1998.

Nourani, C. F. "Functorial Consistency," May 1997, AMS 927, Milwaukee, Wisconsin, 1997. Abstract number 927–03–29.

Nourani, C. F. "Functorial Fragment Definable Models," February 2000. ASL Annual Meeting, March 2000. BSL vol. No.

Nourani, C. F. "Functorial Metamathematics," Maltsev Meeting, Novosibirsk, Russia, November 1998. math.nsc.ru/conference/malmeet/thesis.htm

Nourani, C. F. "Functorial Model Computing," FMCS, UBC Mathematics Department, Vancouver, Canada, June 2005.

Nourani, C. F. "Functorial Model Theory and Generic Fragment Consistency Models," October 1996, AMS-ASL, San Diego, January 1997.

Nourani, C. F. "Functorial model theory and infinitary language categories, September 1994," Association for Symbolic Logic, San Francisco, January 1995, see Association for Symbolic Quarterly, Summer 1996, for recent abstract.

Nourani, C. F. "Functorial Model Theory, Generic Functors and Sets," January 16, 1995, International Congress, Logic, Methodology, and Philosophy of Science, Florence, Italy, August 1995.

Nourani, C. F. "Functorial Models and Generic Limit Functors," April 1996. The abstract number is: 918–18–1508. AMS Contributed Paper, January 1997, San Diego.

Nourani, C. F. "Functorial Models, Admissible Sets, and Generic Rudimentary Fragments," Summer Logic Colloquium, 1997, Leeds.

Nourani, C. F. "Functorial Projective Set Models," ASL Summer Logic Colloquium, Sofia University from July 31 August 5, 2009.

Nourani, C. F. "Higher Stratified Consistency and Completeness Proofs," http://www.logic.univie.ac.at/cgi-bin/abstract/show.pl?new=e049a2efe0c1a4b7a3ddaa11a75d8152; http://www.math.helsinki.fi/logic/LC2003/abstracts/

Nourani, C. F. "Positive Infinitary Forcing and Word Problems, "815th Meeting of the AMS, November 1984", San Diego, Proc. AMS Notices; and the Annual Meeting of ASL, Anaheim, CA. January 1985, Proc. Journal of Symbolic Logic.

Nourani, C. F. (2005a). "Fragment Consistent Algebras, July 2005.

Nourani, C. F. (2005b). Functorial Model Computing," FMCS, UBC Mathematics Department, Vancouver, Canada, June 2005.

Nourani, C. F. (2005c). "Positive Categories and Process Algebras," Draft outline August 2005, Written to a perfunctory Stanford CS project. Abstract published at Paris France:

Nourani, C. F. (2006). "Functorial Generic Filters," July 2005, ASL, Montreal, 2006.

Nourani, C. F. (2011). Filters, Fragment Consistent Models, and Stratified Toposes: Preliminaries. Joint Mathematical Conference of the Austrian Mathematical Society at the Donau-Universitat Krems together with the Catalan, Czech, Slovak, and Slovenian Mathematical Societies, September 25–28, 2011.

Nourani, C. F. (2013). Product Models on Positive Process Algebras World Congress on Universal Logic, www.unilog.org/cont4.html; Department of Philosophy, Federal University of Pernambuco, Brazil.

Nourani, C. F. 1977, "Functorial Models, Admissible Sets, and Generic Rudimentary Fragments, March 1997, Summer Logic Colloquium, Leeds, July 1997. BSL, vol. 4, no.1, March 1998. www.amsta.leeds.ac.uk/events/logic97/con.html

Nourani, C. F. 1984 "Positive Forcing and Complexity," Written at SLK, Manchester, England, July 1984.

Nourani, C. F. 1984, "Forcing, Nonmonotonic Logic and Initial Models," Logic Colloquium, Manchester 1984.

Nourani, C. F. 1995, "Functorial Model Theory, Generic Functors and Sets," January 16, 1995, International Congress, Logic, Methodology, and Philosophy of Science, Florence, Italy, August 1995.

Nourani, C. F. 1997, "Functorial Consistency," May 1997, AMS 927, Milwaukee, Wisconsin, 1997. Abstract number 927–03–29.

Nourani, C. F. 2000, "Generic Limit Functorial Models and Toposes," 15th Summer Conference on General Topology and Its Applications July 26–29, at Miami University, Oxford, 2000.

Nourani, C. F. 2005, "On Certificates and Models," June 2005. Memo to Stanford TCS.

Nourani, C. F. 2005, "Positive Categories and Process Algebras," Draft outline August 2005, Written to a perfunctory Stanford CS project.

Nourani, C. F. 2005, Fragment Consistent Algebraic Models, July 2006 Abstract and presentation to the Categories Oktoberfest, U Ottawa, October 2005.

Nourani, C. F. 2006, "A Sound and Complete AI Action Logic, June 2005," Preliminary version Proceedings MFCSIT-06, August 2006, Cork, Ireland. D. Kozen. 1990, "On Kleene algebras and closed semirings." In, B. Rovan, editor, Mathematical Foundations of Computer Science 1990, volume 452 of Lecture Notes in Computer Science, pages 26–47, Banska Bystrica, 1990. Springer-Verlag.

Nourani, C. F. 2007, Functorial models and positive realizability, ASL, Florida, March 2007.

Nourani, C. F. 2008, "Positive Realizability on Horn Filters," European Summer Meeting of the Association for Symbolic Logic Colloquium '08, Bern, Switzerland, July 3–July 8, 2008.

Nourani, C. F. 2009, Positive omitting types and fragment consistency, Gödel Society conference, Brazil, 2009.

Nourani, C. F. 2011, Filters, Fragment Consistent Models, and Presheaves. Spring Southeastern Section Meeting, Program by Day AMS Sectional Meeting Program by Day. March 4, 2011–00:23:34 Georgia Southern University.

Nourani, C. F. 2011, "Fragment Consistent Kleene Models, Fragment Topologies, and Positive Process Algebras," Algebraic Topological Methods in Computer Science (ATMCS) III www.lix.polytechnique.fr/~sanjeevi/atmcs/program.pdf

Nourani, C. F., Moudi, R. M. "Fields, Certificates, and Models (Preliminary report), AMS, Clairemont, May 2008.

Nourani, C. F. Fragment Consistency on Functorial Models, AMS, San Francisco, April AMS Reference: 1018–18–90.

Nourani, C. F. Functional generic filter, ASL, Montreal, May 2006. Functorial models and positive realizability, ASL, Florida, March 2007.

Nourani, C. F. Functorial ω-Chain Models, 1998, Holiday Mathematics Symposium, La Cruces, NM, Proc. Editor May Jharke. January 1999.

Nourani, C. F. Functorial Admissible Models, 1996–97 Annual Meeting of the Association for Symbolic Logic, Cambridge, Massachusetts, March 22–25, 1997.

Nourani, C. F. Functorial Model Theory, Generic Functors and Sets, "International Congress, Logic, Methodology, and Philosophy of Science, Florence, Italy, August 1995.

Nourani, C. F. Functors Computing Models on Hasse Diagrams, The Preliminary brief, Single-page abstract Announced at Mini Conference on Topology and Computing, U Southern Maine, 1998.

Nourani, C. F. Functors on V. Memo to D. Scott, A. Blass. April 2005 and 2011, respective, Berkeley.

Nourani, C. F. The Incredible String Models, Brief Versions Accepted at WOLLIC, June 1996, and Kosice, July 1996.

Nourani, C. F., "Functorial Model Theory and Infinite Language Categories," September 1994, Presented to the ASL, January 1995, San Francisco. ASL Bulletins 1996.

Nourani, C. F., "Computable Functors And Generic Model Diagrams—A Preview To The Foundations, December 1996.

Nourani, C. F., "Functorial Model Theory and Generic Fragment Consistency Models," October 1996, AMS-ASL, San Diego, January 1997.

Nourani, C. F., "Functorial Models Defined on Initial Ordered Structures—A Preview," March 1997. Brief presented at FMCS UBC, Canada.

Nourani, C. F., 'Functorial Model Theory, Generic Functors and Sets," January 16, 1995, International Congress, Logic, Methodology, and Philosophy of Science, Florence, Italy, August 1995.

Nourani, C. F., "Admissible Models and Peano Arithmetic," ASL, March 1998, Los Angeles, CA. BSL, vol.4, no.2, June 1998.

Nourani, C. F., "Computable Functors And Generic Model Diagrams A Preview To The Foundations, December 1996.

Nourani, C. F., "Computable functors and generic models diagrams," December 1996. ICM, Berlin. Functorial Computability and Generic Definable Models, International Congress, Mathematicians, Berlin, August 18–27, 1998.

Nourani, C. F., "Descriptive Computing—The Preliminary Definition," Summer Logic Colloquium, July 1996, San Sebastian Spain. See AMS April 1997, Memphis.

Nourani, C. F., "Forcing with Universal Sentences and Initial Models," University of Pennsylvania, 1981–82, Annual Meeting of the Association for Symbolic Logic, December 1983, Proc. in Journal of Symbolic Logic.

Nourani, C. F., "Forcing With Universal Sentences," 1981, Proc. ASL, 1983, vol. 49, and AMS 1985, vol.6., no.2. 1971.

Nourani, C. F., "Fragment Consistency on Functorial Models," (A Preliminary) AMS, San Francisco, April 06 Reference: 1018–18–90.

Nourani, C. F., "Fragment Consistent Algebraic Models," Brief Presentation, Categories, Oktoberfest, Mathematics Department, U Ottawa, Canada, October 2005.

Nourani, C. F., "Functorial Admissible Models," November 26, 1996, ASL, MIT, Cambridge, March 1997.

Nourani, C. F., "Functorial Computability and Generic Definable Models," International Congress, Mathematicians, Berlin, August 18–27, 1998.

Nourani, C. F., "Functorial Computability and Initial Computable Models," May 1997, AMS 927 Milwakee, Wisconsin, 1997. Abstract number 97T-68–191. Volume 18, No. 4, p. 624.

Nourani, C. F., "Functorial Consistency," AMS 927, Milwaukee, Wisconsin, Special Session on Applications of Model Theory to Analysis and Topology, Abstract number 927–03–29, May 1997.

Nourani, C. F., "Functorial Model Theory and Generic Fragment Consistency Models," October 1996, AMS-ASL, San Diego, January 1997.

Nourani, C. F., "Functorial Model Theory and Infinitary Language Categories," Proc. ASL, January 1995, San Francisco. See BSL, Vol.2, Number 4, December 1996.

Nourani, C. F., "Functorial Model Theory and Infinite Language Categories," September 1994, Presented to the ASL, January 1995, San Francisco. ASL Bulletins 1996.

Nourani, C. F., "Functorial Models and Generic Limit Functors," April 1996. The abstract number is: 918–18–1508. AMS Contributed Paper, January 1997.

Nourani, C. F., "Functorial Models and Generic Limit Functors", April 1996, Proc. AMS, San Diego, January 1997.

Nourani, C. F., "Functorial Models and Infinitary Godel Consistency," March 4, 1999. Goedel Conference. 5th Barcelona Logic Meeting and 6th Kurt Godel Colloquium.

Nourani, C. F., "Functorial Models and Infinite Language Categories," ASL, Wisconsin, Spring 1996. BSL, Vol.2, Number 4, December 1996.

Nourani, C. F., "Functorial models and positive Realizability," ASL Gainsville, Florida, March 2007. Functors and generic sets, AMS Spring 1996 issue. Re: 96T-03–54 is the abstract number.

Nourani, C. F., "Functorial Models, Admissible Sets, and Generic Rudimentary Fragments," March 1997, Summer Logic Colloquium, Leeds, July 1997. BSL, vol. 4, no.1, March 1998.

Nourani, C. F., "Functorial Models, Horn Products, and Positive Omitting Types," Savoie, France, June 5, 2012.

Nourani, C. F., "Functorial Projective Set Models," Summer Logic Colloquium, Sofia, Bulgaria. July 2009

Nourani, C. F., "Functorial String Models," ERLOGOL-2005: Intermediate Problems of Model Theory and Universal Algebra, June 26–July 1, State Technical University/Mathematics Institute, Novosibirsk, Russia.

Nourani, C. F., "Functors Computing Models On Hasse Diagrams," Topological Computing Foundations, Mini Conference Maine, April 1997.

Nourani, C. F., "Generic functors and generic sets," AMS Spring 96 issue Re: 96T-03–54 is the abstract number. 1996.

Nourani, C. F., "Generic Functors, and Generic Sets," AMS, Burlington, August 1995.

Nourani, C. F., "Generic Limit Functorial Models and Toposes," TOPO2000, August2000, Oxford, Ohio. atlas-conferences.com/c/a/e/u/04.htm. AMCA: Summer Conference on Topology and its Applications, Oxford, Ohio, 2000.

Nourani, C. F., "Higher Stratified Consistency and Completeness Proofs," April 2003, Sumer Logic Colloquium, 2003, Helsinki, August 14–20.

Nourani, C. F., "IFLCs and Grothendeick topology: Preliminaries," February 8, Accepted to PSSL, Glasgow, May 06, 2006.

Nourani, C. F., "Infinitary Language Categories, Generic Functors, and Functorial Model Theory," ASL www.math.ucla.edu/~asl/bsl/0803/0803–006.ps

Nourani, C. F., "Infinite Language Categories, Limit Topology, and Categorical Computing", 1994, Brief Abstract, MFCS, Boulder, Colorado, June 1996.

Nourani, C. F., "Positive Omitting Types and Fragment Consistency Models," XIV Encontro Brasileiro de Lógica Godel Society Conference April 2006.

Nourani, C. F., "Positive Realizability Morphisms and Tarksi Models," Summer Logic Colloquium, Wroclaw, Poland July 2007

Nourani, C. F., "Slalom Tree Computing," 1994, AI Communications, Vol. 9. No.4, December 1996, IOS Press, Amsterdam.

Nourani, C. F., "The Connection Between Positive Forcing and Tree Rewriting," Proc. Logics In Computer Science Conference (LICS) and ASL, Stanford University, July 1985, Proc. Journal of Symbolic Logic.

Nourani, C. F., "The Solution Set Theorem For Categories With Initial Models," European Category Theory Conference, Tours, France, July 1994.

Nourani, C. F., (1982). "Two part paper On Types, Induction, and Inductive Completeness," Second Workshop on Theory and Applications of Data Types, University of Passau, Passau, West Germany.

Nourani, C. F., (2006). "Functorial Generic Filters," July 2005, ASL, Montreal.

Nourani, C. F., 1982 "Two part paper On Types, Induction, and Inductive Completeness," Second Workshop on Theory and Applications of Data Types, University of Passau, Passau, West Germany.

Nourani, C. F., 2006, Positive Omitting Types and Fragment Consistency Models, XIV Encontro Brasileiro de Lógica XIV Brazilian Logic Conference, Celebrating Kurt Gödel's Centennial (1906–2006) Brazilian Logic Society and Association for Symbolic Logic, April 24–28.

Nourani, C. F., Descriptive Computing, Summer Logic Colloquium, San Sebastian, Spain, July 1996. See AMS April 1997 Abstracts.

Nourani, C. F., Fragment Consistent Algebraic Models, 2005 Brief Presentation, Categories, Oktoberfest, Mathematics Department, U Ottawa, Canada October 2005.

Nourani, C. F., Functorial Computability and Generic Definable Models, International Congress Mathematicians, Berlin, August 18–27, 1998.

Nourani, C. F., Functorial Consistency, May 1997, AMS 927, Milwaukee, Wisconsin, Special Session on Applications of Model Theory to Analysis and Topology, 1997. Abstract number 927–03–29.

Nourani, C. F., Functorial Fragment Definable Models, February 2000. ASL Annual Meeting, March 2000 BSL.

Nourani, C. F., Functorial Generic Filters, July 26, 05 ASL, Montreal, May 06 BSL Vol. 2, September 2007.

Nourani, C. F., Functorial generic filters. (2006). Annual Meeting of the Association for Symbolic logic-Addendum, Universitedu Quebec, Montreal, Quebec, Canada May 17–21, 2006. Bulletin of Symbolic-logic Volume 13, Number 3, Sept. 2007.

Nourani, C. F., Functorial Model Computing, FMCS, Mathematics Department, University of British Columbia, Vancouver, Canada, June 2005

Nourani, C. F., Functorial Model Theory and Generic Fragment Consistency Models, October 1996, AMS-ASL, San Diego, January 1997.

Nourani, C. F., Functorial Model Theory and Infinitary Language Categories, Proc. ASL, January 1995, San Francisco. BSL, Vol.2, Number 4, December 1996.

Nourani, C. F., Functorial Models and Infinite Language Categories, ASL, Wisconsin, Spring 1996. BSL, Vol.2, Number 4, December 1996.

Nourani, C. F., Functorial models and positive realizability. Association For Symbolic Logic, Annual Meeting University of Florida. Gainesville, Florida. March, www.aslonline.org/files/ann07program.ps BSL Vol. 2, September 2007.

Nourani, C. F., Functorial Models, Admissible Sets, and Generic Rudimentary Fragments, Summer Logic Colloquium, Leeds, England, August 1997.

Nourani, C. F., Functorial String Models, ERLOGOL-2005: Intermediate Problems of Model Theory and Universal Algebra, June 26–July 1, State Technical University/Mathematics Institute, Novosibirsk, Russia.

Nourani, C. F., Functorial ω-Chain Models, 1998, Holiday Mathematics Symposium, January 1999, La Cruces, NM, Proc. May Jharke, editor.

Nourani, C. F., Generic Limit Functorial Models and Toposes, TOPO2000, August2000, Summer Conference on Topology and its Applications Oxford, Ohio, atlas-conferences.com/c/a/e/u/04.htm AMCA: 2000.

Nourani, C. F., Positive Omitting Types and Fragment Consistency Models, XIV Encontro Brasileiro de Lógica, Godel Society Conference, April 2006.

Nourani, C. F., Positive Realizability Morphisms and Tarksi Models Summer Logic Colloquium, Wroclaw, Poland July 2007.

Nourani, C. F., Th. Hoppe 1994, "GF-Diagrams for Models and Free Proof Trees," Proceedings the Berlin Logic Colloquium, Humboldt University Mathematics, May 1994, Universitat Potsdam, Germany.

Nourani, C. F., "Functorial Model Theory, Generic Functors and Sets," January 16, 1995, International Congress, Logic, Methodology, and Philosophy of Science, Florence, Italy, August 1995.

Nourani, C. F., "Spring ASL March 97, MIT. BSL, V ol.3, 1997. Platek, R. Foundations of recursion theory, Doctoral dissertation and Supplement, Stanford, 1966.

Nourani, C. F." Functorial Model Computing," FMCS, UBC Mathematics Department, Vancouver, Canada, June 2005.

Nourani, C.F "Functors and Model Computation with Hasse Diagrams," Topology and Computing, University of Northern Maine, Mini Conference, Maine, April 1997.

Nourani, C.F "Generic Limit Functorial Models and Toposes," 15th Summer Conference on General Topology and Its Applications July 26–29, at Miami University, Oxford, 2000.

Peter Arndt, Chris Kapulkin Homotopy Theoretic Models of Type Theory, Technical Report, 2012.

Phoa, W. Relative computability in the effective topos. Mathematical Proceedings of the Cambridge Philosophical Society, 106, 419–422, 1989.

Pitts, A. M. The Theory of Triposes. PhD thesis, Cambridge University, 1981. available at http://www.cl.cam.ac.uk/~amp12/papers/thet/thet.pdf.

Pitts, A. M., Conceptual completeness for first-order intuitionistic logic an application of categorical logic, Annals of Pure and Applied Logic, vol. 41 (1989), pp. 33–81.

Platek, R., Foundations of recursion theory, Doctoral Dissertation Supplement, Stanford, 1966.

Pratt, V. R. 1990, "Action Logic and Pure Induction," Invited paper, Logics in AI: European Workshop JELIA '90, ed. J. van Eijck, LNCS 478, 97–120, Springer-Verlag, Amsterdam, NL, Sep, 1990. Also Report No. STAN-CS-90–1343, CS Dept., Stanford, Nov. 1990..

Ramsey, F. P. The foundations of mathematics, Proceedings of the London Mathematical Society, vol. 25, pp. 338–384, 1925.

Robert Goldblatt (1984) Topoi, the Categorical Analysis of Logic (Studies in logic and the foundations of mathematics, 98). North-Holland. A good start. Reprinted 2006 by Dover Publications, and available online at Robert Goldblatt's homepage.

Robinson, A. 1965, Introduction to Model Theory and the Metamathemtics of Algebras, 2nd edition, North Holland, Amsterdam. (first edition 1963).Robinson, A, 1956

Robinson, A., "Infinite Forcing in Model Theory," Proc. 2nd Scandinavian Logic Symposium, edited by, J. Fenstad (North Holland) Amsterdam, 317–340.

Roman Murawski, *Undefinability of truth. The problem of priority: Tarski vs Go̎del, History and Philosophy of Logic,* vol. 19, pp. 153–160, 1998.

Runde, Volker; *A Taste of Topology (Universitext)*, Springer; 1st edition (July 6, 2005). ISBN 0–387–25790-X.

Saunders Mac Lane and Ieke Moerdijk (1992) *Sheaves in Geometry and Logic: a First Introduction to Topos Theory.* Springer Verlag. More complete, and more difficult to read.

Schiemer, G. "Carnap on External Axioms, Completeness of The Models, and Categoricity," ASL-2012, Vol 5, Number 4, December 2012.

Schroder. E. 1895, "Vorlesungenuber die Algebra der Logik (Exakte Logik). Dritter Band: Algebra und Logik der Relative." B. G. Teubner, Leipzig, 1895.

Scott, D., "Outline of a Mathematical Theory of Computation," Technical Monograph PRG-2, Oxford University Computing Lab., 1970, Proc. 4th Annual Princeton Conference on Information Sciences and Systems, 1970, pages 169–176.

Steel, J. R. HODL(R) *is a core model below* Θ, this Bulletin, vol. 1, pp. 75–84, 1995.

Steen, Lynn, A., Seebach, J. Arthur Jr.; *Counter examples in Topology*, Holt, Rinehart and Winston (1970). ISBN 0–03–079485–4.

Stephen Willard, (1970) *General Topology*, Addison-Wesley Publishing Company, Reading Massachusetts. *(Provides an introductory review of filters in topology.)*

Stephen Willard, *General Topology*, (1970) Addison-Wesley Publishing Company, Reading Massachusetts.

Steven Awodey, Andrej Bauer. (2008). Sheaf Toposes for Realizability (pdf), Archive for Mathematical Logic 47, 5, 465–478.

Stone, M. H., "The representation theorem for Boolean algebras," Trans. Amer. Math. Soc. 40 (1936), 37–111.

Strong, H. R. (1968) Algebraically generalized recursive function theory. IBM J. Research and Development 12–465–475.

Tarski, A. Mostowski, R. M. Robinson, Undecidable Theories, North-Holland, 3rd edition, 1971.

Thomas Scott Blyth (2005). Lattices and ordered algebraic structures. Springer. p. 151. ISBN 978–1-85233–905–0.

Thoralf Skolem Einige Bemerkungen zur axiomatischen Begrundung der Mengenlehre, Matematikerkongressen i Helsingfors den 4–7 Juli 1922, Den femte skandinaviska matematikerkongressen, Redogorelse, Akademiska-Bokhandeln, Helsinki, reprinted in, 1970 pp. 137–152; translated in van Heijenoort. pp. 290–301, pp. 217–232. 1967.

van den Dries, L., "Tame topology and o-minimal structures," London Math. Soc. Lecture Note Series 248, Cambridge Univ. Press, Cambridge, 1998.

van den Dries, L., Classical Model Theory of Fields. Model Theory, Algebra, and Geometry MSRI Publications, Volume 39, 2000.

van Oosten, J. Partial Combinatory Algebras of Functions. Notre Dame Journal Formal Logic, 52(4), 431–448, 2011.

van Oosten, J. Realizability: an Introduction to its Categorical Side, volume 152 of Studies in Logic. North-Holland, 2008.

Vaught, R. (1958). Prime Models and Saturated Models, Notices American Mathematical Society, 5. 780.

Vaught, R. (1961). Denumerable Models of Complete Theories, Infinitistic Methods, Pergamon, London, 303–321.

Wagner, E. G. (1969). Uniformly reflexive structures. Trans. American Math. Society 144–1–41.

Willard, Stephen (2004). *General Topology*. Dover Publications. ISBN 0–486–43479–6.

William Lawvere, F., Robert Rosebrugh (2003). *Sets for Mathematics*. Cambridge University Press. Introduces the foundations of mathematics from a categorical perspective.

William Lawvere, F., Stephen, H. Schanuel (1997). *Conceptual Mathematics: A First Introduction to Categories*. Cambridge University Press. An "introduction to categories for computer scientists, logicians, physicists, linguists, etc." (cited from cover text).

Wolfgang Burr. (1999). Concepts and Aims of Functional Interpretations (ps format), in: Löwe and Rudolph (eds), Foundations of the Formal Sciences, Kluwer, pp. 205–218.

Yankov, V. A. (2001). "Brouwer lattice", in Hazewinkel, Michiel, Encyclopedia of Mathematics, Springer, ISBN 978–1–55608–010–4.

Zermelo, E. (1930). Uber Grenzzahlenund Mengenbereiche: Neue Untersuchungen uber die Grundlagen der Mengenlehre, Fundamenta Mathematicae, vol. 16, pp. 29–47.

Zermelo, E. (1967). Untersuchungen uber die Grundlagen der Mengenlehre I, Mathematische Annalen, vol. 65, pp. 261–281, translated in van Heijenoort, pp. 199–215.

INDEX